BIASED SIGNALING IN PHYSIOLOGY, PHARMACOLOGY AND THERAPEUTICS

BIASED SIGNALING IN PHYSIOLOGY, PHARMACOLOGY AND THERAPEUTICS

Edited by

Brian J. Arey

ELSEVIER

AMSTERDAM • BOSTON • HEIDELBERG • LONDON
NEW YORK • OXFORD • PARIS • SAN DIEGO
SAN FRANCISCO • SINGAPORE • SYDNEY • TOKYO
Academic Press is an imprint of Elsevier

Academic Press is an imprint of Elsevier
525 B Street, Suite 1800, San Diego, CA 92101-4495, USA
32 Jamestown Road, London NW1 7BY, UK
225 Wyman Street, Waltham, MA 02451, USA

British Library Cataloguing-in-Publication Data
A catalogue record for this book is available from the British Library

Library of Congress Cataloging-in-Publication Data
A catalog record for this book is available from the Library of Congress

ISBN: 978-0-12-411460-9

For information on all Academic Press publications
visit our website at http://store.elsevier.com/

Typeset by MPS Limited, Chennai, India
www.adi-mps.com

14 15 16 17 18 10 9 8 7 6 5 4 3 2 1

Dedication

This book is dedicated to my wife Tiffany, my sons James and Alexander and my stepson Nicholas. You have been my inspiration, motivation and refuge; I am eternally grateful.

> ...I will conclude with an hypothesis which – in its simplest message argues that biological communication consists of a complex meshwork of structures in which G proteins, surface receptors, the extracellular matrix, and the vast cytoskeletal network within cells are joined in a community of effort, for which my life and those of my colleagues is a metaphor.
>
> **Martin Rodbell 1994 Nobel Lecture**

Contents

Preface

As a beginning to this book, I would like to take a brief moment to explain the motivation and objectives of this text. The ultimate objective of the book is to provide a single but comprehensive source of information (for both personal and educational purposes) concerning signal transduction as it is understood in the present with the addition of a personal vision on receptor signaling.

The first chapter is purposefully basic and intended to provide both a general historical backdrop and description of basic pharmacological principles as a launching point to the more detailed descriptions of other topics provided by my co-authors. As the focus of the book is on common themes in receptor signaling across classes of receptors, there are natural redundancies apparent from chapter to chapter. With limited space, we have attempted to provide a comprehensive view of the field including understanding how we can model and assess signaling bias in the practical sense. I am motivated to put these topics together in order to provide a more complete view of how we envision receptor signaling and to put it in the context of evolution and a more universal view of receptor signaling. That is, despite clear and significant differences between classes of receptors, there are also universal principles to their function that have been selected and carried forward throughout evolution.

As this book is published, it occurs to me that the concepts of biased signaling are just coming to the popular forefront of the minds of the research world. However, in my mind, the contents of this book are a culmination of my thoughts and my journey in science.

There are several key moments in the development of these thoughts on signal transduction and the potential of multiple signaling outputs provided by activated receptors. The first being the observations made by my fellow graduate student, Tom Burris, in the late 1980s that dopamine could not only inhibit prolactin secretion from isolated anterior pituitary cells through Gi activation but also stimulate it through a Gs mechanism, and the discussions that followed within the lab that included Marc Freeman (my graduate advisor) and Béla Kanyicska. My exposure to research in nuclear receptors and the development of selective estrogen and progesterone receptor modulators at Wyeth also had a profound impact on my views since it was through this research that I came to the realization that cell background and associated proteins provided another level of regulation and possibilities for the development of better therapeutics. Lastly, the observation that naturally occurring ligands could also induce signaling bias in collaboration with Francisco López (on glycoprotein hormone receptors) and the development of synthetic biased agonists to the FSH and calcium-sensing receptors have rounded out this view.

In compiling this resource, I have sought out fellow authors who are well-respected experts in their domains and who have had significant impact on their research fields for the topics they provided. I would like to thank them for their commitment to this project and for the outstanding chapters they have written. It is my sincere hope that you, the reader, will find this book useful and compelling.

Brian J. Arey M.S., Ph.D.
Department of Cardiovascular Drug Discovery Biology, Research and Development, Bristol-Myers Squibb Company, Hopewell, NJ, USA

Biased Signaling in Physiology, Pharmacology and Therapeutics is accompanied by a website featuring discussion questions, summaries, full color images and additional resources compiled by the authors. To access these companion resources, please visit http://booksite.elsevier.com/9780124114609.

About the Editor

Brian J. Arey received the Bachelor of Science degree in biology from Saint Louis University where as an undergraduate he performed research on estrogen and progesterone receptors and their interactions with chromatin. He then moved to Florida State University where he received both Master's and Doctorate degrees studying neuroendocrine physiology with Dr. Marc Freeman.

In 1992, his work in graduate school led to an National Institutes of Health (NIH) Postdoctoral Fellowship in the NIH Center for Reproductive Sciences at Northwestern University, where he studied the expression and signaling mechanisms of the mouse and rat prolactin receptors.

While at Wyeth in the mid-1990s, Dr. Arey collaborated to develop some of the first described allosteric agonists, antagonists and partial agonists to the follicle-stimulating hormone receptor (FSHR). As part of this research he was the first to demonstrate that the FSH receptor could activate multiple G-protein signaling pathways. In addition, he proposed the concepts of biased agonism/conformational dynamics as a natural physiological phenomenon in relation to glycoprotein hormone receptors.

Dr. Arey ultimately moved to Bristol-Myers Squibb where he and his colleagues developed allosteric modulators to a number of other receptors including novel allosteric modulators to the calcium-sensing receptor (CaSR). Throughout his tenure in pharmaceutical research, he has contributed to the discovery or development of multiple marketed medicines. Dr. Arey writes reviews and is requested to speak internationally on the topics of allosteric activation of GPCRs and the role of biased signaling.

List of Contributors

Andrew Alt Lead Discovery and Optimization, Research and Development, Bristol-Myers Squibb Company, Wallingford, CT, USA

Brian J. Arey Department of Cardiovascular Drug Discovery Biology, Research and Development, Bristol-Myers Squibb Company, Hopewell, NJ, USA

Barbara Bosier Institute of Neuroscience, Group of Neuropharmacology, Université Catholique de Louvain, Brussels, Belgium

Thomas P. Burris Department of Pharmacological and Physiological Science, Saint Louis University School of Medicine, St. Louis, MO, USA

Colin A. Flaveny Department of Pharmacological and Physiological Science, Saint Louis University School of Medicine, St. Louis, MO, USA

Emmanuel Hermans Institute of Neuroscience, Group of Neuropharmacology, Université Catholique de Louvain, Brussels, Belgium

James Herrington Department of Exploratory Biology and Genomics, Research and Development, Bristol-Myers Squibb Company, Wallingford, CT, USA

Terry Kenakin Department of Pharmacology, University of North Carolina School of Medicine, NC, USA

Douglas J. Kojetin Department of Molecular Therapeutics, The Scripps Research Institute, Jupiter, FL, USA

Luciano Mueller Leads Discovery and Optimization, Research and Development, Bristol-Myers Squibb Company, Princeton, NJ, USA

Guillermo G. Romero Department of Pharmacology and Chemical Biology, University of Pittsburgh School of Medicine, Pittsburgh, PA, USA

Liang Schweizer Leads Discovery and Optimization, Research and Development, Bristol-Myers Squibb Company, Princeton, NJ, USA

Laura A. Solt Department of Molecular Therapeutics, The Scripps Research Institute, Jupiter, FL, USA

John Watson Lead Discovery and Optimization, Research and Development, Bristol-Myers Squibb Company, Wallingford, CT, USA

CHAPTER

1

An Historical Introduction to Biased Signaling

Brian J. Arey

Department of Cardiovascular Drug Discovery Biology, Research and Development, Bristol-Myers Squibb Co., Hopewell, NJ, USA

OUTLINE

INTRODUCTION

Throughout human history, we have been intimately associated with our environment. As part of that environment, we contribute to the efficient ebb and flow of energy through the system. Similar to others within the environment, mankind has evolved and adapted to their surroundings, utilizing them as tools to prosper and continue the spread of the species. Just as we utilized stone for tools to cut wood or kill prey, so we also had an intimate knowledge and use of surrounding substances (most often plants) for medicinal purposes. Indeed, through millennia, the knowledge of medicinal substances was handed down and has grown. It wasn't until the modern era (ca 200 years) that we have harnessed

B. Arey (Ed): *Biased Signaling in Physiology, Pharmacology and Therapeutics*
DOI: http://dx.doi.org/10.1016/B978-0-12-411460-9.00001-X

the technical skill and understanding to develop synthetic medicinal substances. It is perhaps not a coincidence that the ability to develop synthetic medicines occurs in history simultaneously with the development of our understanding of physiology and pharmacology.

The concept of biological activity of endogenous and exogenous substances on the human condition is likely as old as the species itself. Study of these substances and the recording of them can be traced back as far as the Egyptian Empire where the oldest known record of pharmaceutical substances can be found within the Kahun Egyptian papyrus (Figure 1.1) that dates to ca 2000 B.C.[1] This text is found within a compendium of papyri dealing with many aspects of Egyptian life and is primarily a veterinary text, but also includes reference to gynecological issues, treatments, and midwifery. A more comprehensive text, the Ebers Medical papyrus dates to approximately 1550 B.C. and is a cumulative resource of medicinal treatments for many ailments.[1]

Despite the existence of various texts throughout the history of civilization, the knowledge of therapeutic substances was primarily passed down through oral history for millennia. Perhaps the first comprehensive modern text dealing with the use of medicinal substances which received legal distinction was the *Dispensatorium* from Valerius Cordi (ca 1546) that dealt with the specific synthesis of medicinal preparations for treatment of diseases and therefore represents the first modern pharmacopeia.[1]

For the majority of human existence the use of therapeutics to treat disease was focused on using natural remedies from plants and substances readily found within our environment. However, the modern era of synthetic therapies was realized through the slow development of technological and scientific advances that laid the groundwork for better

FIGURE 1.1 The Kahun Egyptian papyrus. The use of medicinal substances is probably as old as man himself. The earliest known written text discussing medicinal substances is found in the Kahun papyrus from Egypt. *Reproduced with permission of the Petrie Museum, University of London.*

understanding of biochemistry and cellular biology. It began with the realization of receptive substances within cells that could act as mediators of exogenous stimuli. The receptor theory was ushered in by two men, working independently but with a similar vision. The basis of their hypotheses was borne from a desire to understand how exogenous substances could impact cellular function. John Langley[2] and Paul Ehrlich[3] came to a similar conclusion but from different directions. Although Ehrlich is most often associated with the concept of receptive substances, it was Langley that perhaps had the clearer vision of the nature and utility of receptors.

Ehrlich was keenly interested in infectious disease and utilized some of the cutting edge techniques in the form of histology that the period had to offer. In the late nineteenth century, the germ theory was also gaining ground, and numerous reports had been published claiming that bacteria were able to produce anti-bacterial substances that would inhibit the growth of other bacterial species. In using the newly emerging histological techniques of the time, Ehrlich hypothesized that each cell contained a specific mixture of receptive substances that would bind the stains (e.g., methylene blue) with which he was working. He hypothesized that the bacteria that were taking up methylene blue contained "side chains" that interacted with the dye and allowed it to bind to the bacteria.

Langley was a neuroscientist interested in understanding the function of neurons and the neuromuscular junction. Initially he was interested in understanding the function of some of the paralytics available at the time. However, over time this research evolved into understanding the autonomic control of muscles and salivary gland secretions. Ultimately, this led to the study of the effects of various "poisons" on the neuromuscular junction.[4,5] His work was aided by other researchers of the time, including the previous work of Ehrlich and his theory of side chains, and the rapidly improving field of histology, especially the work of Ramón y Cajal that helped to elucidate the relationship of neurons to muscle fibers at the motor end-plate.[5] A key observation by one of his previous students, Thomas Elliot, helped to solidify his theory by demonstrating that application of exogenous substances (for example, extracts of the adrenal gland [adrenalin]) induced effects on the muscle similar to that elicited by electrical stimulation of sympathetic nerves. In his now famous publication, Langley not only hypothesized the existence of a receptive substance in muscle cells for adrenalin released by the sympathetic nerve, but also generalized this hypothesis to include the action of other compounds and other cell types:

> So we may suppose that in all cells two constituents at least are to be distinguished, a chief substance, which is concerned with the chief function of the cell as contraction and secretion, and receptive substances which are acted upon by chemical bodies and in certain cases by nervous stimulation. The receptive substance affects or is capable of affecting the metabolism of the chief substance.[2]

ISOLATION AND CHARACTERIZATION OF RECEPTORS

The study of pharmacology as a focus of research traces its origins to the work of Buchheim and Schmiedeberg during the mid to late nineteenth century and developed independently of the receptor theory being studied by Langley and Ehrlich. Bucheim and Schmiedeberg were focused on the understanding of the relationship between chemical structure and biological activity, thus laying the foundation for study of different compounds and assessing their biological effects. For this reason, they are recognized as the fathers of pharmacology.[6] The study of pharmacology was a natural evolution from physiology, just as physiology evolved from the study of anatomy,[7] and it was Bucheim's vision of understanding how chemicals induced effects on tissues that would be extremely important to the development of improved therapeutics for clinical use.

Pharmacology can be defined in the modern world as "the science of drugs, their sources, appearance, chemistry, actions and uses".[8] This broad definition includes the understanding of synthesis, effects, structure–activity relationships, molecular interactions, metabolism and distribution, and therapeutic uses. It also reflects the evolution and expansion in biological knowledge of pharmacology that has occurred over the last 150 years (Table 1.1). We now realize the utility of understanding the principles of drug action on and off their intended target and this is due in large part to the early work of Bucheim and Schmiedeberg.

Bucheim and his student, Schmiedeberg, championed the study of chemicals on biological function against overwhelming doubt from the prevailing scientific community of the time. If Bucheim was responsible for the vision of the importance of understanding the action of chemicals on physiology, Schmiedeberg was responsible for executing this vision and for developing the experimental data to convince others that this new field of study was of immense value. In his 46 years at the University of Strasbourg, Schmiedeberg was incredibly prolific providing numerous important observations that led to the fundamental understanding of pharmacology, including his scientific approach to experimental design. Within the context of Bucheim and Schmiedeberg, and Ehrlich and Langley, it is easy to understand how isolated visions were coming together in the late nineteenth century to give rise not only to pharmacology but also the link between pharmacology and receptor theory. These visions led to a revolution in understanding how biological systems interact with the external environment and how this could be used to not only understand disease but also to treat it. This period of time represents not only one of the key points in human history for understanding physiology, but also for developing therapeutics to correct pathophysiology and ultimately improve the quality of human life. It set in motion a cascade of new discoveries that led to the development of new therapeutics and founded the basic principles that are used in drug discovery today.

Despite the fact that pharmacology allowed for the characterization and profiling of receptors using functional and binding assays, isolation of receptors as separate protein entities remained elusive for almost another 100 years. For this reason, the nature of membrane-bound receptors remained a hotly debated issue within the world of biochemistry and pharmacology. It wasn't until the early 1970s that the first receptor protein was finally purified and characterized by Jean-Pierre Changeux through his isolation of the nicotinic acetylcholine receptor in the electric eel.[9,10] Changeux was an important figure not only for this discovery but also because the roots of his science were in allosteric interactions of proteins. He played a key role in the field of enzymology in which he had hypothesized that enzymes could be acted upon by sites distal from their active sites thus introducing the fundamental principles of protein conformational dynamics.

Changeux hypothesized that enzymes *in situ* were fluid structures. In the absence of substrate, the three-dimensional structure of enzymes was not static but was changing within its microenvironment. Interaction of the enzyme with its substrate was able to

TABLE 1.1 Some Key Discoveries in the History of Pharmacology

Year	Investigator	Observation
1872	Oswald Schmiedeberg	"Father of modern pharmacology" created and led the first research institute devoted to pharmacology
1878	Paul Ehrlich	Proposed that there are specific chemical characteristics of cells that allow them to bind dyes
1878	John Langley	Suggested that pharmacological agents form complexes with cells
1897	Paul Ehrlich	Proposed that cells contain "side chains" for binding toxins
1905/1907	John Langley	Theorized the existence of "receptive substances" in cells and their utility for "chemotherapy" (i.e., therapeutics)
1933	Alfred Clark	Proposed the receptor occupancy theory
1948	Raymond Ahlquist	Suggested the existence of two adrenalin receptor subtypes (α and β)
1956	Robert Stephenson	Proposed the concept of "spare receptors" and efficacy
1964	Pierre Changeux	Developed the allosteric theory of protein regulation in enzymes
1970	Pierre Changeux	Isolated and characterized the nicotinic acetylcholine receptor and suggested an allosteric mechanism of toxins
1972	Martin Rodbell	Demonstrated the existence of a GTP-dependent transducer protein and proposed the theory of "signal transduction"
1980	Alfred Gilman	Isolated the first G protein and determined its regulation of adenylate cyclase
1983	James Black & Paul Leff	Proposed the two-state model of receptor activation dependent upon receptor conformation
1984	Herman Kuhn	Identified β-arrestin activity to modulate rhodopsin activation
1986	Robert Lekfowitz & Richard Dixon	Cloned and expressed the first receptor (β-adrenergic receptor), predicted the seven transmembrane structure of GPCRs
1995	Hinrich Gronemeyer	Solved the first crystal structure of a nuclear receptor ligand binding domain
2000	Krzysztof Palczewski & Masashi Miyano	Reported the first crystal structure of a GPCR (bovine rhodopsin)

stabilize the enzyme into its active conformation. Furthermore, he demonstrated that the product of the enzyme reaction had the ability to interact allosterically to stabilize inactive conformation(s) of the enzyme. These observations were important to his purification of the nicotinic acetylcholine receptor. Thus, his observations laid the foundation in many ways for our current understanding of receptor–ligand interactions and how we now view receptor signaling across all classes of receptors.

At the turn of the twentieth century, there was evolution of thought in all fields of science including chemistry and physics. As these disciplines made new advances, some physiologists pushed to apply some of their techniques, most notably mathematical modeling, to the systems they studied.[7] Archibald Hill was a key member of this movement and worked to apply mathematical models to pharmacological data in order to better quantify the relationship between compound concentration and biological effect.[11] Indeed, his application of mathematics to model pharmacological data bears his name (the Hill equation) and can be found as an integral part of further refinement of modeling of pharmacology that took place in the following years. This includes the application of the mathematical principles of Hill that can be found within the modeling of enzyme kinetics that was published shortly thereafter by Michaelis and Menten[12,13] and in the equation developed by Langmuir to describe the saturable nature of gas adsorption by various substrates (e.g., metal).

The Hill equation is derived from the law of mass action at equilibrium for reversible chemical reactions first proposed by Guldberg and Waage, and later by van Hoft. This model simply states that for simple chemical reactions, the rate of forward and reverse reactions are not only dependent upon the concentration of reactants but also on their affinity, or chemical attraction, for each other. Applying this idea to pharmacology, Hill proposed that this equilibrium model would apply for chemicals and their association with proteins in or on a cell. He was particularly interested in understanding the effect of oxygen partial pressure on binding to hemoglobin. This model was applied by Langmuir to receptor–ligand interactions several years later in response to the development of the receptor theory of drug–receptor interactions by A.J. Clark.[7,14]

The Hill equation is shown in Figure 1.2. Reaction 1 represents the interaction between receptor, *R*, and ligand, *L*. For the sake of this equation, the concentration of *R* is considered constant, whereas the concentration of *L* can be variable but in excess to the concentration of *R*. This relationship implies that the concentration of *R*, and therefore the affinity of *R* for *L*, is rate limiting: Where $[L_nR]$ refers to the concentration of the ligand–receptor complex; $[R_0]$ is the total receptor concentration (receptor number); $[L]$ is the concentration of free ligand (for experimental purposes this is considered the concentration of ligand used); k_1 and k_2 are the rate constants of the forward (association between *L* and *R*) and the reverse (dissociation of *L* and *R*) reactions, respectively; K_d is the equilibrium dissociation constant of the receptor–ligand complex and is equal to k_1/k_2; *n* originally referred to the number of ligand binding sites on the receptor and is also referred to as the Hill coefficient or Hill slope factor; K_A is the concentration of ligand at which half of the receptors are occupied (if $n = 1$, it equals the K_d). K_A serves as a measure of affinity of the ligand for the receptor such that the smaller the K_A, the greater the affinity. The quotient, $[L]^n/([L]^n + (K_A)^n)$, is referred to as the fractional receptor occupancy.

Alfred Clark utilized the Hill equation in the development of his receptor occupancy theory. Specifically, he utilized the Hill equation to model his theory that the concentration

$$nL + R \underset{k_2}{\overset{k_1}{\rightleftharpoons}} L_nR \qquad \text{Rxn 1}$$

$$[L_nR] = [R_0] \cdot \frac{[L]^n}{[L]^n + K_d} = [R_0] \cdot \frac{[L]^n}{[L]^n + (K_A)^n} \qquad \text{Eq 1}$$

FIGURE 1.2 The Hill Equation. This equation applied the law of mass action of chemicals to the interaction between ligands and their receptors in or on cells.

of a ligand was proportional to the effect it produces. Most importantly, he hypothesized that a ligand's effect on a cell was due to an excess of ligand acting upon a limited number of receptors on/in the cell. Thus, by the Law of Mass Action, the concentration–response relationship should follow a simple hyperbolic function as had been shown previously for the adsorption of gases to a metal;[4] that is, in the presence of a limiting number of receptors, the system is saturable. This resulted in the first evaluation of concentration–response curves in context of the effect of a ligand on a living tissue. In his first description of this concept, Clark studied the ability of atropine to block the effect of acetylcholine on isolated cardiac ventricular muscle strips (Figure 1.3[15]). In this use of the Hill equation, $[L_nR]$ is replaced by effect of a specific concentration of a ligand (Yobserved) and $[R_0]$ is replaced by the maximal effect achievable by the ligand in the system being studied (Ymax). In Figure 1.3, Clark's data of the effect of atropine is reproduced from his original paper.[15] It is important to understand that Clark's theory did not model the existence of intracellular mediators (signal transducers) that could amplify the signal of receptor binding, since the existence of these mechanisms was not known at the time, and is therefore a simplistic model of receptor pharmacology, especially as it pertains to agonist responses. This led to frequent inconsistencies between the mathematical modeling of the interaction of ligand and receptor and the observed responses.

As a result of these inconsistencies, and with increasing knowledge of agonist–receptor signal transduction, more sophisticated mathematical models were developed that took into consideration a two-step model of efficacy: signal binding (formation of a physical receptor–ligand complex that serves as an activation stimulus) and signal transduction (an amplified intracellular response elicited by the newly formed complex). James Black developed the operational model of pharmacology based on this two-step process that is still used today to model *E*/*c* (agonist concentration, [A]) curves.[16] However, as we will see later, recent advances in understanding of ligand and receptor interactions as it pertains to signal transduction has led to newer and more detailed mathematical models to describe

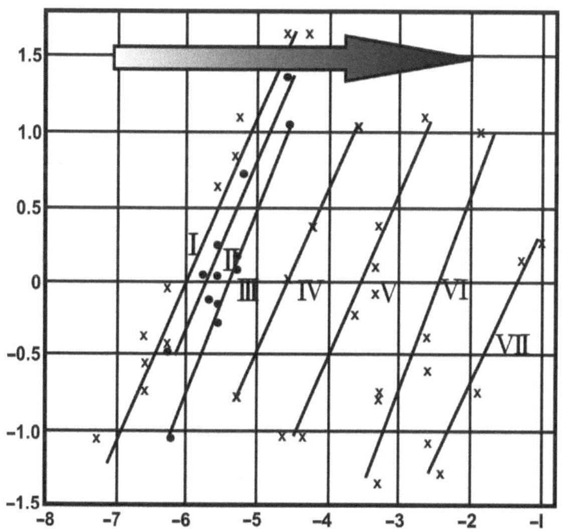

FIGURE 1.3 The effect of increasing concentrations of atropine to inhibit acetylcholine effects on isolated muscle (redrawn from[15]). Alfred Clark studied the effects of increasing concentrations of atropine on acetylcholine-induced muscular contraction in his landmark paper. He noted that as the concentration of the atropine increased (arrow) there was a rightward shift in the ability of acetylcholine to stimulate contraction.

these relationships to efficacy (see Chapter 3). These models are used today to accurately estimate the key characteristics of ligands on biological systems: potency and efficacy.

The understanding of receptors in the relationship of activation of biochemical changes within a cell was slow to develop following the development of the receptor and occupancy theories. Despite the demonstration of "receptive substances" and emerging pharmacological profiles of receptors to many ligands utilizing isolated tissues, the understanding of how the receptor–ligand complex actually induced an intracellular biochemical response was unknown. This was, in large part, due to the prevailing thought of the time that the cell was too complex a system to separate into workable pieces. This consensus prevailed until the 1950s when scientists began to develop ways to better understand the functional systems within cells. Again, development of new technologies and better techniques of isolating cells and cell components led to a new era in understanding of physiology and cell biology. This was a particularly productive time in cell biology and biochemistry, as noted by the many important discoveries of this time including metabolic enzyme pathways inside cells (e.g., Edwin Krebs) and the solution of the structure of DNA (Rosalind Franklin, James Watson and Francis Crick). It is important to note that until this period, and the research of Sutherland, the intracellular mechanisms altered by receptor binding to its ligand to elicit an effect were unknown.

Earl Sutherland was the first to isolate a signaling molecule, cyclic adenosine monophosphate (cAMP). Sutherland was interested in understanding the intracellular mechanism of epinephrine to stimulate glucose production in the liver. His initial work in this regard demonstrated that epinephrine stimulated activity of glycogen phosphorylase to produce glucose from glycogen. In a series of exquisite experiments, Sutherland and colleagues were able to isolate the membrane and cytosolic fractions of liver cells and showed that a heat stable compound was produced from epinephrine exposure to isolated liver membranes that, upon reintroduction to the cytosolic fraction, could induce activation of glycogen phosphorylase activity and increase glucose concentration. Working with his colleagues Ted Rall and Leon Heppel, Sutherland was able to isolate cAMP, show that it was the result of cyclization of ATP, and that addition of cAMP could mimic the effects of epinephrine treatment of liver cells on glucose production.[17,18]

Perhaps more importantly though, Sutherland realized the full magnitude of these observations and hypothesized the existence of "primary messengers" and "second messengers" (Figure 1.4). In Sutherland's view, the primary messenger is represented by the ligand binding to its receptor and the second messenger is represented by the activation or liberation of an intracellular mediator that was responsible for the observed biochemical and functional changes in the cell. These observations would ultimately give rise to the understanding and study of signal transduction. However, Sutherland's model of receptor signaling did not explain how the second messenger (in this case cAMP) was activated following receptor–ligand binding. In their view, Rall and Sutherland envisioned that each receptor not only served the purpose as a ligand binder but also as the liberator of the intracellular mediator; in this case cAMP.[19–21]

The term "signal transduction" was first coined by Martin Rodbell who was instrumental in uncovering the fundamental aspects of transferring the signal of ligand binding to its elicited biological response.[22] Rodbell utilized the concepts championed by Sutherland and adapted them based upon his unique view of cells. Rodbell viewed living cells and tissues as complex, natural computers that were comprised of discriminators, transducers, and amplifiers. Receptors represented the discriminator and signaling molecules such as cAMP represented the amplifier (Figure 1.4). The existence of a transducer molecule that lay between the activation of adenylate cyclase and the bound receptor had been implied. The adipocyte with which Rodbell was studying responded to several different hormones and receptors that could induce production of cAMP. If each receptor was the liberator of the signaling molecule, cAMP, then the response to exposure of the adipocytes to multiple hormones simultaneously should be additive. Rodbell's group had shown it was not additive, suggesting that each receptor could activate a single pool of intracellular adenylate cyclase.[20,21,23] Through Rodbell's careful and elegant studies, he was able to identify the first transducer, which he termed G proteins due to their activity which required guanine triphosphate (GTP). In a series of seminal papers published in 1972, Rodbell demonstrated biochemically the existence of a transducer molecule that lay between the activated receptor and the stimulation of cAMP production by adenylate cyclase.

The discovery itself was serendipitous, in that Rodbell was studying the ability of ATP to uncouple the event of ligand binding to liberation of cAMP. In the process he also studied the ability of other naturally occurring nucleotides to do the same and found that GTP could uncouple signaling at concentrations three orders of magnitude less than ATP. He assumed correctly that the ability of his preparation of ATP to uncouple cAMP production must be due to contamination of the ATP with minute levels of GTP. Based upon his data, he postulated that the receptor–ligand complex stimulated activation of another membrane-associated protein that, in turn, activated adenylate cyclase to convert ATP into cAMP. The isolation of the first G protein from cell membranes was identified a few years later by Alfred Gilman who was strongly influenced by Rall and Sutherland.

Gilman and his colleagues utilized cultures of a mutant immortalized cell line, the S49 cells thought to be devoid of adenylate cyclase (so-called cyc(-) cells). In their experiments, they attempted to study the interactions between the adrenergic receptor and adenylate cyclase by reconstituting receptor-mediated activation of adenylate cyclase activity through combining membrane protein extracts between cyc(-) cells that possessed or didn't possess receptors. Through their careful design of complex and technically challenging

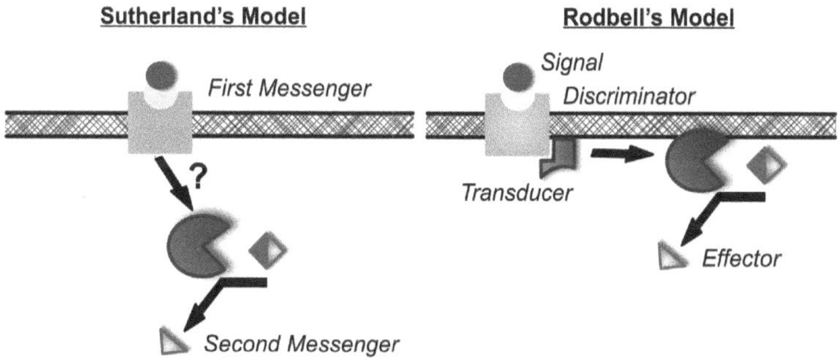

FIGURE 1.4 A comparison of Sutherland's and Rodbell's models of receptor-activated signal transduction. Sutherland's model isolated the first signaling molecule, cAMP, but did not understand that a transducing protein lay between the receptor and its production. Rodbell identified the existence of G proteins and envisioned receptor signaling similar to a computer where the ligand is the signal, the receptor is a discriminator, the G protein is the transducer, and cAMP is the effector.

experiments, Gilman and colleagues ultimately observed the existence of an additional membrane-bound protein that was required for receptor-activated adenylate cyclase activity and which required GTP.[24] Subsequently, they were able to also isolate and purify the subunits that constitute the G_s G protein. Interestingly, it was Rodbell who also hypothesized that receptors may have the ability to activate multiple signaling pathways simultaneously.[25–27]

The true structure of the G protein-coupled receptor itself would not be determined until 1986 when Richard Dixon and Robert Lefkowitz cloned and expressed a functional β_2-adrenergic receptor (epinephrine receptor, a GPCR), revealing its serpentine nature containing seven transmembrane domains, an extracellular amino-terminus, and intracellular carboxy-terminus.[28,29]

As noted earlier, the identity of signaling pathways elicited by both G protein-coupled receptors and ligand-activated ion channels occurred contemporaneously with initial important insights occurring in the late 1950s which gained momentum through the next decade until the elucidation of their signaling molecules in the 1970s. Interestingly, identification of the signaling molecules or initial steps following receptor activation preceded the identification of the receptor protein/gene itself.

MECHANISMS OF RECEPTOR FUNCTION

It is important to note that at the time of the elucidation of G proteins, Rodbell and his colleagues knew that the signal of receptor binding to ligand could be elicited by a single effector (cAMP). It is also important to note that these theories of receptor signal transduction applied to only what we refer to as G protein-coupled receptors. We now know that there are other receptor classes (Figure 1.5) and that each class has its own general mechanism for stimulating the production of effectors in order to alter cellular function/behavior. These include receptors that act as transcription factors (nuclear receptors), receptors that have endogenous tyrosine kinase activity (growth factor receptors), receptors that do not have endogenous enzymatic activity but stimulate tyrosine kinase phosphorylation cascades (cytokine receptors), and those that act as ion channels.

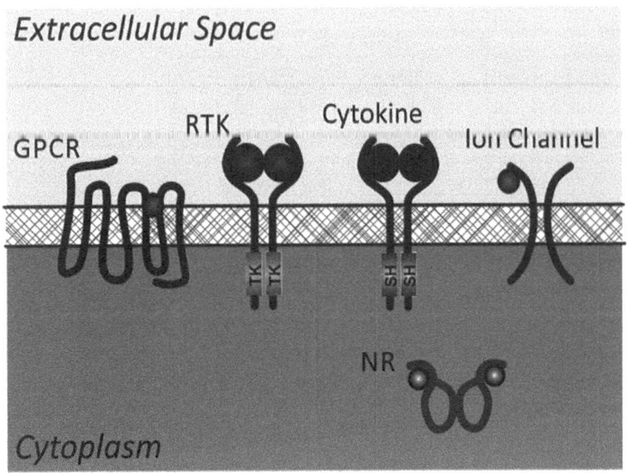

FIGURE 1.5 Receptor classes discussed. The general structure and cellular localization of the different classes of receptors discussed in this chapter are shown. Interaction of each receptor class with its particular ligand types is also shown.

GPCRs were originally thought to signal primarily through adenylate cyclase (cAMP) and phospholipase C (inositol triphosphate, IP_3), but with the identification of other G proteins it is well recognized that these receptors can also activate other signaling pathways. G proteins are heterotrimeric proteins comprised of a determinant α subunit and β and γ subunits that are shared with other $G\alpha$ subunits (Figure 1.6). There are multiple isoforms of G_β and G_γ subunits but each isoform is not expressed in every cell; they have cell-specific gene expression patterns.[30] In addition to G_s (activation of adenylate cyclase), G_i (inhibition of adenylate cyclase) and G_q (activation of phospholipase C), there have been additional G proteins identified over the years including G_o, G_{11}, G_{12}, G_{13}, G_{15}, G_{16}, G_t, G_{olf}, and G_{gust}. Based upon their primary sequence homology and activities, they have been grouped into classes.[30] G_o primarily inhibits adenylate cyclase, similar to G_i, and along with G_i proteins makes up the $G_{i/o}$ class. G_{11} is similar in sequence and activity to G_q and therefore these proteins comprise the $G_{q/11}$ class of G proteins.

Table 1.2 provides a very general outline of the different G protein classes and their activities. In addition to inhibition of adenylate cyclase, some members of the $G_{i/o}$ class of G proteins also activate phosphodiesterases that act to metabolize cAMP. Another member of that class also can inhibit voltage-dependent cation channels, thereby regulating the polarity of the cell membrane. The $G_{12/13}$ class of G proteins acts to stimulate Rho and ERK phosphorylation cascades that are well recognized for modulating cellular growth and proliferation.

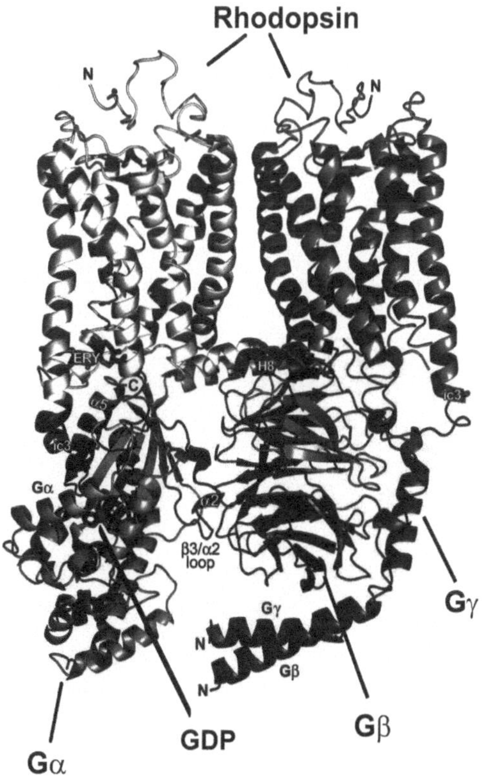

FIGURE 1.6 The crystal structure of a representative GPCR (rhodopsin) in association with G protein. The figure illustrates the areas of interaction between a dimer of rhodopsin and the various subunits of the G protein heterotrimer (α, β and γ) bound to guanine diphosphate (GDP). Note the interactions between intracellular loop 3 of one of the GPCR monomers (IC3) with alpha helix 5 of the G_α subunit and its proximity to the bound GDP. The IC3 of the other GPCR monomer makes contact with helices on the G_γ subunit.

TABLE 1.2 Activities of G Proteins

G Protein Class	# of Isoforms	Primary Signaling Effector(s)[30]
G_s	3	Activation of adenylate cyclase activity, liberates cAMP Calcium channel activation, increased intracellular calcium c-Src tyrosine kinase activation, protein phosphorylation
$G_{i/o}$	8	Inhibition of adenylate cyclase activity, decreases cAMP Activation of phosphodiesterases, decreases cAMP Inhibits voltage-dependent cation channels, c-Src activation
$G_{q/11}$	4	Activation of phospholipase C, liberates inositol phosphate RhoGEF activation, protein phosphorylation
$G_{12/13}$	2	Activation of Rho and ERK kinase pathways, protein phosphorylation
G_T/g	2	Activation of phosphodiesterases
$G_{\beta\gamma}$	6β 14γ	Activation of β-catenin and ERK kinase pathways, receptor internalization

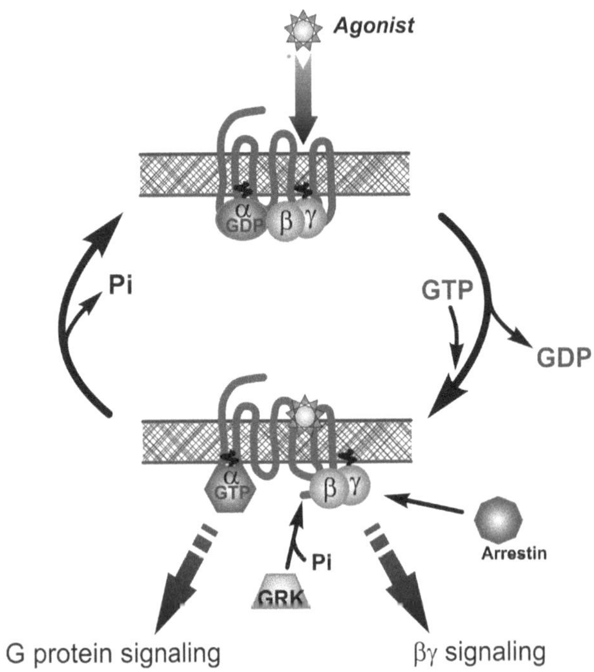

FIGURE 1.7 G protein activation by a GPCR in the presence of ligand. GPCRs act as GTP exchange factors. In the absence of ligand, the receptor binds the heterotrimer of G protein subunits (bound to GDP) through their intracellular loops; the interaction between the G protein subunits does not activate signaling since the $\beta\gamma$ subunits act to inhibit exchange of GDP for GTP in the G_α subunit. When ligand interacts with the receptor, this causes conformational changes to be translated to the G protein heterotrimer that results in exchange of GTP for GDP on the α subunit, thus liberating active α subunit from the $G_{\beta\gamma}$ subunit. Each of the liberated subunits activates their own signaling cascades (Table 1.2). Activation of these cascades also activates G protein regulated kinases (GRK) that phosphorylate the GPCR, thus recruiting the desensitizing and signaling protein β-arrestin. Reassembly of the heterotrimer occurs following hydrolysis of GTP by G_α and is greatly enhanced by the presence of GTPase accelerating proteins (GAPS) such as RGS (regulator of G protein signaling) proteins.

The steps to activation of G proteins upon receptor activation have been described in detail and numerous crystallographic determinations have been made of various G proteins in resting and activated states.[31] G proteins act as GTP exchange factors, whereby they associate with GPCRs as an intact heterotrimer comprised of one G_α subunit, one G_β subunit, and one G_γ subunit bound to guanine diphosphate (GDP)[19] (Figure 1.7). The signal of ligand activation of the receptor induces the exchange of G_α-bound GDP for GTP (Figure 1.7). This exchange results in the breakdown of the heterotrimeric complex of G protein into two subunits: the free GTP-bound G_{α} subunit and the $G_{\beta\gamma}$ dimer. Since the protein–protein interaction between the β and γ subunits is so strong, and their proper protein folding that occurs during protein synthesis requires this interaction,[32] they are generally never found as monomers in the cytoplasm under physiological conditions.[31] Interaction of the GTP-G_{α} with effector enzymes associated with the local signaling complex in the cellular membrane requires hydrolysis of GTP to GDP. Once this occurs, the GDP-G_{α} subunit then recycles back to its original heterotrimeric complex with the $G_{\beta\gamma}$ dimer.

In addition to activation of signaling cascades by activated Gα subunits following GPCR ligand binding, the $G_{\beta\gamma}$ subunits also activate other signaling pathways. $G_{\beta\gamma}$ subunits can activate ion channels in the cell membrane, such as those for potassium, sodium, or calcium. In addition, activated $G_{\beta\gamma}$ subunits also can stimulate ERK kinase pathways. They can also activate pathways more commonly associated with G_q-stimulation, through activation of phospholipase and phosphoinositide-3-kinase enzymes.

Unlike GPCRs, growth factor receptors utilize their own innate tyrosine kinase activity to initiate the signaling cascade. These receptors generally bind large protein ligands. For this reason, they are also commonly referred to as receptor tyrosine kinases (RTKs). Both growth factor and cytokine receptors are comprised of an extracellular, ligand binding domain (LBD), a single membrane spanning domain, and an intracellular domain (Figure 1.8). In the case of growth factor receptors, the intracellular domain contains a protein sequence that is a tyrosine kinase, known as the kinase domain (TKD). The TKD contains numerous tyrosine amino acids that are critical to activation of the tyrosine kinase activity of the receptor. Growth factor receptors form dimers upon ligand binding. Dimerization of some growth factor receptors involves the ligand itself acting as a bridge between two receptor molecules. However, this is not the case for all receptors and some growth factor receptors dimerize following ligand binding, but the ligands are not directly involved in the protein–protein interactions stabilizing the receptor complex.

Some growth factor receptors are also known to oligomerize in the absence of ligand, suggesting that, at least for some receptors of this class, the ability to form complexes is an inherent property. This has also been found for other receptor classes including GPCRs. Receptor activation is initiated by ligand binding with a stoichiometry of 2:2; that is, two ligands stabilize a dimer of receptor molecules. The signal of ligand binding to RTKs induces activation of the tyrosine kinase domain of one of the receptors that subsequently phosphorylates the kinase domain of the other receptor[34] (Figure 1.9). Under resting conditions (no ligand present) the TKD of the receptor is under the influence of a cis auto-inhibitory mechanism that is induced by the intracellular juxtamembrane region of the receptor. Ligand interaction with the extracellular domain relieves this inhibitory influence allowing for a sequential auto-phosphorylation of key tyrosines within the TKD.

Furthermore, phosphorylation of the TKD of some of these receptors greatly increases the catalytic activity of that TKD, thereby enhancing their ability to phosphorylate the adjoining receptor TKD. The act of phosphorylation of the kinase domains of the receptor dimers then provides a newly formed site of assembly for the recruitment and subsequent activation of other signaling proteins. The recruited proteins have common structural motifs within them that recognize the phosphorylated receptors. Both Src homology domains (SH domain) and phospho-tyrosine-binding domains (PTB domain) recognize the phosphorylated tyrosine within the receptor intracellular region and associate with the receptor in a complex.[35,36]

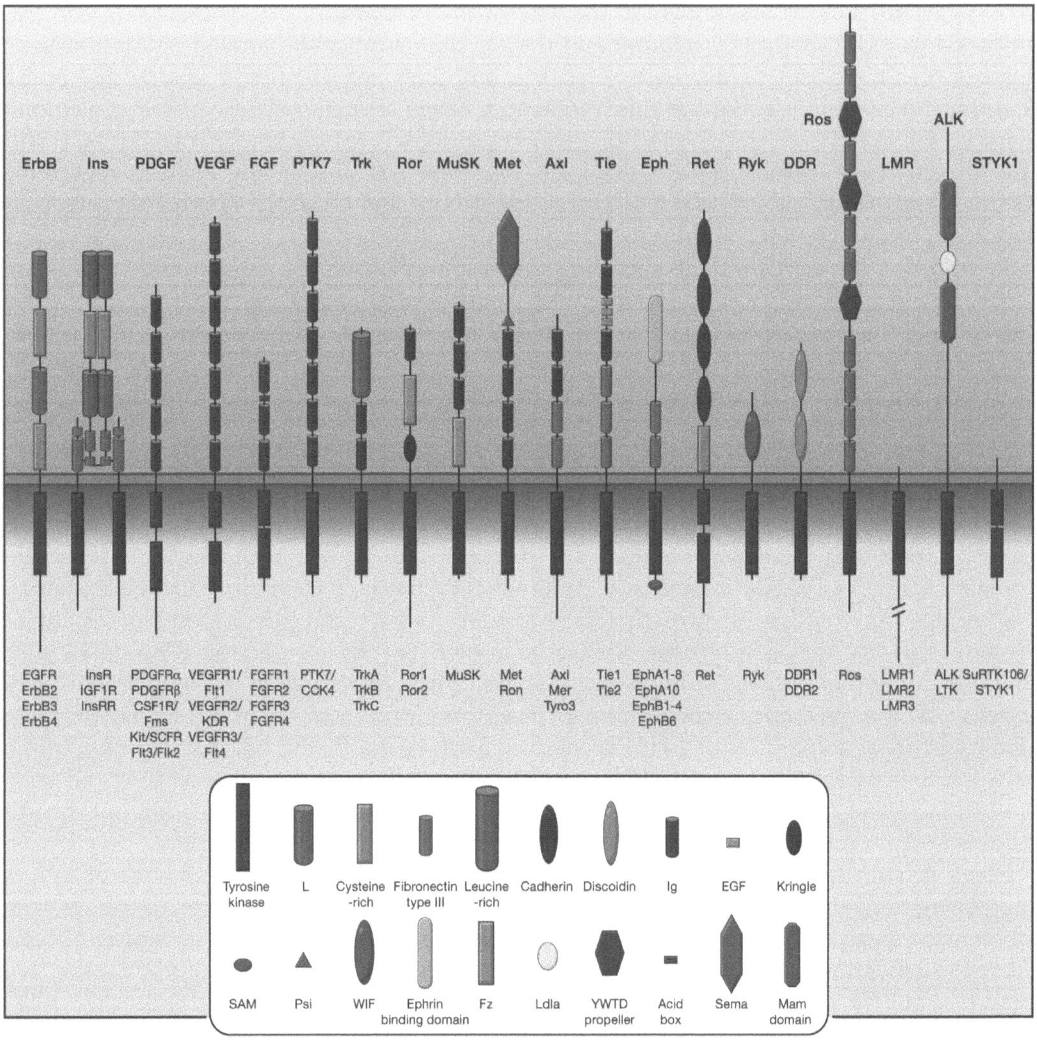

FIGURE 1.8 Comparative structures of the RTK family of receptors.[33] Overview of the general structural motifs present in the RTK family demonstrating the tremendous variability of structure exhibited by this class of receptor. The general structure of the monomers is shown, including the detail of the variable structural protein motifs found within the extracellular ligand binding domains. The various domains are shown as different shapes within each structure. In addition, the intracellular TKD of each receptor is highlighted. Republished with permission from Elsevier Limited, Oxford.

In this way, the ligand-bound, phosphorylated growth factor receptor behaves as a scaffold onto which is built a large macromolecular complex of signaling proteins (Figure 1.9).

Docking proteins associated with activated RTKs include FGF receptor substrate 2 (FRS2α), insulin receptor substrate 1 (IRS1), and the Grb2 (Growth factor receptor bound protein 2)-associated binder (Gab1). These docking proteins contain domains at their amino terminus that allow them to associate with the plasma membrane, thus allowing them to be

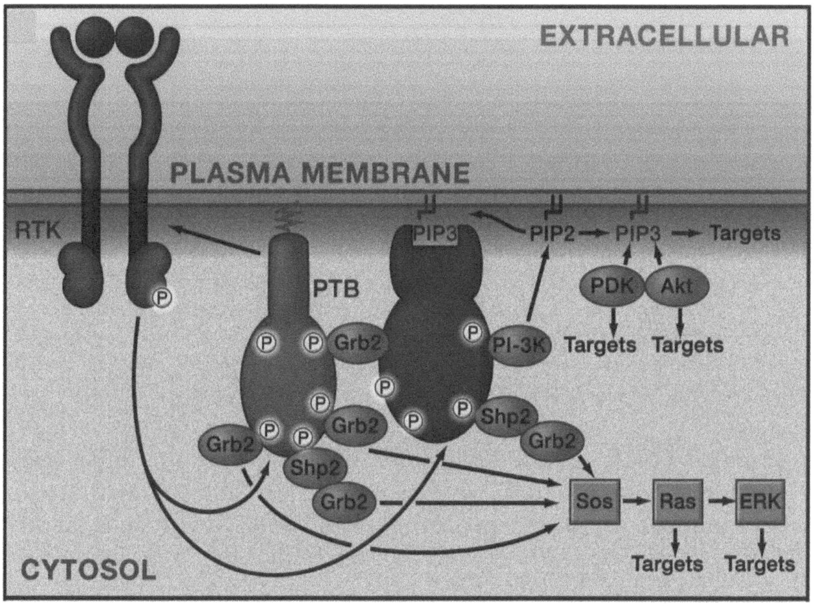

FIGURE 1.9 Signaling mechanism of RTK receptors.[33] The mechanism of ligand-dependent activation of RTK phosphorylation cascades. Ligand stabilization of a receptor dimer leads to activation of the tyrosine kinase activity of the receptor's intracellular TKD domain. Phosphorylation of closely associated docking proteins such as Grb2 and Shp2 acts to recruit other integral membrane protein signaling proteins, such as PTB and PI3K. Recruitment of PTB and PI3K leads to their activation via phosphorylation and this signal is translated by the PTB and PI3K proteins via their phosphorylation of signaling effectors within the cytoplasm (e.g., Sos, Ras and ERK). Reproduced with permission from Elsevier Limited, Oxford.

localized in close proximity to activate receptors (Figure 1.9). They also contain multiple tyrosine residues which are phosphorylated by the bound, activated RTK and are recognized by other signaling proteins such as Grb2, phosphoinositide-3 kinase (PI-3K), and Src homology phosphatase 2 (SHP2) (Figure 1.9). Phosphorylation of FRS1α induces association with both Grb2 and SHP2. These proteins then act to recruit another docking protein, Gab1 that, in turn, recruits additional signaling proteins (e.g., PI-3 K) to its surface via phosphorylated tyrosines. Other proteins can also be recruited.[33] Once bound to the docking site, these recruited proteins become activated to stimulate activity of other enzymes within the cytosol such as Sos, Ras, and ERK. These enzymes induce phosphorylation signaling cascades that regulate numerous biochemical activities of the cell including metabolism, protein synthesis, and gene expression. We can think of the activated RTKs as foci of signaling molecules recruited to interact via recognition of phosphorylated domains within their structures.

Unlike other receptor classes, the nuclear receptors (NRs) are not localized to cellular membranes, rather they are found intracellularly within the cytoplasm or nucleus. Due to their localization, they bind membrane diffusible ligands such as steroids, retinoids, fatty acids, and eicosanoids. Nuclear receptors have a general structure comprised of several domains that perform specific functions of the receptor (Figure 1.10 Schematic of general NR domains). The LBD and a DBD are the best conserved domains across the family of NRs. These two domains are connected by a flexible hinge region. The N-terminus and C-terminus

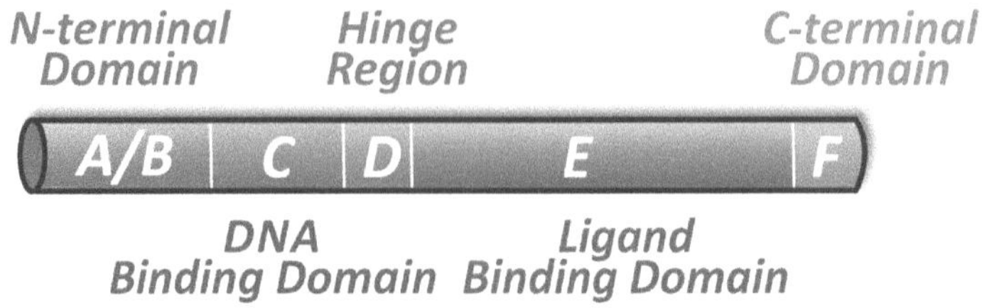

FIGURE 1.10 General structural domains of nuclear receptors. Nuclear receptors are comprised of a DNA binding region and ligand binding region that are connected by the hinge region. NRs generally function as dimers. These dimers, once formed in association with co-activators or co-repressors, bind to promoter regions on genes and regulate gene transcription. The dimers can be either homodimers (e.g., estrogen receptor [ER]) or heterodimers (e.g., RXR and retinoid activated receptor [RAR]). Depending upon the dimer formed, the interactions between the monomers are dependent upon the two partners and have a direct role in the effect of the receptors.

are variable across the family of NRs. As its name implies, the LBD is responsible for recognition and interaction with the ligand. The transcriptional activation and DBDs are responsible for binding and induction of gene promoters. The dimerization domain is the site of interaction between NRs. As can be inferred from their structural domains, NRs recognize membrane permeable ligands and dimerize. Upon activation they interact with specific sequences in the promoter regions of genes to induce or repress gene expression directly as part of an activating or repressing protein complex (Figure 1.10).

Similar to membrane-bound receptors, the NRs also utilize interactions with other proteins to fine-tune their functional activity. In this case, NRs interact in a specific fashion with additional proteins that help to assist them in forming the active transcriptional regulatory complex. These proteins, termed co-activators, recognize the ligand-bound receptor (Figure 1.11). Interactions between the co-activator and the NR induce specific structural changes, such that the complex specifically recognizes and binds to sequences of DNA in the promoters of the regulated genes. Once bound to DNA, co-activator-bound NRs recruit the transcriptional machinery required to drive production of mRNA by RNA polymerase. However, NRs can also repress gene transcription. In this case, the proteins associated with the NR are called co-repressors and in complex with the NR act to inhibit gene transcription once bound to their cognate promoter elements on DNA.

One key feature of co-activators and co-repressors is that they are expressed in a cell-specific fashion. Indeed, the particular repertoire of co-activators and co-repressors expressed in a given cell is specific to that cell type. Pharmacologically and therapeutically, this has been leveraged to produce synthetic ligands that act in a cell or tissue specific manner. Numerous synthetic ligands to NRs have been created that can induce tissue-specific gene regulation and in some cases these ligands can behave as agonists in one tissue and antagonists in another. Such tissue-selective receptor modulators are commonly prescribed drugs in some cases (e.g., tamoxifen, raloxifene, bazedoxifene). This has been utilized for the development of selective estrogen receptor modulators (SERMs) for hormone replacement therapies and for peroxisome proliferator-activated receptor (PPAR) therapies for insulin resistance and hyperlipidemia (e.g., rosiglitazone, fenofibrate). From a pharmacological perspective, this provides the ability to fine-tune drug responses in the context of multiple tissues, thereby producing wanted agonist activity in one tissue or cell type while producing antagonist or null activity in another.

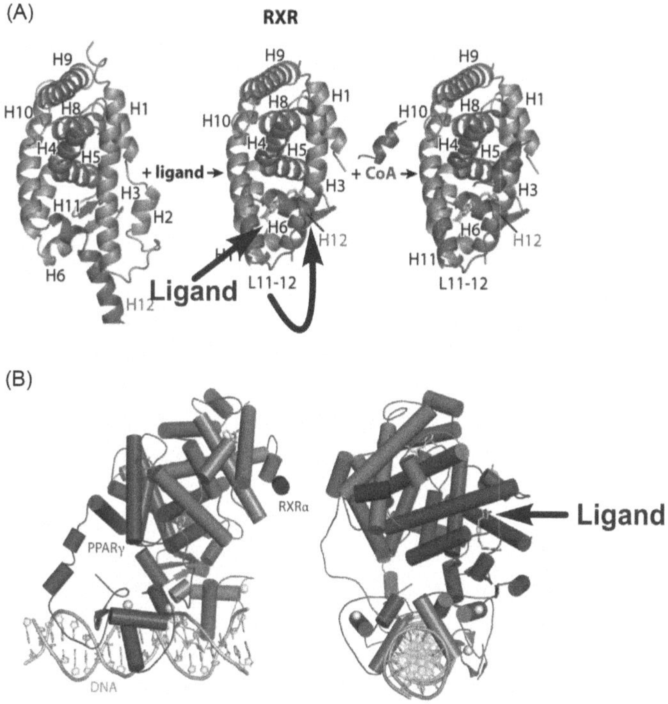

FIGURE 1.11 Crystal structures of an NR in complex with ligand and co-activator.[37] (A) illustrates the crystal structure of an NR (RXR) LBD in the absence of ligand, in the presence of ligand, and in the presence of ligand + co-activator. Note the movement of helix 12 (H12) and the subsequent juxtaposition of H12 and the ligand upon binding. This interaction then leads to the formation of a pocket for co-activator binding. (B) presents the crystal structure of an NR dimer (PPARγ and RXRα) bound to DNA. Note how the DBDs of the dimer wrap around and intercalate into the grooves of the DNA helix. *Adapted and reproduced with permission from Annual Reviews, Palo Alto, CA.*

Many of the mechanistic insights into how NRs work in the absence and presence of ligands and co-regulators have been gleaned from crystallographic analyses of isolated domains from the various NRs; however, Fraydoon Rastinejad and colleagues were able to develop crystals of an intact NR complex.[37,38] This crystal structure revealed key insights into how the NR heterodimer-co-regulator-DNA complex is arranged. They observed that the bound LBD in NRs has interaction with the DBD, thereby suggesting that not only can one achieve cellular/tissue specificity of NR activity via selecting specific co-regulator interaction but one can also actually drive DNA response element selectivity within a cell or tissue via the ligand–LBD conformation (Figure 1.11B). A similar structural relationship has been found for the retinoid X receptor and vitamin D receptor heterodimer in complex with DNA.[39] Therefore, NRs provide excellent opportunities to pharmacologically fine-tune the activity of the receptor with properly designed and directed synthetic ligands.

Ligand-gated ion channel receptors are also unique in their mode of action and signaling properties. Their main function is to modulate the balance of ions across the cell membrane. In the closed state ion channels are impermeable to the flow of ions. Each channel has its own specificity in terms of ion(s) that it is permeable to. Ion channels are found in all living cells and can be classified in several different ways. Generally speaking, ion channels are most often classified by their gating stimulus. Stimuli that induce permeability of channels to ions include membrane voltage (voltage-gated), exogenous ligands (ligand-gated), mechanical stretch (mechanosensitive), and even light (light-gated). For the purposes of this chapter we will focus on ligand-gated ion channels.

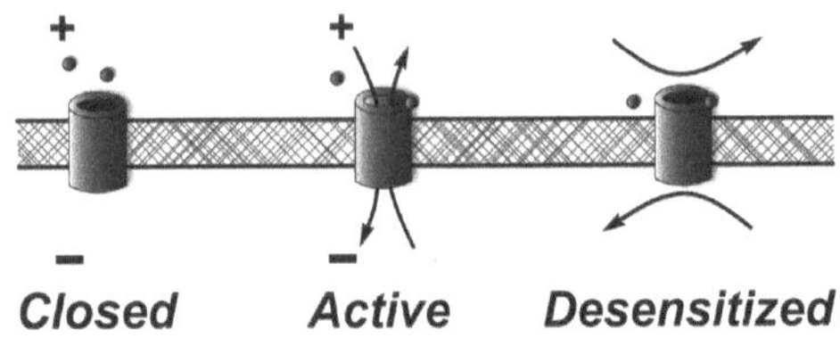

FIGURE 1.12 States of ligand-gated ion channel activation (adapted from[42]). The three states of ion channel activation: closed, active (open), and desensitized. Under resting conditions, the LGIC is closed and impermeable to the passage of ions, thus maintaining the membrane potential (depicted as + and −). Ligands bind to the channel at regions allosteric to the pore of the channel and induce activation, or opening, of the channel. This allows free passage of ions (spheres) across the membrane, thus reducing the polarization across the membrane. After prolonged exposure to ligand, the ion channel becomes desensitized and is also impermeable to ion flow.

Ligand-gated ion channels were first identified by Changeux.[40] Through his work, he identified that the receptor for acetylcholine (ACh) was an ion channel and demonstrated its sensitivity to nicotine. Some 40 years later, we now know that ligand-gated ion channels are highly and ubiquitously expressed in the central nervous system, as well as other tissues including skeletal muscle, heart, kidney, gut, the reproductive organs, etc. Ligand-gated ion channels are multi-subunit proteins and can be grouped into three distinct gene families: the trimeric cation channels of the P2X family that bind ATP, the tetrameric ionotrophic channels that bind glutamate, and the pentameric cys-loop family of receptors that bind glycine, ACh, gamma amino butyric acid (GABA) or N-methyl-D-aspartate (NMDA). These receptors can also be classified by the type of ions that they are permeable to: anion conducting ion channels (GABAR and glycine (GlyR)) and cation conducting channels (glutamate (GluR), purinergic (P2XR), serotonin ($5HT_3R$), and nicotinic AChR). For each ligand there are also multiple receptor subtypes that it can bind and the receptors can also be classified in this manner. Since each ion channel is comprised of multiple subunits, different subunit combinations brought together give rise to the various subtypes of channel associated with a given ligand.

Like other receptors, the general structure of ligand-gated ion channels can be separated into functional domains. The extracellular domain contains the ligand binding region of the receptor; the transmembrane domains of the different subunits of a given receptor come together to create the pore of the channel and the intracellular domain interacts with various cytoplasmic proteins to help regulate receptor localization and activation.[41,42]

Ion channels function to allow passage of ions across the cell membrane and to assist in regulation of electrical potential (voltage) across the surface of the cell. Under resting conditions (absence of ligand, Figure 1.12, states of ion channel open/close) the pore formed by the ion channel subunits is in a closed state, impermeable to the passage of ions. However, the action of ligand binding to the ligand-gated ion channel opens the pore allowing passage of ion across the cell membrane depolarizing the membrane potential. Prolonged exposure of the receptor to its ligand induces a state of desensitization, in which the receptor pore becomes no longer permeable to ions despite the continued interaction with ligand. Desensitization is a key physiological mechanism that has evolved to protect cells from prolonged or overstimulation by a given ligand. It is not unique to ligand-gated ion channels but has been described in one way or another for all receptor classes. In his original studies of the nicotinic ACh receptor, Changeux described two phases of current flow following exposure of the receptor to ACh: activation, associated with a rapid inward current, followed over time by desensitization or a progressive decline in current in the continued presence of ACh. This led to a basic three-state model for ligand-gated ion channel function: inactivation, activation and desensitization (Figure 1.12).

BASIC PRINCIPLES OF RECEPTOR PHARMACOLOGY

Before we can discuss the more detailed effects of ligands (both natural and synthetic) at their given receptor, we must first have an understanding of the key concepts associated with receptor pharmacology. This is important since ultimately we can utilize knowledge of receptor pharmacology to develop synthetic ligands to act as therapeutics and understand physiology. It is important to note, however, that not all therapeutics interact with receptors. For instance, some drugs are designed to inhibit enzymes (e.g., warfarin, statins) or incorporate into macromolecules (e.g., taxol, doxorubicin), while others are used as chelating agents (e.g., diferiprone) or osmotic agents (e.g., mannitol). Discussion of all of these manifestations of therapeutics is outside the scope of this text, therefore we will focus our discussions in terms of receptor pharmacology.

As alluded to earlier, the intensity of action of a ligand depends upon the concentration of ligand present at and bound to its receptor. This was described effectively by Hill through his derivation of the Hill equation (Figure 1.2) and, as we noted, is based upon equilibrium of chemical reactions. Table 1.3 outlines some key terms associated with receptor pharmacology. It is important to understand these terms as they are the fundamental lexicon that we will use to discuss pharmacological basics.

The essence of pharmacology lies in understanding the effect(s) of a given ligand on a biological system. This can be studied at the molecular, biochemical, cellular, tissue, and whole animal levels. Direct interaction of a ligand with its receptor can be studied molecularly through the use of binding assays in which the ability of a ligand to displace a known, labeled agonist is used. These types of assays have been used to great effect to understand the details of the interactions between a ligand and the receptor. They have historically used a radiolabeled form of the naturally occurring ligand but can also utilize synthetic compounds. Radioactive labels that are placed onto the ligand of interest are usually tritium (H^3), iodine (I^{125}), carbon (C^{14}), or sulfur (S^{35}). These experiments can be performed on isolated cell membranes or cytosol, whole cells, or even tissues and are done so at or near equilibrium of the radiolabeled ligand. The end point of these assays is to observe the ability of the labeled ligand to specifically bind to a preparation known to contain the receptor of interest. Binding of the radiolabeled (hot) ligand can be measured using traditional scintillation counters. However, recent advances in development of fluorescent labels have allowed for newer non-radioactive binding assays that utilize sensitive fluorescent readers to measure the bound, labeled ligand.

Regardless of the method used to measure the label, the principles are the same. The general assumption in these types of studies is that the assay matrix (e.g., cell membranes or tissue) contains a finite number of receptors. This implies that the matrix has a saturable number of binding sites such that upon increasing concentrations of ligand, the availability of binding sites decreases. This concept can be seen in Panel A of Figure 1.13 in which bound ligand is plotted against the concentration of total ligand added to the assay. Upon incubation of the receptors with increasing concentrations of labeled ligand more label is bound to the matrix (Total binding, the sum of labeled ligand bound to unsaturable sites

TABLE 1.3 Key Terms Used in Pharmacology

Term	Definition
Agonist	A ligand (natural or synthetic) that binds a receptor and activates its biological effect on the cell. In the case of synthetic ligands these often mimic the effect of the natural ligand
Antagonist	A ligand that binds a receptor and blocks its ability to respond to an agonist
Inverse agonist	A ligand that reverses the basal (or constitutive) activity of a given receptor leading to an observed opposite effect
Partial agonist	A ligand that activates a receptor's biological effect but not to the same degree as the physiological ligand
Super agonist	A ligand that activates a receptor's biological effect to a greater degree than the physiological ligand
Efficacy	The degree to which an agonist or antagonist induces the biological effect of a receptor
Potency	The concentration of a ligand that induces a given degree of efficacy. Most commonly the IC_{50} or EC_{50} is used to determine and compare ligands
EC_{50}	The concentration at which an agonist induces its half-maximal effect
IC_{50}	The concentration at which an antagonist induces its half maximal effect
B_{max}	The estimated number of total available receptors (binding sites) present
E_{max}	The maximal effect induced by a ligand in a given assay
Orthosteric site	The site on the receptor that the naturally occurring agonist binds
Allosteric site	A site on the receptor that some ligands bind that is not the orthosteric site
Affinity	The chemical attraction of ligand and receptor for each other
Dissociation constant (K_d)	A measure of affinity of a receptor for a ligand
Specific binding	In equilibrium binding assays, the amount of measurable binding attributed to the receptor–ligand complex
Non-specific binding	In equilibrium binding assays, the amount of measurable binding due to interactions other than those involving ligand and receptor

in the matrix [low affinity] + labeled ligand bound to saturable sites [high affinity] in the matrix). However, it does not follow a strict linear relationship; there is a concentration of free ligand that is reached in which less bound ligand is measured than would be assumed by a linear relationship. This is the result of the binding of the ligand to saturable "Specific" sites (see Table 1.3), which follows a similar pattern with a rapid increase (calculated as the difference between the Total Binding signal − the Non-specific binding signal) measured before reaching a plateau in which there is obtained a concentration of free ligand that no longer induces an elevation in specific bound ligand.

In contrast, non-specific binding in these assays follows a linear relationship. The direct, saturation binding assay is the cornerstone for studying receptor interactions with ligands and the relationships derived from it are used to estimate the equilibrium dissociation

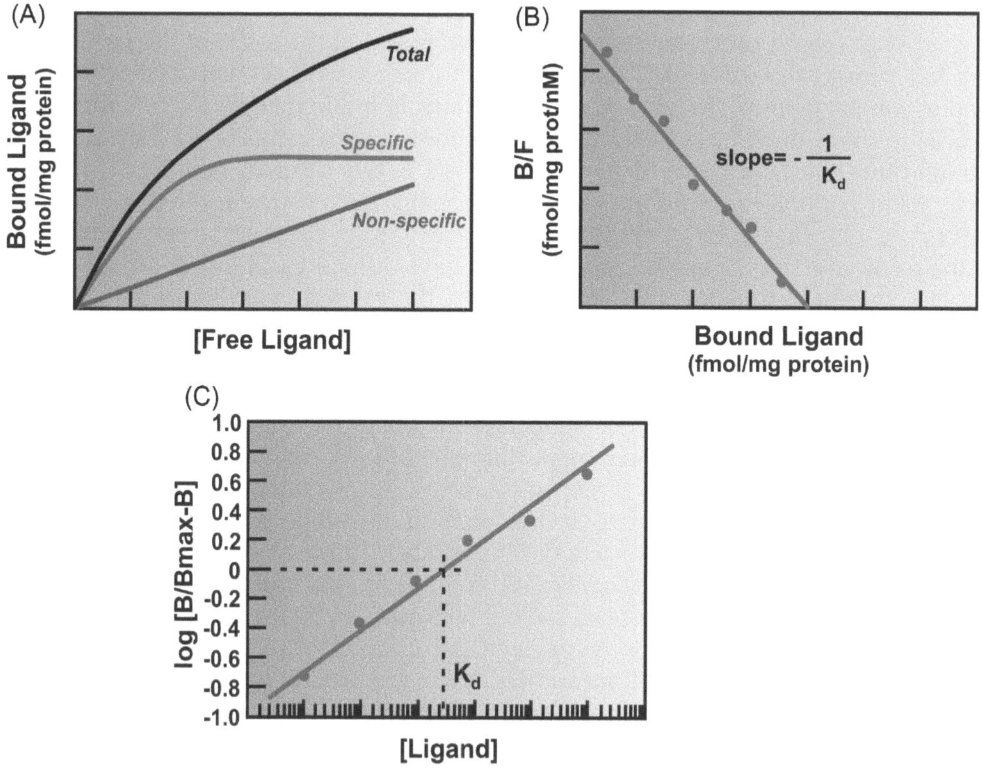

FIGURE 1.13 Equilibrium binding curves. (A) schematically depicts typical curves from an equilibrium binding experiment. As the concentration of radiolabeled free ligand is added to the experiment, the amount of Total binding is increased. Total binding is the sum of specific and non-specific binding of the radiolabeled ligand. Specific binding follows a simple hyperbolic function while non-specific binding is linear. (B) illustrates hypothetical analysis of the data in (A) by the method of Scatchard. Plotting the relationship between the bound/free ligand against the concentration of bound ligand produces a line that can be fitted. The negative reciprocal of the slope of the line estimates the equilibrium dissociation constant (K_D). The total number of binding sites is estimated as the x-intercept. (C) illustrates plotting of the data from (A) using the method of Hill. The slope of the line generated estimates cooperativity of the system. The equilibrium dissociation constant can be estimated as the concentration of ligand at which the log[B/Bmax − B] is zero.

constant (K_d) and receptor binding density (B_{max}). The critical parameters of Total and Non-specific binding are measured by utilizing tubes containing increasing concentrations of labeled ligand in the presence (non-specific) or absence (total) of an excess of an unlabeled competitor. This competitor can be the unlabeled ligand itself or another unlabeled compound thought to bind to the same site on the receptor.

Figure 1.13B and C details some of the common analyses from competition binding experiments. By studying the relationship of ligand binding to receptor we can begin to get an understanding of the details of their interaction and appreciation for the biophysical associations between the two. Figure 1.13B depicts the Scatchard plot proposed by George Scatchard[43] which uses data from the equilibrium binding experiment to plot the ratio of bound/free ligand per mg protein versus the concentration of bound ligand per mg protein. By fitting the resultant curve by linear regression, the equilibrium dissociation constant (K_D) can be estimated as a negative reciprocal of the slope of the line. This constant has proved to be a critical parameter that describes the attraction or affinity of the receptor for the ligand. The total number of available binding sites (B_{max}) is estimated by the x-intercept of the line. In its most simplistic form, the data points will be aligned along a straight line that is easily fit by linear regression. In this case, the binding data suggest a single set of non-interacting sites. However, in many circumstances the data points align in a curved plot that is fitted by non-linear regression. In these instances the data points are the result of two or more binding sites that have different affinities for the ligand or of two sites that have a cooperative association. If the cooperativity is positive (binding of one site promotes the binding at a second site) the curve will turn upward; if the cooperativity is negative the curve will turn downward.

The Hill plot[44] (Figure 1.13C) mentioned earlier can also be utilized with equilibrium binding data in order to determine cooperativity. The Hill plot is drawn by plotting the log of labeled ligand concentration versus the ratio log $[B/(B_{max} - B)]$. [Ligand] represents the concentration of free ligand, B represents the amount of ligand bound, and B_{max} is the maximum number of binding sites. In this analysis the slope of the line fitted by linear regression is called the Hill coefficient or Hill number. In the absence of cooperativity, the Hill coefficient is 1.0. Positively cooperative systems have a Hill number greater than 1.0 while negatively cooperative systems have a Hill number of less than 1.0. Using the Hill plot, the K_D can be estimated by the value of the abscissa when log $[B/(B_{max} - B)] = 0$. In practice, both the Scatchard analysis and Hill plots are used in understanding the interactions of ligands with their receptors. The Hill plot is used extensively to understand the interaction of synthetic compounds with receptors.

By adding an unlabeled competitor that is not the natural ligand, we can begin to understand the pharmacological nature of the compound at the receptor. For instance, by performing receptor binding assays we can divulge whether or not the compound is an orthosteric or allosteric ligand (Table 1.3). An orthosteric ligand will generally be able to displace the labeled, natural ligand from binding to its receptor since the two ligands will be competing for binding at the same site on the receptor. However, not all ligands bind to a receptor at the same site. Ligands that bind to sites other than the natural ligand are said to be allosteric (see Conformational Dynamics and Biased Signaling section below for more detail).

In addition to studying the direct interaction of the ligand with its receptor, we can also study the biological effects of receptor activation. Indeed, it is these analyses that help us understand the potential therapeutic impact of synthetic ligands. Measuring the biological impact of ligands on a biological system allows us to see the phenotypic effects of a ligand that cannot be derived from binding experiments since binding relationships simply tell us that the two entities interact. Biological assays can utilize any measurable effect of the ligand on the cell, tissue, or animal. The most efficient means of studying the pharmacology of ligands is performed using cellular or biochemical assays. Most often, these are signal transduction assays, utilizing the known signaling mechanisms of the receptor to assess the effect of the ligand on the biological system. However, any reasonably consistent end point of a cell, tissue, organ, or animal that is known to be associated with receptor activation can be utilized to understand the pharmacology of a ligand. Indeed, the nature of the effect of a ligand on the biological system of interest is of paramount importance.

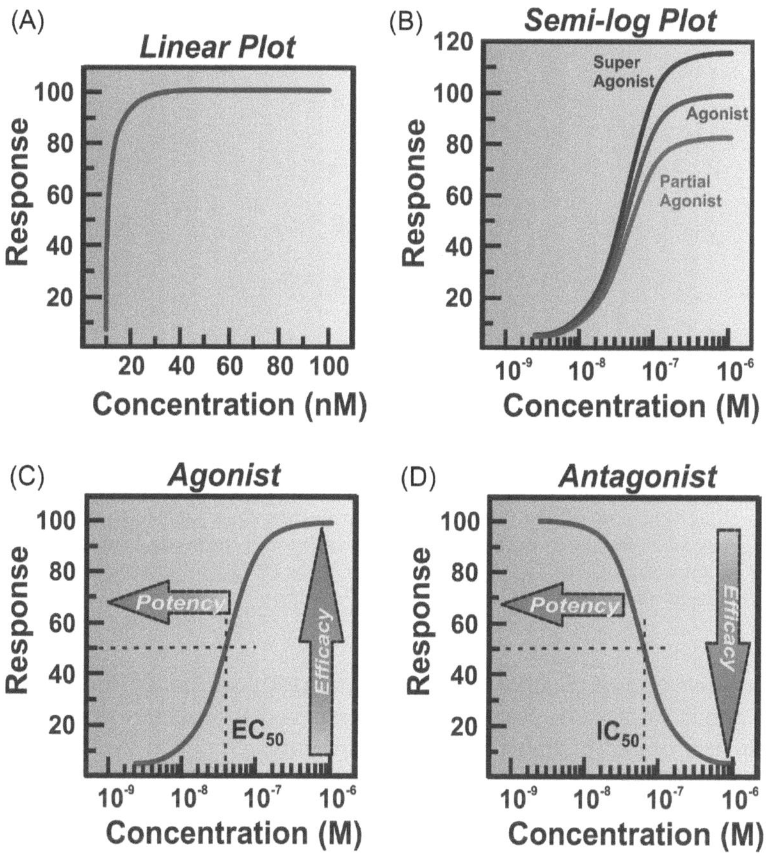

FIGURE 1.14 Concentration–response curves for agonists. Plotting of CRCs (Response vs Concentration) in a linear plot produces a hyperbolic curve. Plotting the data in a semi-log plot reveals the typical sigmoidal curve. Shown are a full agonist, a partial agonist, and a super agonist. Both agonists and antagonists are characterized functionally by potency and efficacy. The arrows in (C) and (D) illustrate that for agonists improved efficacy is related to a larger response, while antagonist efficacy is improved with greater inhibition of the agonist response. For both types of compounds potency is improved as you lower the concentration required to induce a half-maximal effect (EC_{50} for agonist, IC_{50} for antagonist).

Figure 1.14 depicts some of the key aspects of understanding the pharmacological effects of ligands. Synthetic ligands that mimic the effects of the naturally occurring ligand, for example on signal transduction, are defined as agonists. Compounds that block, or antagonize, the effects of a naturally occurring ligand are said to be antagonists. The nature of a compound as an agonist or antagonist is determined by studying the effects of the compound on the biological end point in a concentration–response experiment. With most ligands the biological effects elicited will occur over two to three orders of magnitude of concentration. If one plots the concentration versus effect of an agonist on a linear scale, the result is a hyperbolic curve as shown in Figure 1.14A. As you can see, not much detail is evident in such a plot. However, by plotting these data in a semi-log fashion, where the concentration of ligand is plotted on a

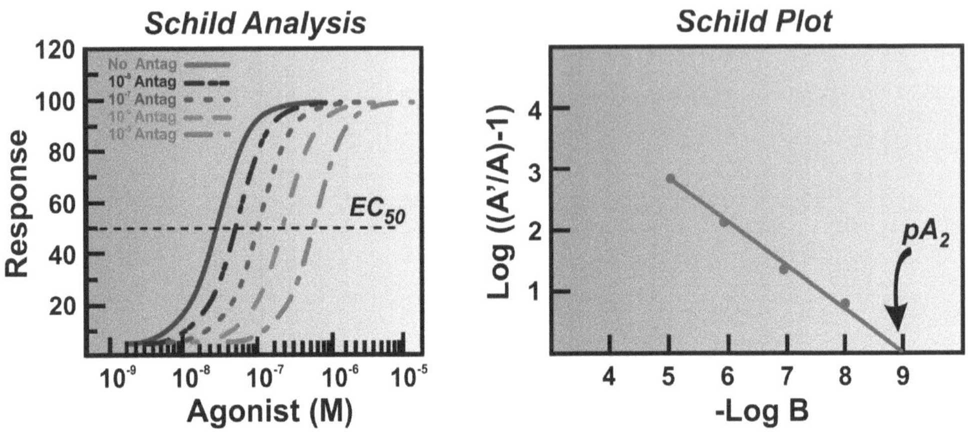

FIGURE 1.15 Schild analysis of antagonists. Antagonist affinity for a receptor can be estimated by the method of Schild. The left panel illustrates the effect of increasing concentrations of a competitive antagonist on the CRC of an agonist. Note the rightward shift of the agonist CRC with increasing antagonist concentration. By calculating the ratio of agonist activities across the concentrations of antagonist (right panel), a relationship is developed that can be fitted by linear regression. The x-intercept estimates the affinity of the antagonist for the receptor that is represented as the negative log (pA_2).

log scale, you derive a sigmoidal concentration–response curve (CRC) for an agonist (Figure 1.14B and 1.14C). These CRCs are often referred to as dose–response curves if the experiments are performed in whole animal experiments.

In terms of pharmacological activity, compounds are generally assessed by two major parameters: potency and efficacy. Potency is defined as the biological effect elicited by a compound by a given concentration. In most cases, ligand potency is described in terms of the half-maximal effect or EC_{50}. The efficacy of a ligand is described as the maximal effect elicited by the ligand in the biological system being studied. In terms of developing therapeutics, the goal is most often to improve potency by lowering the EC_{50} concentration and maximize efficacy. In some cases, synthetic compounds can be produced that have efficacy far exceeding that of the natural occurring ligand (Figure 1.15). These compounds are referred to as super agonists. Furthermore, some receptor ligands (both natural and synthetic) do not induce full activation of the receptor's signaling capacity. These ligands are referred to as partial agonists. And lastly, some ligands act to inhibit the basal signaling capacity of the receptor in the absence of native ligand and are termed inverse agonists.

Compounds that are antagonists are also described in terms of their potency and efficacy. Due to their inherent nature as blockers of the natural ligand's effect, antagonists are usually studied in competition (or antagonist) mode in which the ability of increasing concentrations of the antagonist to inhibit the effect of a chosen concentration of agonist (e.g., EC_{50}) is studied (Figure 1.14D). In the case of antagonist, the potency is often estimated and represented as the concentration of compound that produces its half-maximal inhibition (IC_{50}). Frequently we also study the effect of increasing concentrations of antagonist to shift the CRC of the agonist, as this can provide a more detailed understanding of the interactions between receptor and antagonist. These plots were first described by Arunklashana and Schild to describe the affinity of an antagonist for its receptor, but have been adapted to the study of competitive ligand binding assays.[45] In Schild analysis, a series of CRCs of an agonist is run in an assay; each one is performed in the presence of a particular concentration of antagonist in such a way as to cover the concentration–response curve of the antagonist.

Figure 1.15 depicts an example of a Schild analysis[46] of a hypothetical agonist in the presence of its antagonist. As is evident, with increasing concentrations of antagonist, the CRC of the agonist is shifted to the right. This is due to an increasing requirement of agonist to overcome the antagonist blockade of the receptor as the antagonist concentration increases. Since the ability of the agonist to overcome the presence of the antagonist is dependent upon the affinity of the receptor for the antagonist, one can use the ratio of agonist efficacies across the antagonist concentrations to create the Schild plot. Figure 1.15 shows a plot calculated for our hypothetical agonist deriving ratios of EC_{50} in the presence of a particular concentration of antagonist (A′) to that in the absence of antagonist (A). Plotting the log $([A'/A]-1)$ versus the $-\log B$ (concentration of the antagonist), we derive a plot that can be fit by linear regression. If the slope of the line is approximately -1, then the antagonist is competitive and the antagonist and agonist bind to the same or overlapping receptor sites. If the slope of the line is greater or lesser than -1 the antagonist is non-competitive. The intercept of the linear regression line at the abscissa is equal to the pA_2, which is an estimate of the equilibrium dissociation constant for the antagonist. It is called pA_2 because this point is the $-\log$ of the concentration of antagonist that requires a two-fold increase in agonist concentration to reach 50% efficacy.

In Clark's original modeling of receptor activation, he surmised a direct, linear relationship between the number of receptors bound by ligand and the observed biological effect. That is to say that the K_d of a ligand and its EC_{50} should be approximately equal (taking into account experimental variability). However, it became clear in the years after Clark's description that this was not always the case and dramatic differences could be observed between the K_d of ligand and its observed biological potency, with biological potency being less than the observed K_d. In some cases, the differences were multiple orders of magnitude.[47]

These observations were made clearer by Robert Stephenson when he proposed that a linear relationship need not exist between receptor occupancy and biological effect.[48] Stephenson proposed the concept of spare receptors (or receptor reserve) and that maximal efficacy could be achieved by activation of only a fraction of available receptors. He hypothesized that this was dependent upon the specific interactions between ligand and receptor. Studies by other investigators ultimately supported this notion.[49–51] Thus, the fractional receptor occupancy represents the calculated amount of total receptors occupied by ligand. When the concentration of ligand is large (much greater than the K_d of the ligand), the fractional receptor occupancy will be 1.[47]

It should be noted that plotting the fractional receptor occupancy as a function of efficacy on a linear scale produces a simple hyperbolic function similar to the CRC. When plotted in a semi-log fashion, it appears as a sigmoidal curve. A ligand that has great efficacy requires few receptors in order to achieve a maximal response. In real-world application the appearance of large receptor reserve (e.g., in recombinant cell lines overexpressing a given receptor) can have profound effects on measured potencies as well as perceived signaling properties of a receptor (see Chapter 3).

CONFORMATIONAL DYNAMICS AND BIASED SIGNALING

In the context of the historical background given earlier, recent advances in development of synthetic ligands for receptors of various types have had a profound impact on our understanding of receptor pharmacology and ultimately physiology. Classical receptor theory (prior to 1990) held that receptors possessed the capacity to bind many different ligands but that they possessed only a single active conformation that would induce agonist binding into functional signaling. To put this in terms of enzymology, receptors had only one potential active state. This model suggests that efficacy is an inherently determined function of a receptor and that potency was determined by the ability of a given ligand to induce the active state of the receptor. This suggests, in a way, that the receptor is simply a passive player in the process, acting as a conduit to relay the message to a cell that a given ligand has arrived at the cell. However, advances in our understanding of receptor conformational dynamics reveal that receptors of all classes are both inherently active three-dimensional structures, but also that they behave as ligands themselves in terms of their ability to associate with signal transducing effectors; that is, the three-dimensional shape of the receptor–ligand complex plays an active role in determining signaling activation through its affinity for specific association with signaling proteins.

These insights were drawn from the study by Changeux and his colleagues[40,52–57] of L-threonine deaminase, the enzyme that catalyzes the synthesis of isoleucine from L-threonine. Their work began in hopes of understanding how isoleucine, the product of the reaction, could inhibit the activity of the enzyme. Through their studies, they found that the inhibition of enzyme activity by isoleucine did not fit the well-accepted Michaelis model of competitive inhibition. Their explanation for this was that the regulatory isoleucine bound to a site on L-threonine deaminase that was distinct from the catalytic site of the enzyme. The term "allosteric" (from the Greek words *allos* meaning "other" and *stereos* meaning "space") was coined by Changeux's thesis advisor (Jacques Monod) to describe the non-overlapping nature of the regulatory factor with the active site of the enzyme. Their model (the MWC model, so named for the first letter of each author's surname) explained simultaneously the concept of allosteric modulation of proteins by regulatory effectors, but also introduced the concepts of conformational fluidity of proteins (Table 1.4).

TABLE 1.4 The Four Postulates of Protein Interactions Championed By J.P. Changeux

Key Principles of the Structure of Complex Proteins[40]
1. Proteins are symmetrically organized from subunits into a cooperative structure in which the tertiary organization is constrained by the quaternary structure.
2. Protein quaternary structures exist in few distinct conformations (relaxed <– > constrained) in thermal equilibrium in absence of a regulator.
3. Ligands shift the equilibrium of the protein structure between conformations, selectively stabilizing the one for which it has the highest affinity.
4. Conformational transitions take place in a concerted fashion for the entire protein.

Of course, the full impact of such theories was not fully appreciated at the time despite the excitement that surrounded these observations. Indeed, not until these concepts were translated to explain the mechanisms of drugs interacting with the nicotinic acid acetylcholine receptor were these theories applied to receptor pharmacology.[40] Nevertheless, it is easy to see now how these concepts lay the foundation for modern receptor pharmacology and physiology, and indeed James Black and Paul Leff eventually translated these ideas into their famous operational model of receptor activation.[16]

Based on their studies, Changeux hypothesized that allosterism can exist only within the context of a protein that exists as a continuum in which interactions that affect one part of the molecule can be transmitted and impact the other parts of the molecule. What's perhaps more insightful was that he hypothesized that the complex proteins had a defined quaternary structure that restricts the overall movement of the protein, but that the tertiary structure of the protein was a fluid structure in the absence of interaction with exogenous forces.[40,58] He suggested that the protein was within a thermal equilibrium of constantly changing conformations within the constraints of its quaternary structure. His theory proposed that interactions with other proteins or ligands acted to stabilize the protein into a given conformation thus affecting its functional activity.

The MWC model of allostery is also referred to as the Concerted Model of allostery and has been shown to provide an accurate estimation of cooperative thermodynamics in some systems, such as in hemoglobin.[59,60] Simultaneously with the publication of Monod, Wyman and Changeux's theories on conformational dynamics of proteins, another group of scientists developed a competing theory to explain conformational-dependent effects of ligands on proteins. The theory by Koshland, Nemethy and Filman (the so-called KNF theory) was an extrapolation of Linus Pauling's early explanation of the cooperative nature of oxygen binding to hemoglobin.[60,61] In this view of allostery, multiple conformations of proteins exist independently within the same quaternary structure of the protein. The KNF model proposes that the switch in conformation of one subunit (part) due to stabilization by a ligand promotes or eases the free energy required for other subunits to adopt that conformation. The KNF model is also referred to as the Sequential Model of allostery.

Changeux utilized the principles of the MWC model to eventually demonstrate that ligands activating the nicotinic ACh receptor (an ion channel) were allosteric.[9,10,58] He theorized that in the absence of ligand, the receptor was capable of freely flowing in equilibrium between conformational states and that it had a certain probability of occupying any one of these states at any given time (Figure 1.16A[42]). In the presence of the ligand, the receptor bound to the ligand and the interaction between ligand and receptor stabilized the receptor in a given conformational state. At a minimum, he envisioned three conformational states of the receptor: Closed (impermeable to ions), Active (or open, permeable to ions) and Desensitized (also impermeable to ions) (Figure 1.16A), but it is easy to envision a multitude of conformations possible for any receptor. In this model, the ligand acts allosterically to stabilize the open state of the ion channel since it does not have a direct interaction in the channel pore (analogous to an enzyme active site).

We now know that receptors exist in a form of dynamic equilibrium of ever-changing conformations (Figure 1.16B).[62,63] Since some of these states require more free energy (ΔG) to attain, the receptor does not occupy these states equally. Therefore, one can think of a

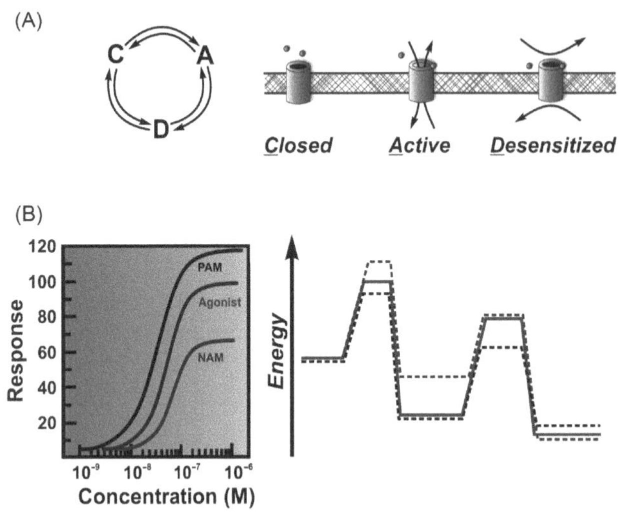

FIGURE 1.16 Conformational dynamics of receptors. (A) shows the equilibrium of states for the nicotinic ACh receptor proposed by Changeux: Closed 'C', Activated 'A', and Desensitized 'D'. The receptor is in dynamic equilibrium of conformations that are affected by ligand. The agonist binds to the receptor and acts to stabilize the active conformation of the receptor, thus promoting the downstream signaling events. Allosteric modulation of ion channels or other receptors acts to affect the energy barriers required for conformational changes by the orthosteric ligand to occur (B). The effect is to reduce or enhance agonist efficacy and/or potency.

ligand as an allosteric or orthosteric integrator that affects the transition energy between conformations of the receptor.[42,58] In the case of an agonist, the conformation stabilized (or preferred) is one that is associated with receptor activation of the downstream signaling apparatus within the cell. Antagonists, on the other hand, stabilize conformations of the receptor that do not induce activation of the signaling apparatus. This model can also be employed to other receptor classes and there are examples of conformational changes within all receptor classes today.[64] Further application of this model to pharmacology suggests that super agonists are capable of stabilizing conformations of the receptor that are more efficient in coupling to signaling pathways than the naturally occurring (or orthosteric) ligand. Partial agonists stabilize conformations of the receptor that also couple with the signal transduction apparatus but do so less efficiently than the orthosteric ligand, or activate multiple signaling pathways that compete.[65,66]

However, receptors are seldom (if at all) found within the context of the cellular milieu as isolated proteins. Rather, they are found associated with other proteins including other receptors in some cases. In the case of membrane-bound receptors, other proteins associated with the inner leaflet of the membrane and integral membrane proteins act as scaffolds to build complex protein complexes which can contain receptors, signaling proteins, and down stream effectors.[42] Nuclear receptors are also found as complexes with other proteins within the cytoplasm, including co-activators and co-repressors, and within the nucleus as larger transcriptional regulatory complexes. If receptors are found in the larger context of support proteins within the cell, it is not hard to imagine that these support proteins can have a significant impact on the properties of the receptor, including its three-dimensional conformation as well as its freedom to move in and out of conformations. These proteins can be considered as either permanently associated with the receptor or transiently associated with the receptor (Figure 1.17).[42]

In some cases, receptors are comprised of separate subunits that are brought together during secondary protein processing. In the case of ion channels this assembly occurs

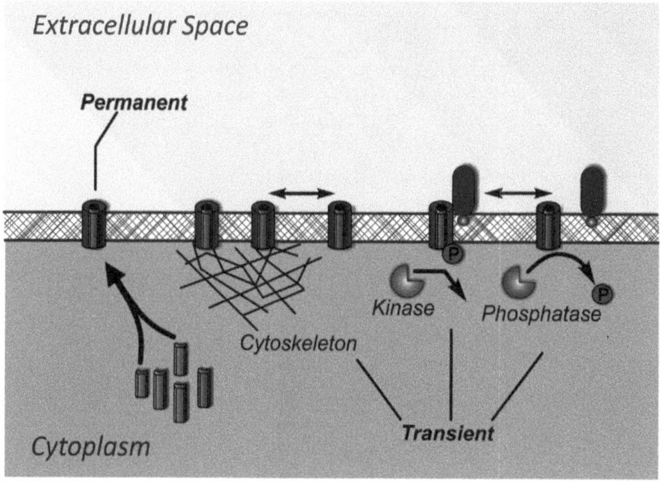

FIGURE 1.17 Accessory proteins can affect receptor function. Receptors are seldom found within a cell as isolated structures. Rather, they are found in complex with other associated proteins that can impact their conformational fluidity and therefore affect receptor responses. The receptor is under the influence of both permanent and transient associated proteins. An example of permanent associated proteins is the accessory subunit for the ACh receptor. Transient accessory proteins can include the cytoskeleton or extracellular matrix proteins, integral membrane proteins, and regulatory proteins such as kinases and phosphatases.

within the Golgi apparatus of cells following synthesis of the proteins in the endoplasmic reticulum to form the complete channel. For example, the nicotinic ACh receptor is comprised of four subunits that comprise the general structure of the receptor and the pore through which the ions flow; however, they also contain a fifth subunit that is added in the construction of the ion channel in the Golgi that is referred to as an accessory subunit.[42,67] In this case, the accessory subunit has been shown to modulate properties of the receptor including response to nicotine, the channel kinetics, and even subcellular localization. It is a permanently associated accessory protein to the receptor.

A similar association between certain GPCRs and single membrane spanning proteins called receptor activity modifying proteins (RAMPs) can be found. Understanding of the interactions between these proteins has shed new light on receptor function and how receptor conformation can be modulated physiologically. RAMPs are known to induce ligand specificity of the calcitonin and calcitonin-like receptors. In this system, two related GPCRs, the calcitonin receptor and the calcitonin-like receptor (CLR), can bind to several hormone ligands (calcitonin, amylin, calcitonin gene-related protein [CGRP] and adrenomedullin). There are three RAMP proteins currently known: RAMP1, RAMP2, and RAMP3. All three of these proteins can associate with both the calcitonin and calcitonin-like receptor. In the case of CLRs, RAMPs become associated with the receptor during secondary protein processing and have been shown to facilitate transport of the receptor to the cell membrane, thereby implying they are permanently associated proteins (e.g., act as chaperone proteins).[68,69]

When associated with the calcitonin receptor, RAMP1, 2 or 3 induce receptor selectivity for the hormone amylin, thereby creating three separate receptor subtypes for amylin (AMY1, AMY2, and AMY3 receptors). When associated with the calcitonin-like receptor, each receptor–RAMP complex has its own ligand specificity. RAMP1 associated with the calcitonin-like receptor induces a receptor with specificity for CGRP; RAMP 2 associated with the calcitonin-like receptor creates a receptor with selectivity for adrenomedullin (AM1 receptor). RAMP3 associated with the calcitonin-like receptor creates a receptor

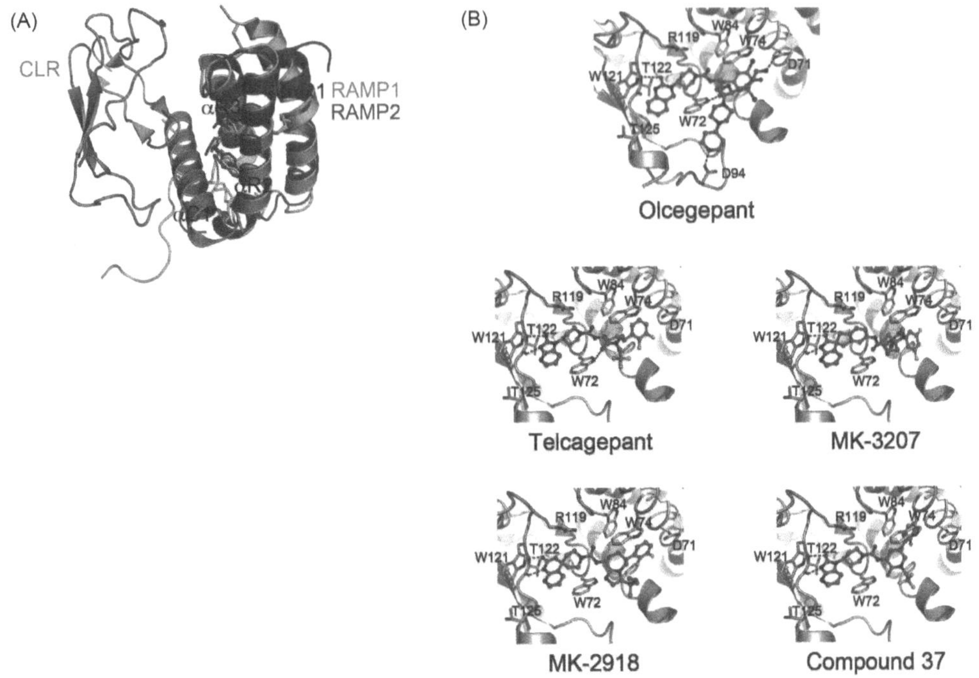

FIGURE 1.18 Crystal structure of the CLR with RAMP proteins and in presence of various antagonists.[71] (A) depicts the crystal structure of the N-terminus of CLR with either RAMP1 or RAMP2. The two RAMPs bind similarly to the CLR. (B) illustrates the crystal structure of the CLR in complex with RAMP in the presence of various antagonists to CGRP. The antagonists bind to the CLR–RAMP complex in an overlapping fashion, bridging the area of interaction of the two proteins (adapted from Ref. 71). *Reproduced with permission from Elsevier, Oxford.*

with dual selectivity for both CGRP and adrenomedullin (AM2 receptor). The calcitonin receptors have long extracellular domains that bind their relatively large protein ligands.

Recently, the three-dimensional structure of the CGRP receptor extracellular domain in complex with RAMP1 and various selective antagonists was generated by x-ray crystallography.[70,71] This structure revealed the key sites of the interaction between the CGRP and RAMP1 extracellular domains. More importantly, these crystals revealed how the RAMP proteins can determine ligand selectivity. Figure 1.18A illustrates the solved three-dimensional structure of the calcitonin-like receptor (CLR) in complex with either RAMP1 or RAMP2. The models derived from the crystal structures show that the extracellular domains of the two proteins form heterodimers in solution and that both RAMP1 and RAMP2 bind CLR in a similar fashion. This is perhaps not surprising since this receptor is known to interact with both RAMP proteins. However, additional sites of interaction between these receptors and their associated RAMP proteins exist at other points, including between the transmembrane domains of the GPCR and the RAMP proteins. Indeed, this has been found for a close relative of CLR, the secretin receptor and its association with

RAMP3.[72] Using these structures it is easy to see how RAMP proteins can affect the three-dimensional structure of the GPCR and thereby stabilize, enable, or restrict movement of the receptor in space. If one takes into account the significant interactions between the two proteins at the extracellular domain, the transmembrane domain, and intracellular domains, this provides a large surface area of contact for the RAMP to affect function of the GPCR.

Figure 1.18B reveals the interactions between some highly selective antagonists to the CLR-RAMP1extracellular domain heterodimer (CGRP receptor).[71] These structures reveal in great detail important interactions between the antagonist, the CLR, and RAMP1.[71] From a general perspective it is clear that the antagonists make significant interactions with amino acid side chains from both the CLR and RAMP1 (Figure 1.18B). These molecules can be broken up into three separate receptor binding components: a CLR binding region; a binding region at the interface of the CLR and RAMP1; and a binding region that interacts with a hydrophobic pocket formed between the CLR and RAMP1.[71] In addition, these structures reveal the intimate interactions between ligand and receptor, illustrating that these partners have a significant spatial impact on each other. Indeed, ligands often possess flexibility within their structures and interaction with the binding site on the receptor can also stabilize specific conformations of the ligand. Drug discovery efforts have focused on optimizing these regions in order to produce more potent antagonists that have suitable pharmacokinetic properties. These types of three-dimensional models have been instrumental in aiding synthesis of new and better compounds and are often used in drug discovery programs for this purpose. They allow drug discovery scientists to envision new substitutions to compounds that could eventually improve the overall quality of the potential therapeutic.

Transiently associated proteins can also be found for all the receptor classes. These interactions usually are associated with the signal of ligand binding and are utilized by the cell to help regulate receptor activation. Examples of these types of associations include tyrosine kinases recruited to growth factor and cytokine receptors, NR co-activators and co-repressors, G proteins and cytoskeletal proteins that act as scaffolds for creation of large multimeric signaling complexes. More detail on these types of interactions can be found in later chapters within this text.

Allosteric binding of a ligand to its receptor is not always mutually exclusive to the binding of the orthosteric ligand and in some cases can act to augment or diminish agonist efficacy.[63,73] These ligands are referred to as allosteric modulators and have been best described for GPCRs. However, given the commonalities of receptor function between classes, it is not hard to envision that similar mechanisms might also be at work for other receptor types. Indeed, allosterism has also been noted for cytokine receptors,[74,75] RTKs,[76,77] ion channels,[40,42] and nuclear receptors.[78] Allosteric modulators act to increase or decrease the sensitivity of the receptor for orthosteric ligands. Therefore, we can distinguish between allosteric modulators and allosteric agonists and antagonists. Allosteric modulators have little or limited effect on the signaling of the receptor alone but have a significant impact on the receptor–signaling apparatus in the presence of another agonist. Those allosteric modulators that promote the effect of an agonist are called positive allosteric modulators (PAMs) and those that act to diminish the effect of an agonist are termed negative allosteric modulators (NAMs). PAMs act to stabilize conformations of the receptor that are thermodynamically more favorable for an agonist to activate the receptor signaling effectors. These compounds lower the energy required by the agonist to induce the active receptor conformation[73] (Figure 1.16B). In contrast, NAMs act to increase the energy required for the agonist to activate the signaling process.[73]

Since the principle of ligand stabilization of conformational fluidity of the receptor leading to effects on receptor signal transduction appears to be common to all receptor classes, it would stand to reason that the properties of allosterics to affect signal transduction of all receptor classes may be equivalent. That is, depending upon the conformation of the receptor stabilized by the allosteric ligand, signaling outputs of the allosteric ligand–receptor complex can take on all or only part of the receptor's full repertoire of signaling capabilities. These compounds can act as modulators, full agonists, partial agonists, super agonists, non-competitive antagonists, and even functional antagonists. This concept that ligands, whether orthosteric or allosteric, can selectively induce conformations of the receptor that selectively activate only part of a receptor's signaling capabilities is referred to as biased signaling. Although the concept of activation of multiple signaling pathways by receptors was first hypothesized by Rodbell in the late 1970s, it was Terry Kenakin, Robert Lefkowitz and others who ultimately illustrated that receptors could activate different patterns of signaling in a ligand-dependent fashion. This type of activity has been best described for synthetic small molecules of NRs and GPCRs, but emerging data recently have found that similar phenomenon can be induced by growth factor receptors and ion channels as well. The role of biased signaling across the various receptor classes will be discussed in detail in the following chapters.

Biased signaling is not restricted to synthetic allosteric ligands but has also been found to be a physiological phenomenon occurring with natural, orthosteric ligands. An excellent example of physiological biased signaling is found in the gonadotropin system.[65,79,80] Gonadotropins (follicle-stimulating hormone [FSH] and luteinizing hormone [LH]) are pituitary hormones that regulate germ cell development. They are found in the blood as a series of glycosylated variants, depending upon the physiologic state of the organism, and bind to cell surface GPCRs to induce their physiological effects.[65] The differently glycosylated variants can be separated based upon their isoelectric points using column chromatography. For many years the physiological roles of glycosylated variants of FSH and LH were studied. Researchers found little effect of isolated variants in receptor binding assays and cellular assays; although it was observed that deglycosylation of FSH and LH produced competitive antagonists. Further study of a recombinant form of FSH produced in insect cells found it behaved as a partial agonist to purified FSH and displayed a bell-shaped CRC.[66]

It was subsequently found that the two phases of the CRC (upward and downward) were due to activation of different G proteins.[66] Interestingly, both the purified and recombinant forms of FSH had similar binding properties. It was later found that these two forms of FSH differed in their complexity of glycosylation; pituitary-extracted FSH having an acidic isoelectric point indicative of a higher content of sialic acid and the recombinant FSH having a more basic isoelectric point suggesting incomplete glycosylation.[81] The pIs of these two proteins were within the range of physiologically identified serum variants and thus were suitable representatives of physiologic FSH. Therefore, fully glycosylated, acidic FSH acts as a full agonist on cAMP production via Gs coupling, but has no effect on Gi signaling. Partially glycosylated, basic FSH acts as a partial agonist biased ligand by conferring both Gs and Gi activation (competing signaling pathways) by the FSH receptor.

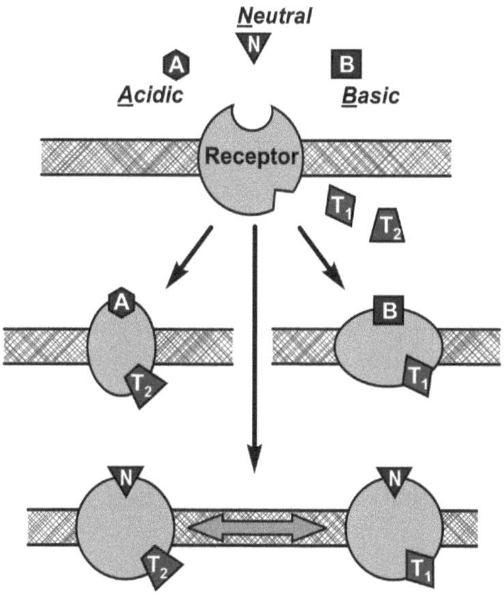

FIGURE 1.19 Schematic diagram of how conformational dynamics leads to biased signaling (adapted from Ref. 65). Using the example of the FSHR, we can understand how conformational dynamics of receptors leads to biased signaling. The FSHR can bind to multiple isoforms of FSH that can be thought of as different ligands. Depending upon the ligand bound, the receptor conformation that is stabilized has its own affinity for the available signal transducers (G proteins, represented here as T_1 and T_2). Some ligand–receptor conformations predominantly activate one signaling pathway, while others can activate different or multiple pathways.

Deglycosylated FSH has little effect on Gs or Gi coupling and therefore behaves as a competitive antagonist.[66,81] These data suggest that one of the physiological roles of circulating glycosylated variants of gonadotropins is to finely regulate the gonadotropic signal from the pituitary as a means of orchestrating the exquisite timing of biochemical signals required for proper gonadal function. Moreover, putting this in context of biased signaling, these data suggest that biased ligands can be found naturally and that the ability of organisms to produce biased ligands provides another level of control over the effects of external signals on cellular function.[65,66,82]

Figure 1.19[65] illustrates a schematic diagram of how conformational dynamics of receptors leads naturally to the evolution of biased signaling as a physiological means of efficiently deciphering complex signals using a limited number of receptors and signaling components.[63,65] In this case we are using a GPCR as an example. Receptors have the ability to bind multiple ligands and these ligands stabilize specific receptor conformations. Once formed, the ligand–receptor complexes can be envisioned as having specific affinities of their own for the potential signaling components. In the case of GPCRs, this would be G proteins or β-arrestin, but for NRs this would be co-activators or co-repressors; for growth factor and cytokine receptors this would be recruited kinases and for ion channels this would be changes in ion selectivity[83] or ERK activation. The signaling effectors that are activated to greater or lesser degrees are dependent upon the stoichiometry between activated receptor and availability of the transducing proteins.

To add another level of fine-tuning to the process, the presence of scaffold and modulating proteins such as RAMPs help to define the potential conformations attainable by the receptor–ligand complex and by localizing receptor and transducing proteins within proximity to one another. So in nature, the cell plays an active role in determining its responsiveness to exogenous signals through coordinated expression of receptor, transducing proteins as well as support proteins. We also know that most natural ligands interact with their receptors in different cells and tissues. Therefore, it would stand to reason that tissue responsiveness to naturally occurring ligands is specific as the cells of these tissues have specific gene expression profiles for these various factors. This is perhaps best described for NRs where this approach has been used extensively to develop tissue specific therapeutics to treat disease and improve quality of life.[84]

Biased signaling by ligands, whether orthosteric or allosteric, has brought along with its discovery a new lexicon to describe the physiological and pharmacological effects of these ligands. Synthetic allosteric compounds have provided a rich resource for exploring the spectrum of functional possibilities provided by signaling bias and in some cases have revealed unforeseen receptor signaling capabilities. Indeed, since all compounds induce efficacy via conformational stabilization, it is highly unlikely that any two ligands could produce identical conformational effects on a receptor. It follows then that any two ligands are highly unlikely to produce exactly the same signaling effects in a given cell.[63]

We have already discussed agonists, partial agonists, and super agonists. Allosteric antagonists and inverse agonists are another class of pharmacological activities. Inverse agonism (Table 1.3) can be visualized in receptor systems where the unoccupied receptor has a measurable level of signaling activation (no ligand present). In the presence of compounds, this basal level of receptor-induced signal transduction can be dose-dependently inhibited by some synthetic ligands. These ligands are referred to as inverse agonists because they induce a negative concentration–response relationship in the absence of agonist. Allosteric antagonists are compounds that act allosterically to block binding or signaling of the orthosteric agonist but do not have a measurable effect on signaling of the receptor on their own. In the purest definition, antagonists (orthosteric or allosteric) do not induce signaling activation. This is termed neutral antagonism. At one time, it was thought that inverse agonism as a pharmacological profile was very rare in comparison to neutral antagonism. However, further study of published antagonists has revealed the opposite to be true: Neutral antagonism is rare but inverse agonism is more common.[63]

PHYSIOLOGICAL AND THERAPEUTIC IMPLICATIONS OF BIASED SIGNALING

It is easy to recognize that conformational fluidity of receptors lends itself to a wide range of possibilities for receptor activation. Indeed, some ligands can produce multiple pharmacological activities. This is evident both in nature (as in the gonadotropin system) and in development of synthetic or therapeutic ligands. From the natural perspective, the ability of receptors to bind to multiple ligands creates efficiencies in terms of requiring synthesis of fewer receptor and ligands. This is obvious among the class I and II cytokine receptors that commonly bind multiple ligands, each with its own cell-specific effects.[85] In addition, cytokines of this family also have the ability to bind multiple receptors.[86,87] This lends itself to a high level of complexity and specificity that can be achieved and required in regulating immune function. In this receptor family, common binding proteins (GP130, β_c and γ_c) form multimeric structures with ligand-specific binding proteins (cytokine α receptors) to provide specific signals to the target cells.[87] In general, the shared cytokine receptors do not have appreciable affinity for their ligands, but in the presence of the cytokine-specific α receptors produce a higher affinity binding complex for their ligands. This allows a limited number of receptors to respond to a large number of potential ligands by utilizing combination of proteins to produce a matrix of unique receptors.

Therapeutically, this type of specificity offers opportunity to develop newer, more specific drugs. Receptors provide a fertile ground for targets of newly discovered therapies. However, this often requires receptors that have clear benefit with little on-target and off-target toxicity. Given that most receptors are pleiotropic within cells and that they are expressed in many cell-types across numerous tissues provides a large challenge to the development of safe and effective drugs. Understanding of the conformational dynamics of receptors in the presence of ligands, as well as understanding of the potential signaling bias by receptors, allows drug discovery scientists to fine-tune the molecules that may potentially become drugs. In some cases, the natural ligand can activate opposing signaling pathways within cells.

Design of therapeutics to selectively activate one type of signaling pattern can be therapeutically useful. Using the gonadotropin system as an example, one major issue with existing fertility treatments is that they rely on purified or recombinant gonadotropins that are agonists. This leads to issues with hyperfertility or overstimulation of follicular development, resulting in heightened risk of multiple birth pregnancies in these women.[88] Understanding of the signaling bias of gonadotropins[65,66,82] has led to the identification and optimization of small molecule allosteric ligands capable of activating a discrete signaling profile induced by the gonadotropin receptor.[65] The thiazolidinone chemotype of allosteric ligands for the FSH receptor demonstrate a spectrum of signaling bias from full agonism, partial agonism, and even antagonism in terms of cAMP production and G_i coupling.[65,81,89] Furthermore, recently van Koppen et al.[90] have reported potent FSH biased ligands that activate G_s, β-arrestin and estrogen production but also act as positive allosteric modulators of FSH enhancing FSH binding and signal transduction activation. Some of these compounds have made their way to clinical trials where they are currently being assessed for safety and efficacy.

Recently, a small molecule biased ligand has been reported for the RTK, fibroblast growth factor receptor (FGFR).[77] This compound was identified in a ligand binding assay but it was found to be extremely weak with an IC_{50} of 2 μM. However, subsequent study of the compound revealed it to be much more potent in *in vitro* cell assays. Further analysis of the compound revealed it to be a selective, allosteric biased antagonist since it was weaker in inhibiting some parts of the FGFR signaling profile than others. This is a significant step forward as most drugs targeting RTKs are antibodies that inhibit binding of orthosteric ligand to the receptor, or antibodies that block receptor dimerization. Some small molecule inhibitors to RTKs exist but these do so by blocking the kinase activity of the receptor.[91,92] Another advantage of a small molecule antagonist for this receptor class would be that biologics (such as antibodies) require parenteral administration due to poor oral bioavailability. Small molecule synthetic compounds provide the opportunity for oral delivery, which can significantly impact patient compliance and make the use of the drug cheaper. Therefore, identification of a specific, small molecule antagonist to the FGFR is the first step toward identification of therapeutically optimized RTK antagonists.

CONCLUSION

To summarize, nature has evolved the capability of biological systems to respond to exogenous factors through a system of ligands, receptor proteins, and transducing proteins that are based upon bidirectional conformational recognition. The most primitive organisms contain receptors and therefore one can assume that the concept of cellular recognition of exogenous signals appears early in the evolution of life. This suggests that the principles of conformation-based recognition of ligands are as equally old. It is not surprising then that many commonalities may have evolved in terms of function of receptors across classes. We have briefly described the history of receptor theory and receptor action, the basics of conformational dynamics and how it leads to biased signaling, and cellular specificity of actions of ligands. Later chapters in this text will focus on biased signaling of specific receptor classes in more detail, but we will also discuss other key aspects of this common physiological mechanism in terms of evolution, cell background, quantitation of signaling bias, and how understanding of biased signaling can provide new opportunities for development of more specific and safe therapeutics.

References

1. Leake CD. *An historical account of pharmacology to the twentieth century*. Springfield, IL: Charles C Thomas; 1975.
2. Langley J. On the reaction of cells and of nerve-endings to certain poisons, chiefly as regards the reaction of striated muscle to nicotine and curari. *J Physiol* 1905;**33**:374–413.
3. Pryull C-R. Part of a scientific master plan? Paul Ehrlich and his receptor theory. *Med Hist* 2003;**47**:332–56.
4. Maehle A-H, Prull C-R, Halliwell RF. The emergence of the drug receptor theory. *Nat Rev Drug Discov* 2002;**1**:637–41.
5. Maehle A-H. "Receptive substances": John Newport Langley (1852–1925) and his path to the receptor theory of drug action. *Med Hist* 2004;**48**:153–74.
6. Koch-Weser J, Schechter P. Schmiedeberg in Strassburg 1872–1918: the making of modern pharmacology. *Life Sci* 1978;**22**:1361–72.
7. Gesztelyi R, Zsuga J, Kemeny-Beke A, Varga B, Juhasz B, Tosaki A. The Hill equation and the origin of quantitative pharmacology. *Arch Hist Exact Sci* 2012;**66**:427–38.
8. Stedman T. *Medical dictionary*. 24th ed. Baltimore, MD: Williams and Wilkins; 1982.
9. Changeux JP, Kasai M, Huchet M, Meunier JC. Extraction from electrical tissue of gymnotus of a protein presenting several typical properties characteristic of the physiological receptor of acetylcholine. *C R Hebd Seances Acad Sci D* 1970;**270**:2864–7.
10. Changeux JP, Kasai M, Lee CY. Use of a snake venom toxin to characterize the cholinergic receptor protein. *Proc Natl Acad Sci USA* 1970;**67**:1241–7.
11. Hill AV. The possible effects of the aggregation of the molecules of haemoglobin on its dissociation curves. *J Physiol (London)* 1910;**40** Proceedings iv–vii.
12. Johnson KA, Goody RS. The original Michaelis constant: translation of the 1913 Michaelis–Menten paper. *Biochemistry* 2011;**50**:8264–9.
13. Michaelis L, Menten ML. Die kinetik der invertinwirkung. *Biochem Z* 1913;**49**:333–69.
14. Goutelle S, Maurin M, Rougier F, et al. The Hill equation: a review of its capabilities in pharmacological modeling. *Fundam Clin Pharmacol* 2008;**22**:633–48.
15. Clark AJ. The antagonism of acetyl choline by atropine. *J Physiol (London)* 1926;**61**:547–56.
16. Black JW, Leff P. Operational models of pharmacological agonism. *Proc R Soc Lond B Biol Sci* 1983;**220** (1219):141–62.

17. Rall TW, Sutherland EW. Formation of a cyclic adenine ribonucleotide by tissue particles. *J Biol Chem* 1958;**232**:1065–76.
18. Sutherland EW, Rall TW. Fractionation and characterization of a cyclic adenine ribonucletotide formed by tissue particles. *J Biol Chem* 1958;**232**:1077–91.
19. Gilman AG. G proteins: transducers of receptor-generated signals. *Annu Rev Biochem* 1987;**56**:615–49.
20. Gilman AG. G proteins and regulation of adenylyl cyclase. Lecture. Medicine NLPa. 1994, Available at: http://www.nobelprize.org
21. Birnbaumer L. The discovery of signal transduction by G proteins: a personal account and an overview of the initial findings and contributions that led to our present understanding. *Biochim Biophys Acta* 2007;**1768**:756–71.
22. National Library of Medicine. Profiles in Science. The Martin Rodbell Papers. [undated]. Available at: http://profiles.nlm.nih.gov/ps/retrieve/Narrative/GG/p-nid/41.
23. Birnbaumer L, Pohl SL, Rodbell M. Adenyl cyclase in fat cells: II. Hormone receptors. *J Biol Chem* 1969;**244**:3477–82.
24. Ross EM, Gilman AG. Resolution of some components of adenylate cyclase necessary for catalytic activity. *J Biol Chem* 1977;**252**:6966–9.
25. Northup JK, Smigel LS, Sternweis MD, Gilman AG. The subunits of the stimulatory regulatory component of adenylate cyclase. Resolution of the activated 45,000 dalton (alpha) subunit. *J Biol Chem* 1983;**258**:11369–76.
26. Northup JK, Sternweis MD, Gilman AG. The subunits of the stimulatory regulatory component of adenylate cyclase. Resolution, activity and properties of the 35,000 dalton (beta) subunit. *J Biol Chem* 1983;**258**:11361–8.
27. Northup JK, Sternweis MD, Smigel LS, Schleifer LS, Ross EM, Gilman AG. Purification of the regulatory component of adenylate cyclase. *Proc Natl Acad Sci* 1980;**77**:6516–20.
28. Dixon RAF, Kobilka BK, Strader DJ, Benovic JL, Dohlman HG, Frielle MA, et al. Cloning of the gene and cDNA for mammalian [beta]-adrenergic receptor and homology with rhodopsin. *Nature* 1986;**321**:75–9.
29. Kobilka BK, MacGregor C, Daniel K, Kobilka TS, Caron MG, Lefkowitz RJ. Functional activity and regulation of human beta 2-adrenergic receptors expressed in Xenopus oocytes. *J Biol Chem* 1987;**262**:15796–802.
30. Wettschureck N, Offermans S. Mammalian G proteins and their cell type specific functions. *Physiol Rev* 2005;**85**:1159–204.
31. Johnson CA, Siderovski DP. Receptor-mediated activation of heterotrimeric G-proteins: current structural insights. *Mol Pharmacol* 2007;**72**:219–30.
32. Higgins JB, Casey PJ. *In vitro* processing of recombinant G protein gamma subunits. Requirements for assembly of an active $\beta\gamma$ complex. *J Biol Chem* 1994;**269**:9067–73.
33. Lemmon MA, Schlessinger J. Cell signaling by receptor tyrosine kinases. *Cell* 2010;**141**:1117.
34. Ullrich A, Schlessinger J. Signal transduction by receptors with tyrosine kinase activity. *Cell* 1990;**61**:203–12.
35. Pawson T. Specificity in signal transduction: from phosphotyrosine-SH2 domain interactions to complex cellular systems. *Cell* 2004;**116**:191–203.
36. Schlessinger J, Lemmon MA. SH2 and PTB domains in tyrosine kinase signaling. *Sci STKE* 2003;**191**:RE12.
37. Huang P, Chandra V, Rastinejad F. Structural overview of the nuclear receptor superfamily: insights into physiology and therapeutics. *Annu Rev Physiol* 2010; **72**: 247–72.
38. Chandra V, Huang P, Hanuro Y, Raghuram S, Wang Y, Burris TP, et al. Structure of the intact PPAR-γ-RXR-α nuclear receptor complex on DNA. *Nature* 2008;**456**:350–6.
39. Orlov I, Rochel N, Moras D, Klaholz BP. Structure of the full human RXR/VDR nuclear receptor heterodimer complex with its DR3 target DNA. *EMBO J* 2012;**31**:291–300.
40. Changeux JP. Allosteric receptors: from electric organ to cognition. *Annu Rev Pharmacol Toxicol* 2010;**50**:1–38.
41. Sobolevsky AI, Rosconi MP, Gouaux E. X-ray crystal structure, symmetry and mechanism of an AMPA-subtype glutamate receptor. *Nature* 2009;**462**:745–56.
42. Araud T, Wonnacott S, Bertrand D. Associated proteins: the universal toolbox controlling ligand gated ion channel function. *Biochem Pharmacol* 2010;**80**:160–9.
43. Scatchard G. The attraction of proteins for small molecules and ions. *Ann N Y Acad Sci* 1949;**51**:660–72.
44. Hill AV. The mode of action of nicotine and curari, determined by the form of the contraction curve and the method of temperature coefficients. *J Physiol (London)* 1909;**39**:361–73.
45. Kenakin TA. *Pharmacology Primer: theory, applications and methods.* Academic Press; 2004.
46. Arunlakshana O, Schild HO. Some quantitative uses of drug antagonists. *Br J Pharmacol* 1959;**14**:48–58.
47. Ruffolo RR. Important concepts of receptor theory. *J Auton Pharmacol* 1982;**2**:277–95.

48. Stephenson RP. A modification of receptor theory. *Br J Pharmacol* 1956;**11**:379–93.
49. Furchgott RF. The pharmacology of vascular smooth muscle. *Pharmacol Rev* 1955;**7**:183–265.
50. Furchgott RF. The use of β-halalkylamines in the differentiation of receptors and in the determination of dissociation constants of receptor–agonist complexes. In: Simmonds NJHaAB, editor. *Advances in drug research.* London: Academic press; 1966. p. 21–55.
51. Nickerson M. Receptor occupancy and tissue response. *Nature* 1956;**178**:697–8.
52. Changeux JP. On the allosteric properties of L-threonine deaminase. I. Methods of studying biosynthetic L-threonine deaminase. *Bull Soc Chim Biol (Paris)* 1964;**46**:927–46.
53. Changeux JP. On the allosteric properties of L-threonine deaminase. II. Kinetics of action of biosynthetic L-threonine deaminase with respect to the natural substrate and inhibitor. *Bull Soc Chim Biol (Paris)* 1964;**46**:947–61.
54. Changeux JP. On the allosteric properties of L-threonine deaminase. III. Interpretation of the inhibitory effect of L-isoleucine steric hindrance or allosteric effect. *Bull Soc Chim Biol (Paris)* 1964;**46**:1151–73.
55. Changeux JP. On the allosteric properties of L-threonine deaminase. IV. The desensitization phenomenon. *Bull Soc Chim Biol (Paris)* 1965;**47**:113–9.
56. Changeux JP. On the allosteric properties of L-threonine deaminase. V. The allosteric transition. *Bull Soc Chim Biol (Paris)* 1965;**47**:267–80.
57. Changeux JP. On the allosteric properties of L-threonine deaminase. VI. General Discussion. *Bull Soc Chim Biol (Paris)* 1965;**47**:281–300.
58. Changeux JP, Edelstein SJ. Allosteric mechanisms of signal transduction. *Science* 2005;**308**:1424–8.
59. Eaton W, Henry E, Hofrichter J, Mozzarelli A. Is cooperative binding by hemoglobin really understood? *Nat Struct Biol* 1999;**6**:351–8.
60. Cui Q, Karplus M. Allostery and cooperativity revisited. *Protein Sci* 2008;**17**:1295–307.
61. Pauling L. The oxygen equilibrium of hemoglobin and its structural interpretation. *Proc Natl Acad Sci* 1935;**21**:181–91.
62. Onaran HO, Scheer A, Cotecchia S, Costa T. A look at receptor efficacy: from the signaling network of the cell to the intramolecular motion of the receptor. In: Kenakin T, Angus JA, editors. *The Pharmacology of functional, biochemical and recombinant systems.* Heidelberg, Germany: Springer; 2002. p. 217–80.
63. Kenakin T, Miller LJ. Seven transmembrane receptors as shapeshifting proteins: the impact of allosteric modulation and functional selectivity on new drug discovery. *Pharmacol Rev* 2010;**62**:265–304.
64. Wrabl JO, Gu J, Liu T, Schrank TP, Whitten ST, Hilser VJ. The role of protein conformational fluctuations in allostery, function, and evolution. *Biophys Chem* 2011; **159**: 129–41.
65. Arey BJ, López FJ. Are circulating gonadotropin isoforms naturally occurring biased agonists? Basic and therapeutic implications. *Rev Endocr Metab Disord* 2011;**12**:275–88.
66. Arey BJ, Stevis PE, Deecher DC, Shen ES, Frail DE, Negro-Vilar A, et al. Induction of promiscuous G protein coupling of the follicle-stimulating hormone (FSH) receptor: a novel mechanism for transducing pleiotropic actions of FSH isoforms. *Mol Endocrinol* 1997;**11**:517–26.
67. Kuryatov A, Onksen J, Lindstrom J. Roles of accessory subunits in alpha4-beta2 nicotinic receptors. *Mol Pharmacol* 2008;**74**:132–43.
68. McLatchie LM, Fraser NJ, Main MJ, Wise A, Brown J, Thomson N, et al. RAMPs regulate the transport and ligand specificity of the calcitonin receptor-like receptor. *Nature* 1998;**393**:333–9.
69. Williams D, Devi LA. Escorts take the lead: molecular chaperones as therapeutic targets. *Prog Mol Biol Transl Sci* 2010;**91**:121–49.
70. ter Haar E, Koth CM, Abdul-Manan N, Swenson L, Coll JT, Lippke JA, et al. Crystal structure of the ectodomain complex of the CGRP receptor, a class-B GPCR, reveals the site of drug antagonsim. *Structure* 2010;**18**:1083–93.
71. Archbold JK, Flanagan JU, Watkins HA, Gingell JJ, Hay DL. Structural insights into RAMP modification of secretin family G protein-coupled receptors: implications for drug development. *Trends Pharmacol Sci* 2011;**32**:591–600.
72. Harikumar KG, Simms J, Christopoulos G, Sexton PM, Miller LJ. Molecular basis of association of receptor-modifying protein 3 with the family B G protein-coupled secretin receptor. *Biochemistry* 2009;**48**:11773–85.
73. Keov P, Sexton PM, Christopoulos A. Allosteric modulation of G protein-coupled receptors: a pharmacological perspective. *Neuropharmacology* 2011;**60**:24–35.

74. Quiniou C, Sapieha P, Lahaie I, Hou X, Brault S, Beauchamp M, et al. Development of a Novel noncompetitive antagonist of IL-1 receptor. *J Immunol* 2008;**180**:6977–87.
75. Matthews EE, Thévenin D, Rogers JM, Gotow L, Lira PD, Reiter LA, et al. Thrombopoietin receptor activation: transmembrane helix dimerization, rotation, and allosteric modulation. *FASEB J* 2011;**25**:2234–44.
76. De Meyts P. The insulin receptor: a prototype for dimeric, allosteric membrane receptors. *Trends Biochem Sci* 2008;**33**:376–84.
77. Bono F, De Smet F, Herbert C, De Bock K, Georgiadou M, Fons P, et al. Inhibition of tumor angiogenesis and growth by a small-molecule multi-FGF receptor blocker with allosteric properties. *Cancer Cell* 2013;**23**:477–88.
78. Buzon V, Carbo LR, Estruch SB, Fletterick RJ, Estebanez-Perpina E. A conserved surface on the ligand binding domain of nuclear receptors for allosteric control. *Mol Cell Endocrinol* 2011;**348**:394–402.
79. Ulloa-Aguirre A, Crépieux P, Poupon A, Maurel M-C, Reiter E. Novel pathways in gonadotropin receptor signaling and biased agonism. *Rev Endocr Metab Disord* 2011;**12**:259–74.
80. Landomiel F, Gallay N, Jégot G, Tranchant T, Durand G, Bourquard T, et al. Biased signalling in follicle stimulating hormone action. *Mol Cell Endocrinol* 2014;**382**:452–9.
81. Arey BJ, Yanofsky SD, Claudia Pérez M, Holmes CP, Wrobel J, Gopalsamy A, et al. Differing pharmacological activities of thiazolidinone analogs at the FSH receptor. *Biochem Biophys Res Commun* 2008;**368**:723–8.
82. Arey B. Allosteric modulators of glycoprotein hormone receptors: discovery and therapeutic potential. *Endocr* 2008;**34**:1–10.
83. Nys M, Kesters D, Ulens C. Structural insights into Cys-loop receptor function and ligand recognition. *Biochem Pharmacol* 2013;**86**:1042–53.
84. Burris TP, Solt LA, Wang Y, Crumbley C, Banerjee S, Griffett K, et al. Nuclear receptors and their selective pharmacologic modulators. *Pharmacol Rev* 2013;**65**:710–78.
85. Ozaki K, Leonard WJ. Cytokine and cytokine receptor pleiotropy and redundancy. *J Biol Chem* 2002;**277**:29355–293558.
86. Renauld J-C. Class II cytokine receptors and their ligands: key antiviral and inflammatory modulators. *Nat Rev Immunol* 2003;**3**:667–76.
87. Wang X, Lupardus P, LaPorte SL, Garcia KC. Structural biology of shared cytokine receptors. *Annu Rev Immunol* 2009;**27**:29–60.
88. Schieve LA, Devine O, Boyle CA, Petrini JR, Warner L. Estimation of the contribution of non-assisted reproductive technology ovulation stimulation fertility treatments to US singleton and multiple births. *Am J Epidemiol* 2009;**170**:1396–407.
89. Yanofsky SD, Shen ES, Holden F, Whitehorn E, Aguilar B, Tate E, et al. Allosteric activation of the follicle-stimulating hormone (FSH) receptor by selective, nonpeptide agonists. *J Biol Chem* 2006;**281**:13226–33.
90. van Koppen CJ, Verbost PM, van de Lagemaat R, Karstens WJ, Loozen HJ, van Achterberg TA, et al. Signaling of an allosteric, nanomolar potent, low molecular weight agonist for the follicle-stimulating hormone receptor. *Biochem Pharmacol* 2013;**85**:1162–70.
91. Tvorogov D, Anisimov A, Zheng W, Leppänen VM, Tammela T, Laurinavicius S, et al. Effective suppression of vascular network formation by combination of antibodies blocking VEGFR ligand binding and receptor dimerization. *Cancer cell* 2010;**18**:630–40.
92. Chung AS, Ferrara N. Developmental and pathological angiogenesis. *Annu Rev Cell Dev Biol* 2011;**27**:563–84.

CHAPTER

2

The Role of the Cell Background in Biased Signaling

Guillermo G. Romero

Department of Pharmacology & Chemical Biology, University of Pittsburgh School of Medicine, Pittsburgh, PA, USA

OUTLINE

BIASED SIGNALING

The concept of biased signaling commonly refers to the selectivity of receptor agonists to differentially activate alternative signaling pathways. This phenomenon has been proposed to result from agonist-selective stabilization of alternative active receptor conformations.[1,2] This conformational model is an extension of the classical

B. Arey (Ed): *Biased Signaling in Physiology, Pharmacology and Therapeutics*
DOI: http://dx.doi.org/10.1016/B978-0-12-411460-9.00002-1

ligand-receptor-transducer-effector model (Figure 2.1). It implies that the activation of selective signaling pathways by different agonists is an intrinsic property of specific receptor–ligand interactions. This operational model has some limitations. Cell surface receptors do not exist in isolation, as they engage multiple cell surface and intracellular interactive partners that modify their function. Thus, the receptor interactome significantly influences signaling outcomes. Furthermore, a "conformational" model applies primarily to multiple conformations that may exist in equilibrium or under steady state conditions. However, receptors participate in multiple processes that are irreversible from a thermodynamic point of view, such as phosphorylation and compartmentalization (i.e., endocytosis). Thus, the classical ligand-receptor-transducer-effector model is a

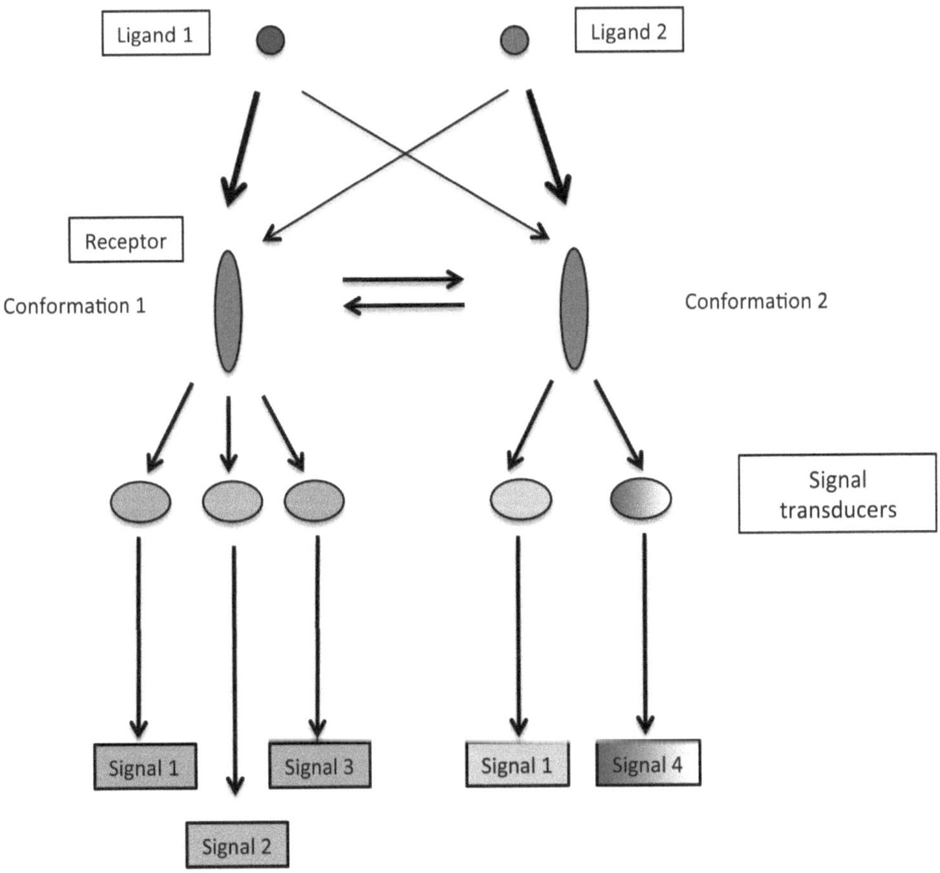

FIGURE 2.1 A simplified classical ligand bias model. A receptor exists in two alternative conformational states. Ligand 1 has higher affinity for Conformation 1, which couples to multiple signal transducers to induce three different signals. Ligand 2 binds Conformation 2, which cannot couple to Signal 2, but it now couples to Signal 4. Ligand 1 and Ligand 2 are biased agonists. Full agonists activate all available downstream pathways.

useful first approximation but its application to real biological systems must be subject to experimental verification.

An example to illustrate the relevance of the cellular physiological background in biased signaling can be found in the parathyroid hormone receptor system. The parathyroid hormone receptor type 1 (PTH1R) is a G protein-coupled receptor (GPCR) that interacts with parathyroid hormone (PTH) and a selected set of PTH analogs. When activated, the PTH1R promotes G-protein-dependent signaling (cAMP production, calcium mobilization) and arrestin-dependent signaling (activation of ERK kinases). The PTH1R ligand D-Trp(12),Tyr(34)-bPTH(7-34) [PTH(7-34)] does not activate cAMP production or calcium release from intracellular stores. However, this ligand induces endocytic traffic of the PTH1R by an arrestin-independent mechanism in kidney distal convoluted tubule (DCT) and ROS 17/2.8 cells but does not in proximal tubule cells.[3] The origin of these distinct behaviors has been linked to the expression of an adaptor protein known as the Na^+/H^+ exchanger regulatory factor 1 (NHERF1), a protein that binds the C-terminus of the PTH1R and modulates its functions.[3] Transfection of NHERF1 to DCT and ROS 17/2.8 cells inhibits PTH(7-34)-induced PTH1R endocytosis. PTH(7-34) does not stimulate β-arrestin translocation to the membrane in DCT, ROS 17/2.8,[3,4] or CHO[5,6] cells independently of the expression of NHERF1. No effects of PTH(7-34) on β-arrestin recruitment and ERK phosphorylation have been detected on multiple cell lines transiently or stably transfected with human PTH1R.[7]

In contrast, a different study reported that PTH(7-34) is a biased agonist that inhibits G protein-dependent signaling while promoting β-arrestin signaling and ERK activation in HEK293 cells.[8] Furthermore, PTH(7-34) reportedly increased bone anabolism in wild-type mice[9] but not in parathyroidectomized rats.[10] In conclusion, the specific responses to PTH (7-34) are exquisitely dependent on the target cells and their environment. This example is by no means isolated. Because of this, the design of robust platforms to detect and study biased agonism requires reliable information regarding crosstalk between signaling pathways and the proteome of the cell models selected.

RECEPTOR INTERACTING PROTEINS: DEFINING THE RECEPTOR INTERACTOME

The idea that cellular receptors function as multimeric signaling complexes has been well established for multiple types of receptors. Nuclear and cell surface receptors interact with multiple proteins that regulate receptor localization, functional partners, and, ultimately, biological function. The yeast two-hybrid system (Y2H) first described by Fields and Song in 1989[11] has probably been the most widely used procedure to detect and identify receptor interacting proteins. This technology is based on the complementation of two separate fragments of a yeast transcription factor. One of the fragments is fused to a selected, specific bait gene (or gene fragment) and the second to the multiple genes included in a prey cDNA library. The fragments of the transcription factors are designed such that the isolated gene products are not functional unless the two fused genes or gene fragments (bait and prey) bind to one another.

Since the Y2H system is based on the interactions between nuclear factors, the methodology has been particularly useful in the detection of nuclear receptor partners. In fact, the technology has been adapted to high-throughput screening analysis of nuclear receptor partners.[12] The results of these efforts, however, highlight the daunting complexity of protein–protein interactions: Albers et al.[12] detected in excess of 1600 interaction pairs using 38 of the 48 known human nuclear receptors as bait in their screening assays. Global analyses based on the Y2H system have been published (see, for instance, ref. 13). Since interacting partners often modify function, these interactome maps have significant implications for the drug discovery process, as they point towards potential new therapeutic targets.

The application of the original Y2H screening methods to the analysis of cell surface receptors also yielded useful novel information. Numerous partners for GPCRs and receptor tyrosine kinases (RTK) were initially detected in early Y2H screens.[14] However, RTKs and GPCRs are membrane proteins, such that traditional Y2H approaches do not work with the full-length protein. Thus, most efforts were concentrated on the study of the intracellular regions of the receptors. Moreover, some receptor–target interactions are dependent on post-translational modifications, such as phosphorylation, that cannot be mimicked in yeast. This led to the development of alternative two-hybrid technologies (including one-hybrid and three-hybrid systems) that can be adapted to cell surface receptors and performed in mammalian cells.[15] Other alternative methodologies have been developed in recent years, including tandem affinity/pull-down methods and alternative methods for the solubilization of the targeted receptor.[14,16–18]

The discovery of multiple interacting proteins that play some role in specific signaling processes has led to models in which the functional units are higher order multimeric complexes containing one or more classes of receptors, scaffolds, signal transducers, and effectors (Figure 2.2). These complexes have been loosely termed signalosomes. The actual signaling outcomes of these signalosomes depend to a large extent on the composition of the complex, the effects of agonists on complex formation and stability, the traffic of the receptor complex, the relative concentrations of the components, and, often enough, the crosstalk with alternative signaling pathways. Figure 2.2 illustrates two common mechanisms by which the cellular environment may cause pharmacological diversity. Figure 2.2A shows some of the potential signaling consequences of receptor heterodimerization. The receptor heterodimer may:

a. enhance the responses of both monomers to their specific agonists;
b. reduce responses to both types of agonists;
c. increase responses to one class of agonist while reducing responses to the other;
d. alter the behavior of antagonists;
e. modify signal specificity;
f. cross-desensitize both receptors;
g. alter receptor traffic; and
h. create a new pharmacological entity that responds to ligands that do not interact with either monomer.

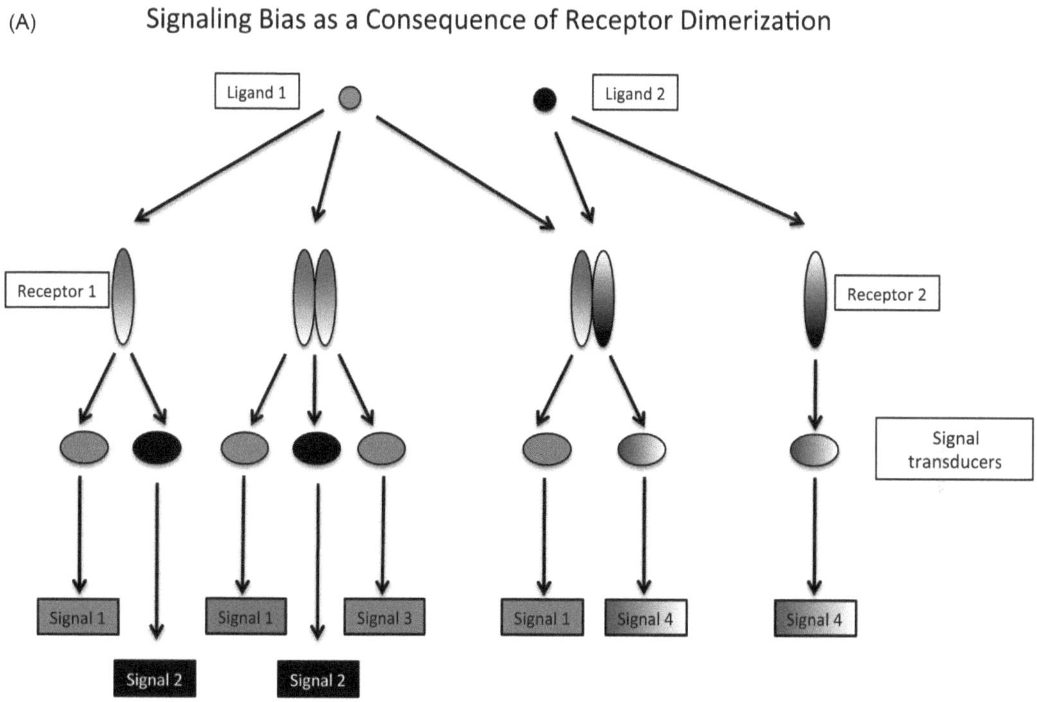

FIGURE 2.2 Two classical examples of the influence of the cellular proteome on biased signaling. (A) Receptor dimerization. Receptor 1 may exist as a monomer, a homodimer, or a heterodimer associated to Receptor 2. Ligand 1 binds Receptor 1 independently of the state of oligomerization, and elicits Signals 1–3 when binding to the monomer or the homodimers. Ligand 2 only binds Receptor 2, and triggers Signal 4 when bound to the Receptor 2 monomer. Either ligand triggers Signals 1 and 4 when bound to the heterodimer. (B) Regulation by accessory proteins. A single class of receptors interacts with alternative accessory proteins, which modulate the affinity of the receptor towards specific ligands and the specificity of the coupling to downstream signaling pathways.

Accessory proteins (Figure 2.2B) can also modify receptor pharmacology in multiple ways. In the following sections, we will examine the role of special classes of scaffolding and accessory signaling proteins in the assembly of signaling complexes and their downstream functions.

It must be noted that signaling outcomes are a function of the levels of expression of at least some of the accessory signaling proteins. This is a topic that has not been studied in sufficient detail because most studies have relied on the overexpression of receptors in cell culture models. There are three main reasons for this. The first one is thermodynamic: Signaling processes require dynamic protein–protein interactions such that signaling complexes must exist in a steady state or equilibrium. Because of mass action, high concentrations of components favor the formation of complexes thus increasing the efficiency of coupling the receptor to downstream signaling events. In contrast, when the concentration of the components is low, complex concentration is reduced, and coupling may be

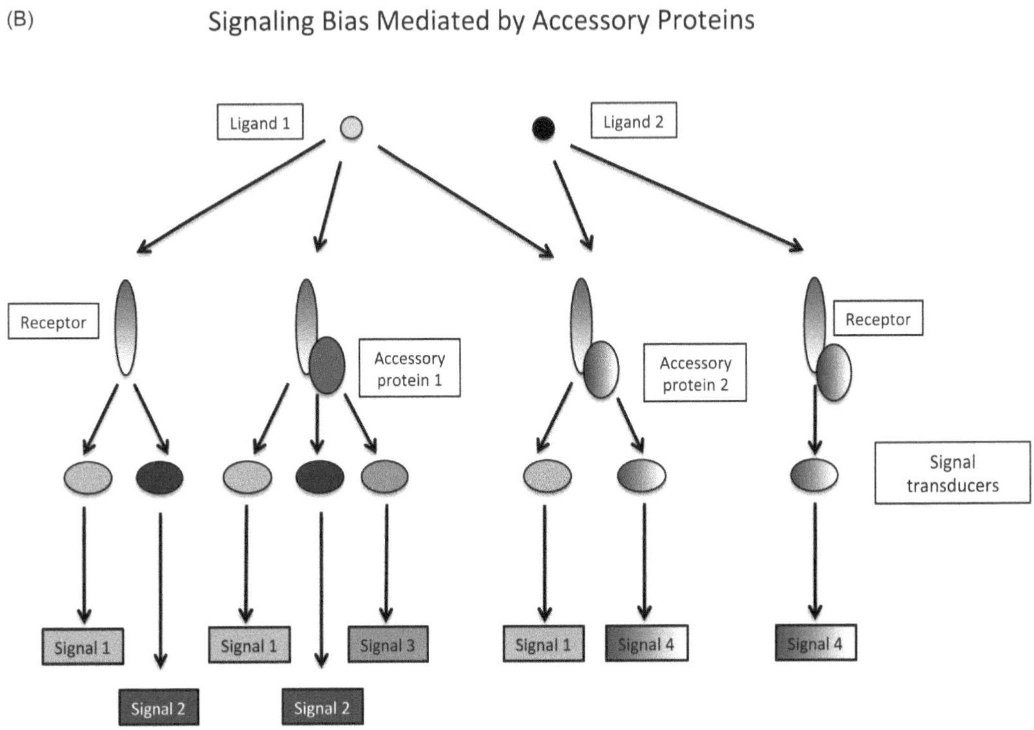

FIGURE 2.2 *(Continued)*

impaired. The second reason is a consequence of the involvement of multivalent scaffolding proteins in the coupling of the signaling cascade. Coupling efficiency is a biphasic function of the concentration of multivalent scaffolds: At low scaffold concentrations, scaffolds enhance coupling, whereas at high concentrations the scaffold interferes with coupling.[19,20]

The third problem is the substantial overlap of binding sites, which indicates that not all protein partners can bind the receptor simultaneously, and that a binding/signaling hierarchy must be established for proper function. For example, a 17-residue sequence of the carboxy-terminal tail of the PTH1R (^{470}IKKSWSRWTLALDFKRK^{486}) reportedly interacts with calmodulin,[21] Gβγ,[22] ezrin,[23] and the scaffold Dvl2.[24] It is evident that this short PTH1R sequence cannot accommodate all four partners simultaneously. Hence, a still uncovered dynamic hierarchy of protein–protein interactions must be established for proper PTH1R function. And finally, when binding sites overlap or when the accessory protein determines relative input to one pathway vs another, the relative concentrations and the affinities for the signal transducers will affect the outcome. For example, the effects of an accessory protein that increases the affinity of a receptor for a specific transducer by 10-fold will be less pronounced in cells where such transducer is expressed at saturating concentrations.

G PROTEIN-COUPLED RECEPTORS

G protein-coupled receptors (GPCRs) constitute the largest family of membrane proteins in mammalian cells.[25] They regulate multiple processes, including metabolism, hormone function, and sensory perception. Reportedly, more than 40% of the drugs currently in the market target GPCR function.[16] GPCRs are characterized by a common structure comprised of seven transmembrane domains with an extracellular N-terminus and an intracellular C-terminus.[25] Agonists have been found to bind to the extracellular domain, extracellular loops or in pockets formed by the juxtaposition of the transmembrane domains. Binding of an agonist induces conformational changes in the transmembrane domains and the intracellular regions. These conformational changes lead to the activation of heterotrimeric G proteins via mechanisms that are the subject of intensive investigation using structural biology approaches. These structural studies have identified a well-defined set of conformational changes that occur upon the binding of activating (agonists) or inactivating (antagonists) ligands.[26–32] The extracellular domain, comprising the N-terminus and the three extracellular loops, is responsible for the initial binding of specific ligands and is characterized by substantial structural diversity among receptor types. By contrast, the intracellular module, which comprises the C-terminus and the three intracellular loops, is highly conserved and undergoes much larger conformational changes upon ligand binding.[29]

The primary mechanism for GPCR signaling involves the coupling of the activated receptor to heterotrimeric G proteins. The classification of heterotrimeric G proteins is based primarily on the structure of their respective α-subunits. Four distinct families of Gα have been identified: α_s, $\alpha_{i/o}$, $\alpha_{q/11}$, and $\alpha_{12/13}$. These G proteins exist as GDP-bound heterotrimers in the basal state. Upon activation, the receptor interacts with the G protein and promotes the exchange of GDP by GTP. GTP binding destabilizes the heterotrimer, and the G protein dissociates into an active α.GTP complex and a βγ dimer. Both α.GTP and βγcomplexes activate distinct signaling pathways (Table 2.1).

Surprisingly, it was soon realized some GPCRs activate multiple classes of α subunits. This result was initially assumed to be a consequence of the expression of multiple receptor subtypes or a consequence of the overexpression of tagged receptor/G protein combinations, but further work demonstrated that a single receptor subtype couples with multiple G proteins. Coupling to multiple G proteins is not uncommon among some GPCR subclasses. For example, the parathyroid hormone type 1 receptor (PTH1R), a member of the secretin family, couples to G_s, G_q, $G_{i/o}$, and $G_{12/13}$ when stimulated with full agonists.[33–38] According to the classical views of GPCR activation, this phenomenon can be explained using a dynamic model in which the ligand-bound, activated receptor fluctuates among multiple conformations that favor interactions with subsets of specific G protein species (Figure 2.1). However, the classical model fails to explain the observation that, although G_s, G_q, $G_{i/o}$, and $G_{12/13}$ are found ubiquitously, the coupling of the PTH1R to these pathways is rather cell-specific. The first clue concerning the origins of this specificity was the discovery of the linkage of the activation of phospholipase C (PLC) signals by PTH1R full agonists to the expression of a scaffolding protein, the Na^+/H^+ exchanger regulatory factor 2 (NHERF2).[39] This discovery was followed by the demonstration that the closely related NHERF1 had a similar effect.[40] Further studies on the direct coupling of PTH1R with

TABLE 2.1 Heterotrimeric G Proteins

Family	Subtype	Main Physiological Effects	Expression
$G_s\alpha$	$G_{s(Short)}\ \alpha$ $G_{s(Long)}\ \alpha$ $G_{s(XL)}\ \alpha$ $G_{olf}\alpha$	Activation of adenylyl cyclase Activation of Src family kinases	$G_s\alpha$: ubiquitous $G_{olf}\alpha$: olfactory neurons and certain CNS ganglia
$G_{i/o}\alpha$	$G_{i1-i3}\alpha$ $G_{o1}\alpha$ $G_{o2}\alpha$ $G_Z\alpha$ $G_t\alpha$ $G_{gust}\alpha$	Inhibition of adenylyl cyclase Inhibition of Ca^{2+} channels Activation of K^+ channels Activation of Src family kinases Activation of cGMP PDE Activation of Cdc42	$G_{i1-3}\alpha$: ubiquitous $G_{o1}\alpha$, $G_{o2}\alpha$: neurons, glia, heart, neuroendocrine organs $G_Z\alpha$: platelets, neurons, chromaffin cells, secretory cells $G_{t1}\alpha$, $G_{t2}\alpha$: retina, taste buds $G_{gust}\alpha$: taste buds, chemoreceptor cells
$G_{q/11}\alpha$	$G_q\alpha$ $G_{11}\alpha$ $G_{14}\alpha$ $G_{15}\alpha$ $G_{16}\alpha$	Activation of phospholipase Cβ Activation of p63-RhoGEF Activation of K^+ channels	$G_{q/11}\alpha$: ubiquitous $G_{15/16}\alpha$: hematopoietic cells
$G_{12/13}\alpha$	$G_{12}\alpha$ $G_{13}\alpha$	Activation of phospholipase D Activation of phospholipase Cε Activation of iNOS Increase E-cadherin-mediated cell adhesion Activation of multiple RhoGEFs	Ubiquitous
$G\beta/\gamma$	β_{1-5} γ_{1-12}	Activation of phospholipase Cβ Inhibition of adenylyl cyclase I; activation of adenylyl cyclase II, IV, and VII Activation of PI3 kinase Activation of K^+ channels Inhibition of Ca^{2+} channels Activation of cJun N-terminal kinase Activation of Src family kinases Activation of protein kinase D Membrane recruitment of GRKs Activation of p114-RhoGEF	Most cells express multiple subtypes $\beta_1\gamma_1$: retinal rod cells $\beta_3\gamma_8$: retinal cone cells β_5: neurons, neuroendocrine cells

heterotrimeric G proteins led to the demonstration that NHERF1 and NHERF2 differentially influence the efficiency of the coupling of the receptor with specific G proteins: NHERF1 promotes coupling with G_q, and NHERF2 enhances coupling to both G_i and G_q.[41]

NHERF1 and NHERF2 are scaffolding proteins of the PSD95/Discs-large/Zo-1 (PDZ) family. They interact with the extreme C-terminus of the PTH1R and other GPCR targets via their two N-terminal PDZ domains.[42] NHERF1 and NHERF1 are multi-domain scaffolding proteins that play a major role in the formation of multimeric complexes.[39,40] Both NHERF1 and NHERF2 promote the formation of complexes that include the PTH1R and PLCβ via direct interactions of their PDZ domains with the C-termini of the receptor and the enzyme.[39,40] Therefore, the original dynamic model of GPCR activation must be revised to include protein partners that modify activity and function. Agonists do not bind to isolated receptors; rather, they interact with multimeric complexes in which the receptor is the critical partner (Figure 2.2). It is important, therefore, to determine the "signaling unit" of each specific receptor, and this requires the identification of partners that regulate coupling to downstream signals. This is especially important for the design of high-throughput screening assays, as many of these rely on events that occur several steps downstream of receptor activation.

GPCR-Associated Protein Complexes and the GPCR Interactome

To date, most efforts to identify GPCR-associated protein complexes (GAPC) have been based on the use of soluble subdomains of the target receptors, in particular the C-terminal domain.[43–45] By far the most widely used approach to identify GAPC candidates is the yeast two-hybrid approach and most of the confirmed GAPC known to date have been identified by this technology. However, other technologies based on affinity chromatography, pull-downs, and alternative global approaches using full-length receptors have become available.[14,16,18,43]

Proteins involved in receptor traffic are commonly identified during these screens and are often not taken into consideration for further functional analyses. This is an acceptable practice as a first approximation, but its drawbacks are obvious; some proteins, like the β-arrestins are both traffic regulators and crucial functional modifiers. The discussion that follows focuses primarily on proteins that have been demonstrated to alter ligand-dependent receptor function.

Two families of proteins that are invariably identified in these screens are the α-subunits of heterotrimeric G proteins and the β-arrestins. The role of these GPCR partners in ligand bias is discussed elsewhere in this book (see Chapter 3) and will not be addressed here. Rather, the discussion that follows focuses on functional partners that modify the coupling of GPCRs to their main downstream targets.

PDZ Proteins

PDZ domains derive their name from the first three proteins in which these domains were identified (Post-synaptic density 95 [PSD95], *Drosophila* Discs-large [Dlg1], and Zonula occludens-1 [ZO-1]). PDZ domains are well-defined structural motifs of around 80–90 amino acids organized in six β-sheets (βA–βF) and two α-helices (αA, αB).[46,47] PDZ proteins—defined as proteins that contain at least one PDZ domain—constitute the most numerous family of intracellular protein scaffolds, with over 200 members of the family identified to date.[47] The targets for these proteins contain, in most cases, somewhat specific carboxy-terminal sequences that depend on the structure of individual PDZ domains, but invariably terminate in a hydrophobic amino acid.[42,47,48] This C-terminal ligand binds to an extended groove between βB and αB, with the C-terminal hydrophobic residue occupying the hydrophobic cavity at the top of the binding groove.

Three classes of PDZ ligands are currently recognized (Table 2.2). PDZ domains that interact with these three classes of ligands differ in their structure: Class I PDZ domains

TABLE 2.2 Classification of PDZ Binding Motifs

Class	Binding Motif
Class I	-[D/E]-[S/T]-X-Φ
Class II	-X-Φ-X-Φ
Class III	-X-[D/E/K/R]-X-Φ
Other	-X-X-C-X-Φ-[D/E]

contain a conserved histidine that coordinates the conserved serine or threonine of the Class I ligands, whereas Class II PDZ domains contain conserved hydrophobic residues at the end of βB to stabilize the binding of the proximal hydrophobic residue.[42,48] Importantly, the interactions of PDZ proteins with their targets are not limited to the C-terminal PDZ ligands. For example, Mahon and Segre[40] identified residues as far as 18 amino acids upstream of the C-terminal ligand as part of the recognition site of the PTH1R by the PDZ protein NHERF1.

Roughly, 80 + human GPCRs terminate in canonical PDZ ligands belonging to Classes I, II, or III. A few other GPCRs that do not contain a recognizable PDZ interacting motif have also been shown to interact with PDZ proteins. In addition, some PDZ proteins interact with binding motifs that are not located at the extreme C-terminus of their GPCR targets. A partial list including only GPCRs with demonstrated association to PDZ proteins is shown in Table 2.3.

PDZ proteins regulate multiple aspects of GPCR function. These include traffic to the plasma membrane, endocytosis, recycling, coupling to specific G proteins, and assembly of downstream signaling pathways. One of the best-studied PDZ proteins is NHERF1, also known as ezrin binding protein of 50 kDa (EBP50). NHERF1 contains four well-defined structural motifs that are essential for its function: a) two N-terminal PDZ motifs (PDZ1, PDZ2); b) a C-terminal Ezrin binding (EB) motif that also interacts with merlin, radixin, and moesin; and c) a canonical Class I PDZ ligand at the very end of the C-terminus. NHERF1 interacts with numerous members of the GPCR family through at least one of its PDZ ligands. Some receptors show significant selectivity for PDZ1 or PDZ2, but others do not.[3,49] NHERF1 (and the closely related NHERF2/E3KARP) play three main roles in the regulation of GPCR signaling: a) scaffolding of signaling complexes; b) regulation of traffic and recycling; and c) conferring signaling bias.

The PDZ protein NHERF1 mediates the formation of multimeric complexes by scaffolding a large variety of partners. For example, NHERF1 scaffolds the formation of a multimeric complex involving the β2-adrenergic receptor (β2-AR) and the cystic fibrosis transmembrane conductance regulator (CFTR), which is required for the regulation of CFTR function by protein kinase A (PKA).[50] In addition, NHERF1 regulates GPCR function by coordinating the assembly of ternary complexes that may or may not include the specific GPCR. For example, NHERF1 assembles a ternary complex containing the renal phosphate transporter Npt2a and the PKA-anchoring protein ezrin.[51] This complex is required for the regulation of phosphate transport by the PTH1R, although there is no

TABLE 2.3 GPCR Regulated by PDZ Proteins

ID	GPCR	PDZ Ligand	Known PDZ Partners
ADA1A	α-1A Adrenergic Receptor	GEEV[c]	nNOS
ADA1D	α-1D Adrenergic Receptor	ETDI[a]	Syntrophin α, β1 and β2
ADRB1	β-1 Adrenergic Receptor	ESKV[a]	SAP97/Dlg1, PSD95, GIPC, CAL, MAGI2, MAGI3
ADRB2	β-2 Adrenergic Receptor	DSLL[a]	NHERF1, NHERF2, Snx27, GIPC, PDZK1
CALCR	Calcitonin Receptor	ESSA[e]	NHERF1 (via RAMP3)
CCR5	Chemokyne Receptor 5	SVGL[b]	NHERF1
CXCR2	CXC Chemokyne Receptor type 2	SVVI[b]	PSD95, NHERF1
D2R	Dopamine Receptor Type 2	ILHC[d]	GIPC
D3R	Dopamine Receptor Type 3	ILSC[d]	GIPC
FZD1	Frizzled-1	ETTV[a]	Dvl1-3, NHERF1, PSD95
FZD2	Frizzled-2	ETTV[a]	Dvl1-3, NHERF1, PSD95
FZD4	Frizzled-4	ETVV[a]	Dvl1-3, NHERF1, MAGI3, PSD95
FZD5	Frizzled-5	LSHV[a]	Dvl1-3[e]
FZD7	Frizzled-7	ETAV[a]	Dvl1-3[e], NHERF1, MAGI3, PSD95, syntenin
FZD8	Frizzled-8	LSQV[a]	Dvl1-3[e]
FZD9	Frizzled-9	PTHL[a]	Dvl1-3[e]
FZD10	Frizzled-10	PTCV[a]	Dvl1-3[e]
GPR124	Orphan Receptor 124	ETTV[a]	SAP97/Dlg1
GPR125	Orphan Receptor 125	HETI[c]	Dvl (in Zebrafish)
κOR	Kappa Opioid Receptor	NKPV[c]	NHERF1
LHCGR	Lutropin-choriogonadotropic Hormone Receptor	YTEC[d]	GIPC
LPAR1	Lysophosphatidic Acid Receptor Type 1	HSVV[a]	GIPC, PDZ-RhoGEF, LARG
LPAR2	Lysophosphatidic Acid Receptor Type 2	DSTL[a]	MAGI3, NHERF2, PDZ-RhoGEF, LARG
mGLUR1	Metabotropic Glutamate Receptor Type 1	HVQL[b]	Tamalin
mGLUR2	Metabotropic Glutamate Receptor Type 2	TSSL[a]	Tamalin
mGLUR3	Metabotropic Glutamate Receptor Type 3	TSSL[a]	Syntenin, Tamalin
mGLUR5	Metabotropic Glutamate Receptor Type 5	SSSL[a]	Tamalin, NHERF2
MT1	Melatonin Receptor Type 1	VDSV[c]	MUPP1
OR2AG1	Olfactory Receptor 2AG1	HSTL[a]	MUPP1

(Continued)

TABLE 2.3 (Continued)

ID	GPCR	PDZ Ligand	Known PDZ Partners
P2Y1	Purinoceptor 1	DTSL[a]	NHERF1, NHERF2
P2Y12	Purinoceptor 12	ETPM[a]	NHERF1
PTGIR	Prostacyclin Receptor	CSLC[d]	NHERF4/IKEPP, PDZK1
PTH1R	Parathyroid Hormone Receptor Type 1	ETVM[a]	NHERF1, NHERF2
5HT2AR	Serotonin Receptor Type 2A	VSCV[a]	PSD95
5HT2BR	Serotonin Receptor Type 2B	VSYV[a]	Syntrophin A1, nNOS
5HT2CR	Serotonin Receptor Type 2C	ISSV[a]	PSD95, MPP3

[a]*Class I PDZ ligand.*
[b]*Class II PDZ ligand.*
[c]*Class III PDZ ligand.*
[d]*Atypical PDZ ligand.*
[e]*Internal (i.e., not restricted to the extreme C-terminal sequence) PDZ ligand.*

evidence that the receptor binds to it. Finally, the interactions of NHERF proteins with the C-terminus of their targets may result in the inhibition of certain functions. For example, the interactions of NHERF1 with the C-termini of a subset of Frizzled receptors (Fzd1, Fzd2, Fzd4 and Fzd7) inhibit canonical Wnt signaling and the activation of β-catenin.[49]

From the point of view of biased signaling, the formation of complexes with specific G proteins induced by NHERF1 and NHERF2 is of the highest interest. The PTH1R activates multiple G protein signaling pathways, including cAMP production, calcium mobilization, and cytoskeletal reorganization. These effects are, however, cell-specific and totally linked to the expression of NHERF1 and NHERF2: In the absence of both NHERF proteins, PTH1R full-agonists signal exclusively through the cAMP pathway and fail to activate calcium transients.[39,40] Initially, it was proposed that NHERF1 and/or NHERF2 acted by assembly of a complex containing the PTH1R and phospholipase C (PLC) β1 or PLCβ3.[40] More recent work has led to an even more fundamental role of NHERF1 and NHERF2 in the coupling of specific G protein pathways. Measuring the binding of GTPγ[^{35}S] to specific G proteins in cells that contain low levels of NHERF1 and NHERF2 (PS-120, CHO), Wang et al.[41] demonstrated that activation of the PTH1R leads exclusively to the activation of the heterotrimeric G protein G_s and signals exclusively via cAMP-dependent pathways. Transfection of NHERF1 and NHERF2 transgenes, however, restored PTH1R catalyzed activation of G_q and release of intracellular calcium.[41] Finally, activation of G_i required specifically the expression of NHERF2.[41] These effects were recapitulated in human embryonic kidney HEK-293 cells, which express both NHERF1 and NHERF2, by transfection with specifically targeted shRNA constructs.

The simplest model that explains these data implies that NHERF proteins are molecular switches that govern the relative magnitude of the input to specific G protein signaling pathways by modulating the assembly of specific signaling complexes. Importantly, this G protein switch phenomenon does not seem to be limited to the PTH1R/NHERF system.

The melatonin receptor MT1 couples to G_i only in the presence of the multiple PDZ domain protein MUPP1.[52] However, PDZ proteins may elicit alternative signaling responses by other mechanisms. For example, in rat osteosarcoma ROS 17/2.8 cells, NHERF1 is required for PTH-induced calcium transients but the process involves entry of calcium to the cell and is completely dependent on PKA activity.[53] This suggests that PDZ proteins regulate convergent signaling pathways by multiple mechanisms.

Receptor Oligomerization

GPCRs have been assumed to exist and function as monomers, and the classical paradigms of GPCR signaling have been built on this assumption. However, there is mounting evidence that many members of the GPCR family form homo- and heterodimers.[54] The functional relevance of these assemblies is still largely unknown. In some cases, dimerization is required for the traffic of specific receptors to the cell surface. For example, the $GABA_{B1}$ receptor (which is the target for most $GABA_B$ ligands) only traffics to the surface of cells that also express the $GABA_{B2}$ receptor (which does not bind $GABA_B$ ligands)[55]; thus, for all practical purposes, the $GABA_{B1}/GABA_{B2}$ dimer is the functional unit. In other cases, oligomerization modulates function by altering ligand binding,[56–60] modifying G protein signaling,[61,62] promoting or disrupting β-arrestin binding,[63,64] and changing endocytic traffic patterns.[63,65] In general, most Family C GPCR are obligatory dimers, while many Family A and some Family B GPCR form oligomeric complexes. These properties have been reviewed frequently in recent years (see[54,55,66–69]).

It is important to point out that, with the exception of Family C receptors, GPCR dimerization/oligomerization is not required for function.[70,71] However, multiple classes of functional units with distinct ligand binding properties and specific downstream signaling pathways may exist in a single cell depending on the types of GPCR expressed and their relative abundances. Given that the analysis of signaling pathways and drug screens is usually done in model cell systems, knowledge of the proteomics of the chosen system is important. As an example, consider the case of HEK293 cells, one of the most popular cell lines for functional analysis and screening tests. HEK293 cells express many endogenous GPCRs, such as the β2-AR.[72,73] The β2-AR interacts with multiple GPCRs, including the α1B-AR, the adenosine A1R, the angiotensin II AT1R, the δ-opioid receptor, and others[74]; it is, therefore, apparent that screening tests performed in HEK293 cells will not necessarily yield the same results as tests done in cells that do not express β2-AR.

The next question is whether or not receptor multimers are valid pharmacological targets. In a recent publication, the IUPHAR (International Union of Basic and Clinical Pharmacology) has advanced three criteria to define the physiological relevance of GPCR multimerization. First, a physical interaction between receptors must be established in native tissues or primary cells. Second, the oligomer must have specific signaling or ligand binding properties distinct from the properties of the individual components. Third, IUPHAR recommends the verification of these observations in a suitable *in vivo* model, either using transgenics or siRNA technology.[75] IUPHAR recommends that at least two of these three criteria must be fulfilled. In general, the functional consequences of heteroligomerization include: increased or decreased ligand affinity, altered trafficking properties, altered signaling to G proteins or β-arrestins, cross-activation by specific agonists, cross-inactivation by specific antagonists, and others. Novel pharmacological properties are not observed in all cases.

A partial list of receptor heterodimeric pairs with specific novel pharmacological properties is included in Table 2.4. Homo-oligomeric structures have been reported as well, but

TABLE 2.4 GPCR Heteromers with Distinct Pharmacological Properties[a]

Receptor Pair[b]	Detection Method	Pharmacological/Physiological Effects	PMID Accession Number
FAMILY A RECEPTORS			
5HT2A–D2	Co-IP, FRET	D2R agonists increase 5HT2AR ligand affinity but decrease signaling. 5HT2AR expression is required for full D2R antagonism.	20831885 21645528
α1A–α1B	Co-IP, FRET	Increased surface expression.	14645668
α1B–α1D	Co-IP	Increased surface expression of α1D.	14736874
α1D–β2	Co-IP	No effect on ligand affinity. Increased surface α1D numbers and more efficient coupling to Ca^{2+} signals.	15615865
α2A-β1	Co-IP	Co-internalization; altered affinity for specific ligands	12529373
α2A–μOR	FRET	μOR ligands attenuate α2A signaling.	18193048
A1–A2A	Co-IP, BRET, FRET	A2AR activation reduces affinity of A1R for agonists; heteromer is a target for caffeine.	16481441
A1–D1	Co-IP	Reduced responses to D1 agonists upon co-activation; antagonistic modulation of one receptor by the other.	10890919
A2A–CB1	Co-IP, BRET	A2A activation is required for agonist-dependent CB1 signaling	17356572
A2A–D2	Co-IP, BRET, FRET	Co-internalization and co-desensitization induced by A2A and D2 ligands.	11872740 12933819
AT1–B2	Co-IP, BRET	Enhanced Ang II binding and signaling.	10993080
AT1–β1 or β2	Co-IP	β-antagonists block AT1 function; AT1 antagonist interferes with β-agonist responses	12963634
AT1–CB1	Co-IP, BRET	Potentiation of AT1 signaling; allows AT1 agonist-dependent activation of Gi; CB1 antagonists block some of the downstream functions of AT1.	21540834
β1–β2	Co-IP	Increased affinity for both subtype-specific agonists; increased signaling for both receptors; reduced basal activity.	16002745
β2–δOR	Co-IP	Agonists for either subtype induce dimer endocytosis.	11134510
β2–κOR	Co-IP	Dimer does not internalize in response to agonists for either subtype; reduced β2-induced ERK activity.	11134510
CB1–D2	BiFC, FRET	Signaling switch from Gi/o to Gs; inhibition of D2 agonist binding by CB1 agonist.	14634050 18262573

(Continued)

TABLE 2.4 (Continued)

Receptor Pair[b]	Detection Method	Pharmacological/Physiological Effects	PMID Accession Number
CB1−δOR	Co-IP, BRET	Cross-desensitization.	22235275 16682964
CB1−μOR	BRET	Reciprocal inhibition.	16682964
CCR2b−CCR5	Co-IP, BRET	Allosteric cross-modulation. Reduced binding of agonist.	16467191
CCR5-μOR	Co-IP	Cross-desensitization.	14729105
CXCR2−δOR	Co-IP, BRET, FRET	CXCR2 antagonists enhance ligand-dependent δOR signaling.	18307412
δOR−κOR	Co-IP	Reduced binding of antagonists and subtype-selective agonists.	10385123
δOR−μOR	Co-IP, BRET	δOR ligands (agonists and antagonists) increase μOR signaling; altered β-arrestin signaling; dimer selective agonists.	15044695 17384143 22860219
D1−D2	Co-IP, BRET	Enhanced signaling by some D2 agonists; signaling switch from Gi to Gq.	21785426 17194762
D1−D3	Co-IP, BRET, FRET	Synergism; D3 activation increases affinity for D1 agonists; altered desensitization/internalization.	18644790 18424554
D1,D2−H3	BRET	H3 agonists inhibit and H3 antagonists potentiate D1 and D2 signaling.	18547596
D2−SSTR5	FRET	Increased D1 agonist binding and activity.	10753124
μOR−SSTR2A	Co-IP	Cross-phosphorylation; cross-desensitization.	11896051
SSTR1−SSTR5	FRET		1071310
SSTR2A−SSTR3	Co-IP	Inactivation of SSTR3. SSTR2A function is normal; reduced receptor internalization.	11134004
V1A-V2	Co-IP, BRET	Prolonged β-arrestin binding to the dimer, reduced recycling of V1A.	14757828
FAMILY B RECEPTORS			
CT−Δe13CT	Co-IP	Complex does not traffic to surface.	15860547
GIP−GLP1	BRET	Ligand induces heteromerization. Dimer shows altered calcium and β-arrestin responses.	22108912
PTH1−Δe14PTH1	Co-IP	Reduced traffic of complex, inhibition of signaling.	20578167
VIP1−VIP2	Co-IP, BRET	Ligand binding reduces oligomerization.	16244179

[a]*This list only includes pairs with novel pharmacological properties. Receptor pairs without distinct, specific functional properties have been excluded.*

[b]*Abbreviations used: 5-HT, serotonin receptor; α, α-adrenergic receptor; A, adenosine receptor; AT, angiotensin receptor; β, β-adrenergic receptor; B, bradykinin receptor; CB, cannabinoid receptor; CCR, CCXR, chemokine receptors; CT: calcitonin receptor; D, dopamine receptor; GIP, gastrointestinal peptide receptor; H, histamine receptor; GLP1, glucagon-like peptide receptor; OR, opioid receptor; PTH1: parathyroid hormone receptor type 1; V, vasopressin receptor; VIP, vasoactive intestinal peptide receptor; Δe13CT, Δe14PTH1: splice variants of the calcitonin and PTRH receptors missing exons 13 and 14, respectively.*

their pharmacological properties are hard to distinguish from those of the monomer state. Moreover, recent study suggests that the dimer, rather than the monomer or higher order oligomeric complexes, is the most abundant state for some of these receptors.[76] The pharmacology of receptor homodimers has been reviewed recently.[77]

Novel links between GPCR oligomerization and human disease have been suggested in recent years. For example, the formation of dopamine D1 and D2 receptor complexes was recently associated with depression. The association between D1R and D2R is increased in samples obtained from patients with severe depression and the administration of a peptide that disrupts the formation of the complex to the prefrontal cortex of rats had anti-depressant effects.[78] Several other physiological and pathological states, like pre-eclampsia,[79] Parkinson's disease,[80] analgesia,[81] schizophrenia,[82] and asthma[83] have been linked to altered GPCR heterodimerization. Thus, GPCR multimeric complexes have emerged as novel, valid drug development targets, suggesting new uses for old compounds and fueling the discovery of novel compounds that target specifically the dimeric/multimeric receptor state.

Several examples of drugs that target receptor dimers have already been reported. For example, 6′-guanidinonaltrindole (6′-GNTI), a drug developed as a kappa-opioid receptor (κOR) agonist, preferentially activates δOR-κOR heterodimers.[84] Moreover, signaling mediated by this compound is blocked by both δOR and κOR antagonists, either *in vitro* or *in vivo*.[84] However, the analgesic effects of the agonist are evident when the drug is administered intrathecally but not upon intracerebroventricular administration, suggesting tissue selectivity that may be exploited in the development of novel analgesics with reduced side effects.[84] A second example is N-naphthoyl-β-naltrexamine (NNTA), a selective activator of κOR-μOR heteromers.[85] This compound is a potent analgesic in mice and a potent activator of downstream signals that express κOR and μOR.[85] However, NNTA is a potent opioid antagonist in cells that only express the μOR.[5] Furthermore, ligands that were traditionally assumed to selectively activate homomeric μOR[morphine, [d-Ala(2)-MePhe (4)-Glyol(5)]enkephalin (DAMGO)], selectively activate δOR-μOR heteromers with substantially greater efficacy than the homomeric μOR.[86] Likewise, the putative selective δOR agonist SNC80 is significantly more active when the δOR is co-expressed with the μOR.[87] The opioid receptor heteromer selectivity of these ligands has been confirmed in transgenic animals deficient in one of the receptor subtypes involved in the formation of the heteromer.[85–87] Thus, heteromers are relevant specific targets for drug development.

The next logical step in the targeting of receptor heteromers is the development of bifunctional ligands. Bivalent ligands may function in at least two ways: interacting with the two orthosteric binding sites of neighboring protomers (orthosteric ligands) or interacting with one primary orthosteric site and an additional allosteric site (dualsteric ligands). Most of the research so far has been focused on the synthesis and evaluation of orthosteric ligands. Several libraries of bivalent ligands targeting oligomeric receptors of different classes have been reported recently.[88–91] Many of these novel ligands stabilize dimeric receptors and have become very useful tools in the investigation of the structure and pharmacology of the dimer. Homobivalent ligands are also of significant interest. For example, homobivalent antagonists for the D2 receptor derived from the linking of two clozapine pharmacophores using spacers of 16–18 atoms have much greater affinity for the receptor (about two orders of magnitude) and potency (about 10-fold) relative to the original pharmacophore clozapine.[92] Importantly, the length of the spacer is critical for the affinity of the homobivalent ligand. Spacers of 16–18 atoms were the most potent/active, with a gradual decrease in activity at shorter and longer spacer lengths.[92]

Co-Receptors and Accessory Signaling Proteins

The pharmacological properties of some subfamilies of GPCR are strongly influenced by the expression of sets of accessory proteins that may become relevant pharmacological targets in the future. An excellent example of a family of accessory proteins are the receptor activity modifying proteins (RAMPs), a class of proteins that enhance diversity in GPCR function, facilitating the creation of binding pockets for alternative ligands, controlling the pharmacology of some GPCR, and regulating cell-surface trafficking and internalization of specific GPCR targets.

The RAMP family includes three single-transmembrane-domain proteins initially shown to regulate glycosylation, traffic, and ligand binding specificity of the calcitonin (CTR) and calcitonin-like (CLR) receptors.[93] RAMP1 and RAMP2 were identified as alternative protein partners for the calcitonin-like receptor (CLR) that modulated receptor traffic and ligand specificity. Both proteins facilitated the proper traffic of the CLR to the plasma membrane; however, when expressed along with RAMP1, the CLR binds to the calcitonin-gene related peptide (CGRP), but when expressed with RAMP2 the activating ligand is adrenomedullin.[93] A third RAMP protein, RAMP3, also chaperones the CLR to the cell surface but enables the CLR to respond to both CGRP and adrenomedullin.[94]

The RAMPs also interact with the CTR, inducing a distinct pharmacological phenotype since the CTR/RAMP complexes bind and respond to amylin.[95,96] CLR/RAMP and CTR/RAMP complexes are presently recognized as specific receptor subtypes. Thus, the new IUPHAR consensus nomenclature is: a) AMY1, AMY2, and AMY3 refer to the CTR/RAMP1, CTR/RAMP2, and CTR/RAMP3 complexes, respectively; b) AM1 and AM2 refer to the CLR/RAMP2 and CLR/RAMP3 complexes; and c) CGRP1 refers to the CLR/RAMP1 complex.[97] The stoichiometry of these complexes remains elusive. Early cross-linking data suggested a 1:1 CLR:RAMP1 ratio,[98] but more recent bimolecular fluorescence complementation measurements suggest that the complexes contain two CLR molecules per RAMP1 monomer.[99] Recent reports have suggested that other members of Family B may interact with RAMPs. Thus, the secretin receptor (SCTR) interacts with RAMP3, but no specific pharmacological properties have been assigned to the SCTR/RAMP3 complex.[100] Likewise, the glucagon receptor, the parathyroid hormone receptors PTH1R, PTH2R, and the vasoactive intestinal polypeptide/pituitary adenylate cyclase-activating peptide receptor (VPAC1R) form complexes with specific RAMPs.[101] Of these, only the VPAC1R/RAMP2 complex shows novel pharmacological properties. The VPAC1R/RAMP2 complex enhances calcium signaling but does not affect cAMP formation.[101]

An important but largely unexplored issue is the role played by the RAMPs in the coupling of downstream signals. RAMPs are not required for CTR traffic to the cell surface or for calcitonin-induced cAMP production. However, the C-termini of the RAMPs markedly modulate the AMY receptor phenotype. RAMP C-terminal truncation does not affect receptor traffic, but markedly reduces ligand affinities and cAMP responses for AMY receptors.[102] Since overexpression of Gs partially rescues this phenotype, this suggests that RAMPs contribute to the coupling of the AMY receptors to G proteins. Moreover, a third protein component, the CGRP-receptor component protein (RCP), interacts with the second intracellular loop of the CLR and is required for effective signal transduction.[103,104] Thus, multiple protein partners regulate the traffic, ligand binding properties, and signaling functions of the CTR/CLR family of receptors.

Another class of GPCR whose function is strongly influenced by proteins co-expressed in the cellular environment is the Frizzled (Fzd) receptor class. In mammals, the Fzd receptor class includes 10 receptors expressed in most organs and tissues. Fzd receptors are the main targets for the 19 Wnt ligands expressed in mammalian systems.[105] Wnt ligands and Fzd receptors modulate a plethora of important biological processes, from early embryonic development to the formation of synapses to tissue repair after injury. Disruption of Wnt signals leads to multiple diseases.[106] Wnt/Fzd signals must be turned on and off at critical moments during embryonic development and tissue homeostasis. This is accomplished primarily by controlling the expression of multiple regulatory proteins that modulate the signaling outcome.

Fzd receptors regulate multiple pathways. The best characterized is the so-called canonical Wnt pathway, which involves the activation of β-catenin-regulated gene transcription. In naïve cells, β-catenin levels are regulated by a destruction complex that includes the protein kinases CK1 and GSK3β and the scaffolding proteins axin and adenomatous polyposis coli (APC).[105] Upon binding to this complex, β-catenin becomes phosphorylated in a specific set of N-terminal residues and further ubiquitylated by the E3 ubiquitin ligase β-Trcp.[105] In the presence of Wnt ligands, Fzd receptors form trimeric complexes including Wnt, Fzd, and the lipoprotein receptor-related protein 5 or 6 (LRP5/6).[107] These complexes activate the adapter protein Dishevelled (Dvl1/2/3) by mechanisms that involve G protein signaling that are still poorly understood. Dvl proteins and axin bind the Wnt/Fzd/LRP complex, which disrupts β-catenin phosphorylation and ubiquitylation. As a result, β-catenin accumulates, traffics to the nucleus, and binds to a subset of transcription factors for the regulation of Wnt responsive gene expression.[105] Activation of the pathway requires the co-expression of multiple signaling proteins, including at least one class of Fzd receptors and one member of the LRP5/6 family.

All Wnt signaling functions that do not depend on β-catenin are grouped under the title of non-canonical Wnt signaling pathways. These include the planar cell polarity pathway, which involves several additional protein partners (vangl/Strabismus, Celsr, Prickle, Daam1, PTK7), and a calcium-dependent pathway that leads to the activation of CamKII, PKC, or calcineurin.[108] Based on the effects of their over/underexpression in *Xenopus* embryonic development or in the transformation of mammary epithelial cells, Wnt ligands were initially classified as canonical and non-canonical.[109,110] Canonical Wnts, such as Wnt-1, -3a, -8, and -8b, induce secondary axis formation in *Xenopus* embryos and cell transformation of human mammary epithelial cells by mechanisms that involve β-catenin stabilization.[109,110] Non-canonical Wnts, such as Wnt-4, -5a and -11, fail to elicit these responses. Moreover, non-canonical Wnts, in particular Wnt5a, inhibit Wnt/β-catenin signaling in various contexts.[111] However, it was soon discovered that Wnt5a is perfectly capable of activating β-catenin responses when co-expressed with Fzd4 receptors.[112] This puzzle was solved with the discovery that the tyrosine kinase Ror2 binds Wnt5a and mediates the inhibition of the Wnt/β-catenin pathway.[112,113]

The expression of alternative receptors to Wnt ligands is not the only mechanism by which cells and tissues regulate canonical vs non-canonical signaling inputs. Some cells and tissues secrete proteins that specifically target different regions of the extracellular domains of LRP5/6 to competitively inhibit Wnt binding, thus preventing the activation of the Wnt/β-catenin pathway without affecting input to non-canonical pathways. Examples of these regulatory proteins are the members of the Dickkopf (DKK) family[114] and sclerostin.[115] Interestingly, however, these secreted proteins bind to very different regions of LRP5/6. The extracellular domain of LRP5/6 comprises four β-propeller motifs organized in two semi-rigid regions.[116] Recent work using specific antibodies has shown that Wnt1, 2, 6, 7a, 7b, 9a, 9b, and 10b (Wnt1 class) interact primarily with β-propeller 1 of LRP6, whereas Wnt3 and Wnt3a (Wnt3 class) interact primarily with β-propeller 3.[117] DKK1 binds to both the first and the third β-propeller regions[116] and inhibits both Wnt1 and Wnt3-class ligands. Sclerostin, however, binds only β-propeller 1 and, therefore, only inhibits Wnt1-class ligands.[117] Thus, the classification of Wnt ligands as "canonical" and "non-canonical" does not reflect intrinsic properties of the ligands; rather, whether a Wnt ligand behaves as "canonical" or "non-canonical" largely depends on the cellular environment. Figure 2.3 depicts a simplified version of the multiple alternative interactions established among Wnt signaling proteins.

Receptor Trafficking

Receptor endocytosis and recycling are differentially regulated by various ligands, independently of their action on the canonical signaling pathways of the receptor. For example, the PTH1R is fully activated by amino-terminal fragments of PTH (PTH[1-34]) and parathyroid-hormone-related peptide (PTHrP[1-36]). Both ligands are regarded as full agonists, since under standard laboratory conditions they maximally activate cAMP production and intracellular Ca^{2+} release.[118] However, the fate of the ligand-bound PTH1R depends on the agonist. This was first detected in washout experiments, in which the target cell is exposed to the ligand for a short period after which the ligand is removed. Although both PTH(1-34) and PTHrP(1-36) induce receptor endocytosis, the PTHrP(1-36)-activated receptor recycles rapidly to the surface, whereas the PTH(1-34)-activated receptor is sequestered in an intracellular compartment that continues generating cAMP.[119] Therefore, the effect of PTHrP(1-36) in the production of cAMP is transient, returning to basal within a few minutes, whereas the generation of cAMP induced by PTH(1-34) is sustained.

Further work suggested the formation of a stable oligomeric complex containing the ligand-bound receptor, β-arrestin, and Gβγ dimers that continues signaling from endosomes for many minutes after the initial exposure of the cells to PTH(1-34).[120] This complex is formed exclusively when using PTH(1-34) and other closely related analogs, but not upon stimulation with PTHrP-derived peptides.[119,120] Little is known about the mechanism of termination of the signal, although the retromer complex—a sorting machinery that regulates retrograde traffic from early endosomes to the Golgi—has been

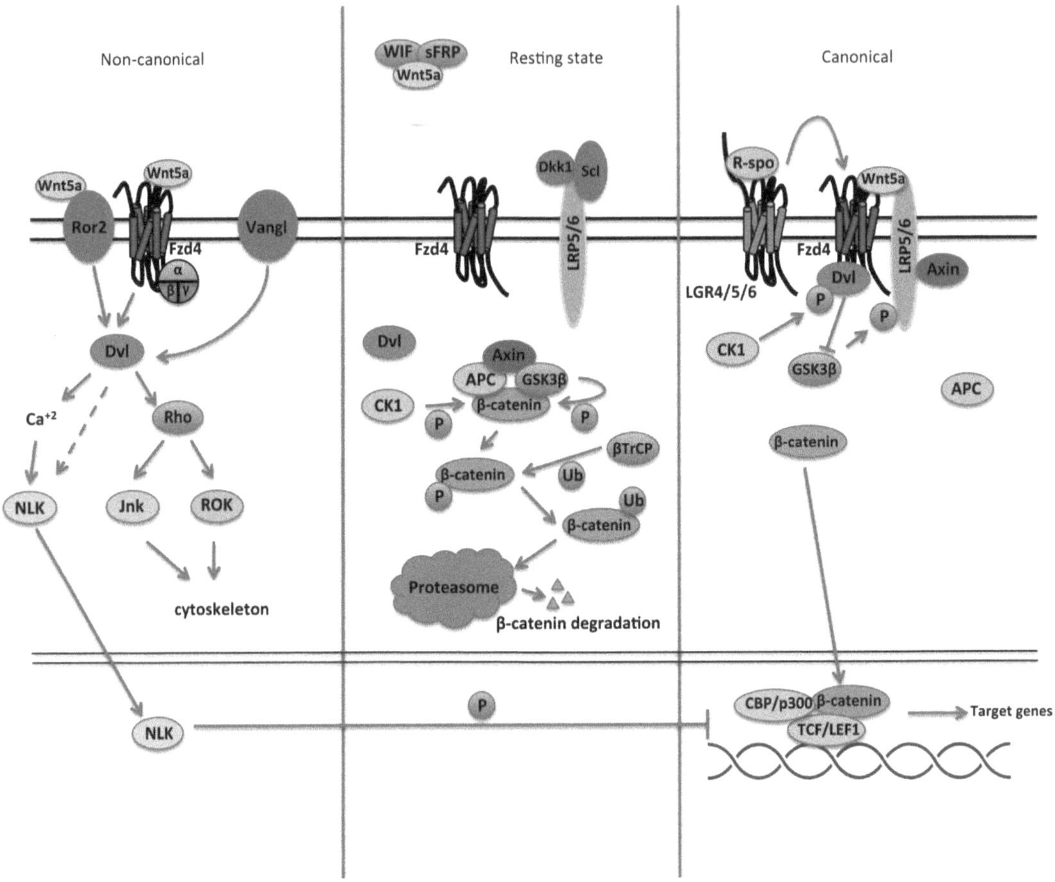

FIGURE 2.3 Alternative Wnt signaling pathways. Wnt5a activates or inhibits β-catenin signaling depending on the cell context. In the resting state, Wnt ligands are sequestered by soluble factors and LRP5/6 associate to the inhibitors Dickkopf 1 (Dkk) and sclerostin (scl). Under these conditions, the axin/APC/Gsk3β/β-catenin complex is stable, which results in the phosphorylation of β-catenin by casein kinase 1 (CK1) and GSK3β. Phosphorylated β-catenin is further ubiquitylated and degraded in the proteasome. Activation with Wnt5a may result in two opposite effects depending on the levels of expression of accessory proteins. In the presence of the tyrosine kinase Ror2 and the planar cell polarity (PCP) factors (such as Van Gogh-like, Vangl), the interaction of Fzd4 with Disheveled (Dvl) leads to the release of calcium from intracellular stores and, in parallel, to the activation of Rho GTPases. The mechanisms involved are poorly understood, but heterotrimeric G proteins seem to be required. This leads to reorganization of the cytoskeleton and activation of the Nemo-like kinase (NLK), which translocates to the nucleus and phosphorylates TCF4/LEF1 and blocks β-catenin signaling. In the absence of Ror2 and the PCP core factors, Fzd4 and LRP5/6 form a complex with Wnt5a, which recruits Dvl to the surface. Further phosphorylation of LRP5/6 by Gsk3β and of Dvl by CK1 leads to the formation of a more stable complex that recruits actin and leads to the inactivation of Gsk3β. This effect is enhanced by a family of secretory proteins, the R-spondins (R-Spo), which bind to a novel family of GPCRs known as LGR4, 5 or 6, which act as modulators of the canonical pathway via mechanisms that remain obscure. Since β-catenin is no longer phosphorylated, it accumulates in the cytosol and eventually moves to the nucleus, where it forms a complex with TCF/LEF transcription factors and the CREB binding protein CBP to direct the transcription of specific target genes.

implicated.[121] Qualitatively similar phenomena have been reported with other GPCRs, although the existence of the receptor-β-arrestin-Gβγ complex remains to be established.[122–124] This phenomenon (signaling from multiple compartments) may explain the quantitative differences between the responses to similar ligands.

One consequence of the activation of signaling pathways in distinct compartments is the spatial–temporal organization of the signal. cAMP signaling is highly organized by multiple scaffolding proteins, such as the A-kinase anchoring proteins (AKAPs). These proteins link PKA to signaling complexes assembled on scaffolds targeted to specific cellular subdomains. These signaling complexes often include, in addition to PKA, other kinases, protein phosphatases, and phosphodiesterases that break down cAMP.[125] This results in remarkable compartmentalization of the signal.[126] A second example of signal compartmentalization in the GPCR signaling pathways is the activation of ERK kinases. GPCR engage multiple mechanisms to activate the ERK pathway. These include typical G protein-mediated events (PKA, PKC), transactivation of the EGFR, direct interaction with tyrosine kinases of the Src family, and focal adhesion-based ERK activation through the protein kinase Pyk2.[127]

In general, the mechanism of ERK activation by GPCR is remarkably receptor- and cell dependent. Even more remarkable is the subcellular organization of the system. For instance, Angiotensin II (Ang II) activates ERK in cardiac myocytes via a mechanism that involves the transactivation of the EGFR.[128] However, although EGF activation is required for Ang II-dependent activated ERK phosphorylation, the fate of the activated ERK is not the same (Figure 2.4). Whereas EGF-activated ERK translocates to the nucleus, ANG II-activated ERK remains in the cytoplasm, attached to β-arrestin on the surface of endosomes.[129] Thus, signaling pathways that share common mechanisms may lead to very distinct biological outcomes. Indeed, whereas activation of the EGFR by EGF has a potent mitogenic effect, the indirect activation of EGFR by Ang II does not.[127]

RECEPTOR TYROSINE KINASES

Unlike GPCRs, which were traditionally considered monomeric signaling units, receptor tyrosine kinases (RTK) have been known to function as dimers almost since the outset. Activation of RTKs requires the phosphorylation of specific tyrosines, which requires close proximity of two RTK monomers. These phosphorylated tyrosine residues anchor specific adapter proteins and tyrosine kinase targets. Signaling bias at various levels has been detected in RTKs. An early example of this is the activation of the EGF receptor (EGFR/HER1/ErbB1) by specific ligands. Two of these, EGF and TGFα, induce similar spectra of functional responses in some assay systems, but very different outcomes in others. The origin of this ligand bias has been linked to the different traffic of the EGF-EGFR and TGFα-EGFR complexes. Both types of complexes are subjected to endocytic traffic, but TGFα-bound receptors are recycled much faster than EGF-bound receptors.[130] In fact, EGF-EGFR complexes are often targeted to the lysosomes, whereas TGFα-EGFR receptors are not.[130] Because of this, TGFα has been considered a more potent mitogenic ligand.[131] However, whereas internalized

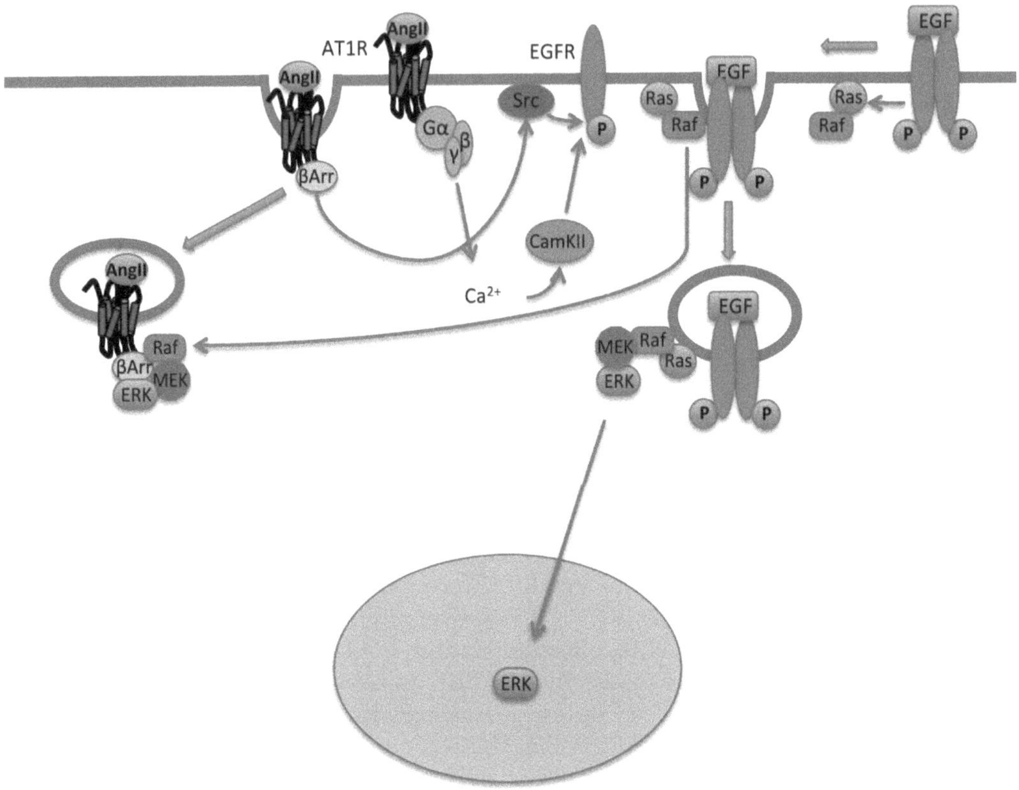

FIGURE 2.4 Two distinct modes of activation of the ERK cascade. Angiotensin II binds the angiotensin type 1 receptor (AT1R) inducing G protein activity, Ca^{2+} release, and calmodulin kinase II (CaMKII) activation. The activated receptor binds β-arrestin and becomes internalized. β-arrestin activates the tyrosine kinase Src, which in turn contributes to the activation of cell surface EGF receptors (EGFR). The EGFR activates Ras, which in turn activates Raf and initiates the signaling cascade leading to the activation of ERK. β-arrestin serves as a scaffold, keeping Raf and the other components of the ERK cascade attached to the surface of endosomes. As a consequence, ERK remains trapped in the cytosol and does not enter the nucleus. In contrast, EGF promotes dimerization, cross-phosphorylation, and internalization of the EGFR, inducing activation of the ERK cascade on the cell surface and the surface of endosomes. However, the binding of ERK to endocytic vesicles is, in this case, only transient. ERK dissociates after activation and enters the nucleus to regulate specific transcriptional events.

TGFα-EGFR complexes rapidly dissociate and recycle to the membrane, EGF-EGFR complexes continue signaling from the endosomal compartment.[132] This leads to a distinct pattern of ERK1/2 phosphorylation and different signaling outcomes.[133]

RTK Signaling Networks

RTK signaling is usually the result of a complex network of protein interactions. These networks include multiple receptors that associate to form heteromeric complexes. These complexes bind multiple ligands to elicit distinct physiological functions.

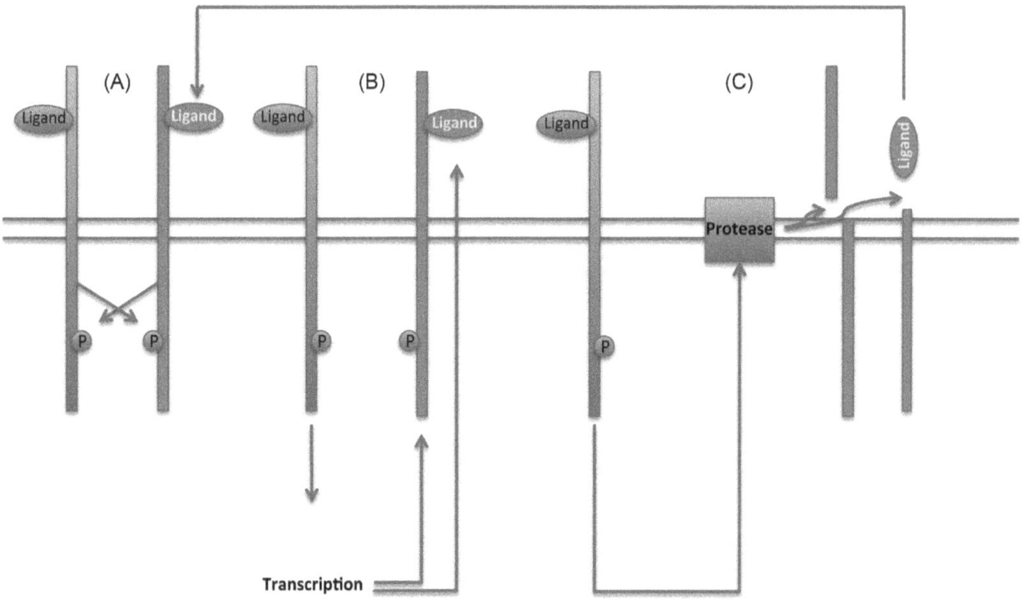

FIGURE 2.5 Receptor tyrosine kinase cross-activation. Three common mechanisms of RTK crosstalk: (A) Cross-phosphorylation as a result of stable or transient heterodimerization. (B) Ligand-dependent transcription of RTK and/or growth factor ligand. (C) Activation of proteolytic enzymes that cleave and activate receptor or release surface-bound ligand.

Cellular responses depend on the nature of the ligands, the structure of the ligand complexes, and the presence of specific accessory proteins. Disruptions of RTK signaling are most often associated with cancer. Thus, a great deal of research has been directed to the discovery of drugs that inhibit receptor function. Most of the drugs currently used in the clinic or in advanced clinical trials are either tyrosine kinase inhibitors or antibodies that target the receptors. There is a scarcity of novel ligands or allosteric modifiers and, in consequence, relatively little is known regarding the influence of the cellular milieu on biased responses. However, there are several examples of signaling bias by natural ligands that regulate RTK signaling networks. Crosstalk between RTK signaling pathways is common and can be a consequence of interactions at different levels (Figure 2.5). RTKs can induce activation of related and unrelated RTK by direct dimerization,[134,135] transactivation by cross-phosphorylation, activation of non-receptor kinases such as c-Src,[136] transcriptional regulation,[137] or activation of cell surface proteases that cleave and activate the receptor or release active growth factor from a cell surface precursor.[138]

RTKs signal to a common set of signal transducers: the Ras/Raf/ERK cascade, the activation of phospholipase Cγ (PLCγ), and the activation of type I phosphatidylinositol-3′-kinases (PI-3K). Table 2.5 summarizes recent studies on the mechanisms of crosstalk among RTKs and between RTKs and GPCR signaling pathways.

TABLE 2.5 Signaling Crosstalk among Receptor Tyrosine Kinases

Receptors[a]	Mechanism	Physiological Consequences	PMID Accession Number
EGFR-HER2	Heterodimerization	Enhanced signaling (mitogenesis, transformation). Altered endocytic traffic of the EGFR.	7565730 8665853
EGFR-HER3	Heterodimerization	Enhanced signaling (mitogenesis, transformation).	7565730 8665853
EGFR-HER4	Heterodimerization	Cross-phosphorylation, increased survival.	7565730
EGFR-AXL	Heterodimerization, transactivation	EGFR transactivates AXL, which becomes ligand independent. Promotes drug resistance, EMT.	23921085
EGFR-IGF1R	Heterodimerization, cross-phosphorylation	Ligand-independent activation, drug-resistant phenotype in some cancers, increased mitogenic activity, increased Akt, ERK phosphorylation.	14593113 17047074
EGFR-cMET	Heterodimerization, transactivation, transcriptional regulation	Enhanced mitogenesis. Phosphorylation at specific tyrosine residues.	10722725 18234969
EGFR-PDGFRβ	Heterodimerization, transactivation mediated by Src	Enhances PDGF-dependent ERK activation.	11533228
FGFR1-PDGFRβ	Heterodimerization, cross-activation	PDGF activates FGFR1. Increased cell proliferation.	19339244
HER2-HER3	Heterodimerization	Enhanced signaling (mitogenesis, transformation).	8665853
HER2-HER3-IGF1R	Heterotrimerization	Enhanced signaling, resistance to anti-HER2 antibodies.	20103628
HER3-cMET	Heterodimerization	Cross-activation.	18180459
IGF1R-IR	Hybridization[b]	Hybrid receptors signal like IGF1R. Increased proliferative responses.	1698059
cMET-PDGFRα	Transcriptional regulation	c-MET increases PDGF, PDGFRα expression. Activation of PDGF signaling pathways.	21496277
cMET-RON	Heterodimerization, transphosphorylation	Signal amplification, transformation	10871856
PDGFR-VEGFR	Dimerization	Dimerization withVEGFR inhibits PDGFR function.	22426130
CROSSTALK WITH OTHER PATHWAYS			
GPCR-EGFR	Release of EGFR ligands. Src-mediated activation.	Mediates ERK activation by many GPCRs.	23992425 16769815 19735466 18372913

[a]*Abbreviations used: AXL: AXL receptor tyrosine kinase; c-MET: Hepatocyte growth factor receptor; EGFR: Epidermal growth factor receptor; HER2, HER3: Heregulin receptors 2 and 3, also known as ErbB2, ErbB3; IGF1R: Insulin-like growth factor 1 receptor; IR: insulin receptor; PDGFR: platelet-derived growth factor receptor; RON: RON tyrosine kinase, also known as macrophage stimulating 1 receptor or PTK8; VEGFR: vascular endothelial growth factor.*

[b]*Hybridization: IGF1R and IR are heterotetramers formed by the proteolytic processing of receptor dimers assembled by disulfide bonding. IGF1R-IR hybrids are receptors containing one IR dimer and one IGF1R dimer.*

The EGF Receptor Signaling Network

The EGF receptor signaling network has been thoroughly studied because of its relationship with oncogenesis. This network encompasses four single-pass tyrosine kinases (HER1-4, also termed ErbB1-4) that homo- and hetero-dimerize to yield 10 different receptor types that interact with a dozen ligands to generate multiple biological outcomes that depend strongly on the engagement of specific accessory proteins.[139,140] Overexpression of specific ErbB gene products can bias dimer formation leading to disruption of signaling homeostasis and disease. The EGFR/HER1 binds seven distinct naturally occurring ligands, including EGF and TGFα. The remaining ligands are: heparin-binding EGF-like growth factor (HB-EGF), betacellulin (BTC), amphiregulin (AR), epiregulin (EPR), and epigen.[140] All seven are made as type I transmembrane proteins that are inserted into the plasma membrane and then cleaved by proteolytic enzymes to release the mature growth factor.[140] HB-EGF, BTC, and EPR also bind and activate HER4. BTC is the only member of the family that binds HER2/HER3 heterodimers, although it does not interact with HER3/HER3 homodimers.[140] A different class of ligands, the neuregulins, bind both HER3 and HER4.[139] There are no known ligands for HER2; amplification/overexpression of HER2 induces constitutive activity often associated with cancer. Furthermore, HER3 has a defective kinase domain.[141] Thus, the consensus is that HER2 and HER3 function primarily as subunits of higher order complexes involving other members of the family.

This idea is strongly supported by experimental evidence. For example, the overexpression of HER2 in mammary carcinoma is accompanied by elevated tyrosine phosphorylation of HER3.[142] Conversely, reduced expression or inhibition of HER2 diminishes HER3 phosphorylation.[143] Likewise, inhibition of HER3 expression in cells that overexpress HER2 reduces PI3 kinase signaling and impairs cell proliferation.[143] These data strongly support a model in which heterodimers consisting of ligandless HER2 and kinase dead HER3 behave as potent oncogenic complexes driving breast cancer cell proliferation.

Although it is clear that ligand binding induces HER dimerization, the formation of dimers in the absence of ligand is controversial. Analysis of the X-ray crystal structure of the extracellular domain of EGFR, HER3, and HER4 has revealed a conserved β-hairpin loop that remains locked in the unliganded receptor but projects away from the receptor:ligand complex.[144] This "finger" mediates the formation of receptor dimers, and is known as the dimerization loop.[144] HER2 is the exception. The ectodomain of HER2 is locked in an "active" conformation that exposes the dimerization loop.[145] Recent work using co-immunoprecipitation methods and spectroscopic approaches has confirmed that all members of the ErbB family can form homo- and hetero-oligomers even in the absence of ligand.[146,147]

Early studies focused on the regulation of EGFR activity by its natural ligands and suggested that all seven natural ligands have very similar effects on the activation of downstream pathways. However, the fate of the receptor is differentially regulated by these ligands. Whereas all EGFR/HER1 ligands activate receptor endocytosis, in HEp2 cells only two (BTC and HB-EGF) retain the receptor in the endocytic compartment and target it to late endosomes and lysosomes.[131] Close to 100% of all EGFR internalized after treatment with TGFα or EPI are recycled to the plasma membrane, whereas AR and EGF produce intermediate responses (≈50% retention).[131] Some elements of these differential effects are a consequence of intrinsic receptor–ligand interaction patterns. Internalized receptors traffic to early endosomes, which are gradually acidified during traffic. TGFα dissociates from EGFR/HER1 at pH 6.5 and EGF dissociates at pH 5.5,[130,131] whereas HB-EGF, BTC, and AR remain associated to the receptor below pH 5.[131]

Most of these observations are consistent with a model of dissociation of the ligand as the interior of the endosome acidifies followed by recycling of the dissociated receptor to the surface; in contrast, receptors that remain ligand-bound traffic to the lysosome where they are degraded. However, AR does not follow this pattern; although AR binding is insensitive to intravesicular pH changes within the physiological range, a large fraction of the EGFR internalized after AR treatment recycles to the surface and less than 30% of EGFR is degraded after AR treatment.[131] The solution to this puzzle is the differential interaction of EGFR-ligand complexes with an intracellular partner, the ubiquitin ligase c-Cbl. Ligands promote the association of EGFR to c-Cbl, which results in ubiquitylation and degradation of the receptor. Whereas c-Cbl forms long-lived complexes with EGFR after activation with HB-EGF, EGF, and BTC, this is not the case when TGFα, EPI, or AR are selected as ligands. As a result, EGFR ubiquitylation is transient after treatment with TGFα, EPI, and AR, but subsistent after EGF, BTC, and HB-EGF. This was linked to differential phosphorylation of specific tyrosine residues in the C-terminus of the receptor.[131] Interestingly, various members of the ErbB/HER family traffic to the nucleus where they mediate a variety of novel functions.[148] Given the substantial differences in the trafficking patterns elicited by EGFR ligands, significant bias in terms of the nuclear functions of the EGFR should be expected. This issue, unfortunately, remains largely unexplored.

One of the landmarks of ErbB/HER signaling is the crosstalk with multiple signaling pathways. Besides heterodimerization involving other members of the HER/ErbB family, ErbB/HER signaling can be strongly influenced by unrelated RTKs. For example, HER2 interacts with the insulin-like growth factor receptor 1 (IGF1R).[135] This interaction is enhanced by either neuregulin-1 (also known as heregulin) or IGF-1 in breast cancer cells.[135] There is also strong evidence of EGFR/IGF1R crosstalk. In COS-7 cells, IGF-1-induced phosphorylation of downstream targets is inhibited by tyrphostin AG1478, an inhibitor of the EGFR kinase activity that has no effects on the autophosphorylation of the IGF-1R, supporting a transactivation model.[149] The mechanism that mediates these effects has not been fully elucidated, but the evidence suggests that IGF-1 activates metalloproteinases and induces the release of EGFR ligands from the cell surface.[149] Likewise, EGFR tyrosine kinase activity is required for the effects of IGF1 on ERK activation and mitogenesis in normal human breast cells and primary cultures of breast epithelium.[134] However, inhibition of the EGFR tyrosine kinase in breast cancer cell lines does not affect IGF1 responses.[134] Therefore, the linkage between EGFR and IGF-1 signaling strongly depends on the cellular environment. The origin of these disparities remains obscure.

Crosstalk of the EGFR network with other tyrosine kinase pathways has been reported. For example, the RTK c-MET (the hepatocyte growth factor receptor) is constitutively phosphorylated in human hepatoma cell lines and the human epidermoid carcinoma cell line A431, which express TGFα, but not in normal human hepatocytes.[150] Constitutive c-Met phosphorylation in A431, HepG2, AKN-1, and HuH6 cells was inhibited by neutralizing antibodies against TGFα and/or EGFR. Furthermore, increase of c-Met phosphorylation by exogenous TGFα in A431 cells was inhibited by neutralizing antibodies against TGFα and/or EGFR and by the EGFR-specific inhibitor tyrphostin AG1478. In tumor cells, c-Met co-immunoprecipitated with EGFR regardless of the presence of ligand.[150]

Other RTK Networks

Receptor oligomerization is a common motif among receptor tyrosine kinases. The platelet derived growth factor (PDGF) receptor family is a classical example. Two distinct type 1 membrane proteins mediate most of the actions of PDGF (PDGFRα and PDGFRβ). These receptors are characterized by five extracellular immunoglobulin loops and a split intracellular tyrosine kinase domain.[151] The natural PDGF ligands are disulfide-linked dimers of four distinct proteins termed PDGF-A, PDGF-B, PDGF-C, and PDGF-D. All ligands form homodimers, but only PFGF-A and PDGF-B are capable of forming heterodimers. PDGF-A homodimers bind exclusively PDGFRα whereas PDGF-B homodimers bind both receptor types.[152] Thus, PDGF-A homodimers will promote the formation of αα receptor homodimers, PDGF-A/PDGF-B heterodimers will bind to αα and αβ complexes, and PDGF-B homodimers will bind all three possible receptor combinations. It is unclear whether dimerization can occur in the absence of ligand. Interestingly, recent data suggest that the transmembrane domains of the PDGFRβ may spontaneously associate forming dimers and higher order aggregates.[153]

PDGF receptors interact with several other RTKs both directly and indirectly. Direct interactions between PDGFRββ and EGFR receptors have been reported.[154] In vascular smooth muscle cells, these interactions result in EGF-independent phosphorylation of the EGFR and increased ERK phosphorylation after treatment with PDGF. Heteromer disruption reduces ERK activation by PDGF.[154] However, cross-activation does not result from phosphorylation of EGFR by PDGFR, and requires c-Src activity.[154] A direct interaction between PDGFRα and VEGFR2 has also been reported. In this case, however, formation of the heteromer inhibits PDGFR function.[155] A different class of crosstalk mediated by transcriptional regulation with no evidence of heteromerization exists between PDGFRα and c-MET.[156] Other examples of RTK crosstalk are the activation of the receptor tyrosine kinases AXL and RON by c-MET.[157] In both cases, heterodimerization causes transphosphorylation and signal amplification, both of which play a major role in transformation and anti-cancer drug resistance.

The extensive crosstalk among RTK signaling networks and the relevance of this crosstalk to cancer therapy has fueled multiple efforts to compile information on the phosphoproteome of normal and transformed cells. Some initial studies have led to rather unexpected results. Although the phosphoproteomes associated with different RTKs may overlap quite substantially, the biological outcomes may be very diverse. For example, the tyrosine phosphorylation patterns of human mesenchymal stem cells (hMSC) treated with EGF and PDGF are nearly identical, but only EGF causes the cells to differentiate along the osteogenic pathway.[158] The most obvious distinction between both phosphoproteomes is the relative activation of the PI3K pathway: Whereas PDGF activated PI3K, EGF did not. Suppressing the PI3K pathway allowed PDGF to promote osteogenic differentiation of hMSC.[158]

These proteomics approaches have been used to analyze differences in the phosphorylation patterns induced by wild type and mutant receptor tyrosine kinases. A recent study using isogenic immortalized human bronchial epithelial cells (HBECs) expressing either WT or two mutants of EGFR (Del E746-A750 and L858R) demonstrated much higher basal levels of tyrosine phosphorylation in the two mutants, with the Del E746-A750 mutant exhibiting a five-fold increase in phosphorylation at the Y727 site.[159] Furthermore, proteomics studies have emphasized the role of the cellular environment in the activation of specific pathways. For example, a recent study examined the phosphotyrosine contents of a cohort of non-small cell lung carcinoma (NSCLC) cell lines and tumor samples directly obtained from patients.[160] Interestingly, the subsets of activated RTKs in NSCLC cell lines differed from those found in tumors. EGFR, c-Met, and EphA2 receptors had the highest levels of tyrosine phosphorylation in NSCLC cell lines, but the discoidin domain receptors (DDR1 and 2) and EGFR formed the largest component of phosphorylated RTKs in tumor specimens. In general, signaling processes in cancer are influenced by cellular context, such as the presence of additional mutations within the same cell, or extracellular stimuli such as matrix and stromal interactions that modulate specific signaling outcomes.

Although sometimes difficult to interpret, proteomics studies have shed important light on the crosstalk between multiple RTK pathways. For instance, a recent study employed quantitative mass spectometry methods to demonstrate crosstalk between a truncated EGFR mutant (EGFRVIII) and c-MET in glioblastoma.[161] Based on this finding, a combined treatment regimen that used a c-MET kinase inhibitor and an EGFR kinase inhibitor was devised. This regimen resulted in enhanced killing of EGFRvIII-expressing cells compared with treatment with either compound alone. EGFR–cMET crosstalk also leads to drug resistance. A recent study reported that over 60% of drug-resistant lung cancer tumors overexpress HGF, the c-MET ligand.[162] This exemplifies the complexity of the crosstalk between independent RTK pathways and the value of a proteomic approach to sort out potential alternative therapeutic strategies.

THE MICROENVIRONMENT

Sixty years ago, Furchgott and Bhadrakom[163] reported that acetylcholine (ACh), an agent widely recognized as a potent vasodilator in whole animal and organ perfusion studies, induced contraction of isolated strips of aortic muscle. This paradox was not resolved until the discovery that the muscle relaxation effects of ACh require a two-cell system: Whereas denuded aortic muscle strips contract in response to ACh, intact aortic preparations that conserve the endothelial layer exhibit relaxation.[164] This led to the discovery of nitric oxide (NO), a critical second messenger that mediates numerous cell responses. Acetylcholine stimulation of endothelial cells induces the production of NO, which diffuses to the vascular smooth muscle cell where it activates a guanylate cyclase. The accumulation of cGMP mediates the relaxation of the muscle cell and supersedes the activation of muscarinic receptors, which are responsible for the contractile response of the isolated muscle cells. Figure 2.6 illustrates how cellular responses to a single ligand are critically influenced by the environment.

Numerous additional examples illustrate the role of the local environment in the regulation of cellular processes. Even systems that are regarded as relatively simple integrate a very large number of signaling pathways. An example of a three-cell system is shown in Figure 2.7. The figure depicts signal integration in the bone. Three primary cells participate in the regulation of bone mass: osteoblasts, osteoclasts, and osteocytes. The interactions among these three cells involve one hormone (PTH), five paracrine/autocrine factors

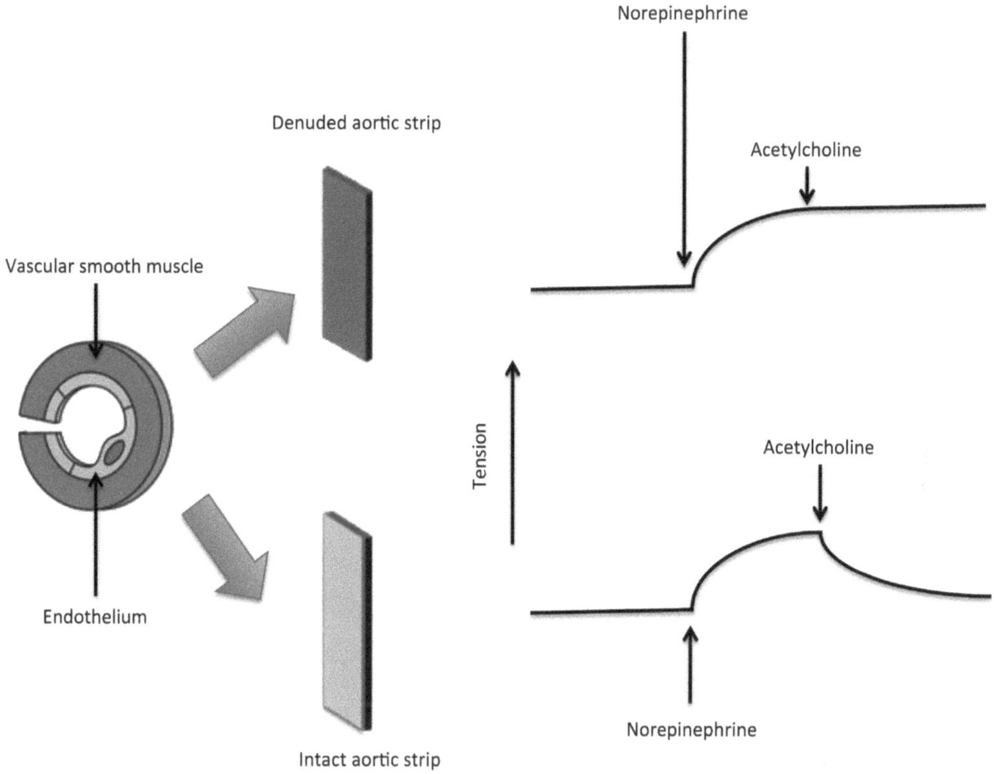

FIGURE 2.6 The aortic strip experiment. Denuded aortic strips contract in response to norepinephrine (NE). Acetylcholine (ACh) cannot reverse this effect. Intact aortic strips, containing endothelial cells, contract in response to NE and relax upon addition of ACh. This is a consequence of the production of nitric oxide (NO) in the endothelium in response to ACh. NO diffuses to the vascular smooth muscle cell and activates cGMP production, which mediates muscle relaxation.

produced locally (BMP, RANK ligand, Wnt3a, Wnt5a, and Wnt10b), three secreted feedback regulatory proteins (DKK1, osteoproteregin, and sclerostin), and at least seven receptors (BMPR1, several Fzd, the Wnt co-receptors LRP5/6 and Ror2, PTH1R, and RANK).

This system performs two main functions: osteoclasts remove bone, whereas osteoblasts deposit new bone material.[165] Osteocytes are derived from osteoblasts and regulate the function of both osteoblasts and osteoclasts. The primary regulator of the system is parathyroid hormone, PTH, which binds to PTH1R receptors present in the osteoblasts and osteocytes. The PTH1R performs multiple functions, including the activation of cAMP production and the activation of β-catenin. PTH function promotes the differentiation of osteoprogenitors into osteoblasts and induces expression of osteocalcin and collagenase-3, two proteins required for the formation of new bone. PTH alone cannot drive the differentiation program, and requires parallel activation of the β-catenin-mediated canonical Wnt

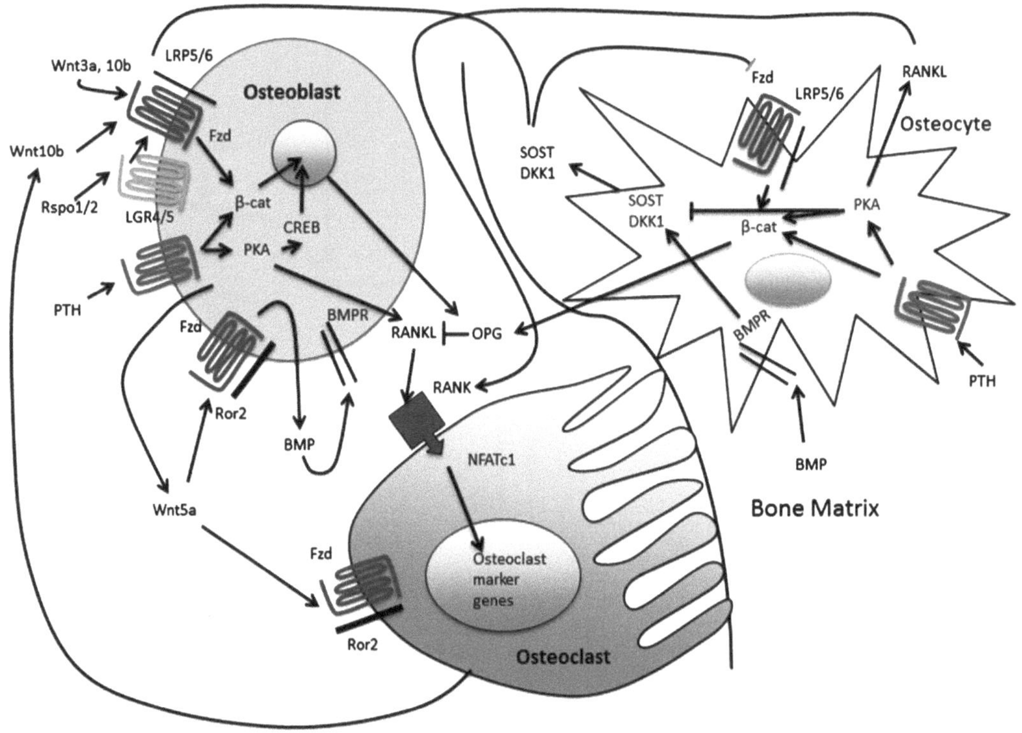

FIGURE 2.7 Bone homeostasis. The osteocyte secretes the Wnt antagonists sclerostin (SOST) and DKK1. The expression of these is repressed by parathyroid hormone (PTH) via PKA and stimulated by bone morphogenetic protein (BMP). Wnt signaling in osteocytes regulates expression of osteoprotegerin (OPG), which binds to the osteoclast differentiation factor RANKL and prevents activation of the RANK receptor. Osteoblasts secrete Wnt5a, which acting via Fzd/Ror2 further inhibits differentiation of osteoclast precursors. Osteoclasts, in turn, secrete Wnt10b, which activates canonical Wnt signaling in osteoclasts. Activation of the PTH1R further increases β-catenin signaling, promoting osteoblast differentiation, increasing local BMP secretion and, subsequently, RANKL expression that will activate osteoclast differentiation.

signaling pathway. One of the osteoblast genes regulated by PTH is RANK ligand (RANKL), a membrane protein that is sometimes shed by the action of metalloproteinases.[166] RANKL binds receptors located on the surface of the osteoclast, a macrophage-derived cell type responsible for the resorption of bone. Activation by RANKL drives the differentiation of osteoclasts.

However, efficient osteoclast differentiation requires inhibition of the β-catenin pathway. This is accomplished by Wnt5a, another protein produced by the differentiating osteoblast, which binds Fzd and its coreceptor Ror2 to block canonical Wnt responses. Differentiating osteoclasts also produce soluble Wnt ligands that potentiate the osteoblast differentiation responses. Bone resorption is indirectly turned off by BMP, which is produced by both osteoblasts and osteocytes in response to PTH and non-canonical Wnt signaling. BMP acts via the BMPR1A receptor and stimulates the secretion of osteoproteregin (OPG), a decoy receptor for RANKL. Secreted OPG binds RANK and prevents its interaction with RANKL. Furthermore, in osteocytes, BMP also regulates the expression of DKK1 and sclerostin (SOST), which bind to LRP5/6 and block Wnt signaling and further osteoblast differentiation. These are by no means the only signals that regulate bone. For instance, osteocytes express mechanosensors that reduce the expression of SOST,[167] and multiple additional GPCRs modulate the process.[168]

CONCLUSION

The responses of cells to specific ligands are critically influenced by the cellular proteome and the cell environment. Signaling processes are integrated in microscopic and multicellular networks that integrate the output of many signals to produce one main biological response. Whereas in the past the analysis of signaling processes traditionally focused on the study of signaling pathways that were often assumed to be linear, it has become increasingly clear that signaling pathways are in reality signaling networks. Biological processes are the outcome of the integration of multiple signals arising from within one cell, from the niche occupied by this cell, and from the whole organism. One of the keys for understanding the phenomenon of biased signaling is the crosstalk between the signaling pathways modulated by specific cell surface receptors. Because of this crosstalk, drugs that bind specific factors may have unpredictable outcomes that must be taken into account when exploring new therapies assumed to be specific. We have only scratched the surface of the complex protein–protein interaction networks that underlie cell signaling cascades. Within this context, signaling bias cannot be viewed exclusively as a bifurcation of the signaling outputs of single receptor–agonist pairs. On the contrary, signaling bias often results from the integration of alternative pathways, the modification of the targeted receptor by scaffolding proteins and other signaling partners, or the influence of processes elicited in other cells that modify the local environment of the target. This has important consequences in the study of the pharmacology of new compounds.

References

1. Kenakin T. The potential for selective pharmacological therapies through biased receptor signaling. *BMC Pharmacol Toxicol* 2012;**13**:3.
2. Kenakin T. Functional selectivity and biased receptor signaling. *J Pharmacol Exp Ther* 2011;**336**(2):296–302.
3. Sneddon WB, Syme CA, Bisello A, Magyar CE, Rochdi MD, Parent JL, et al. Activation-independent parathyroid hormone receptor internalization is regulated by NHERF1 (EBP50). *J Biol Chem* 2003;**278**(44): 43787–96.
4. Sneddon WB, Magyar CE, Willick GE, Syme CA, Galbiati F, Bisello S. Ligand-selective dissociation of activation and internalization of the parathyroid hormone (PTH) receptor: conditional efficacy of PTH peptide fragments. *Endocrinology* 2004;**145**(6):2815–23.
5. Wheeler D, Sneddon WB, Wang B, Friedman PA, Romero G. NHERF-1 and the cytoskeleton regulate the traffic and membrane dynamics of G protein-coupled receptors. *J Biol Chem* 2007;**282**(34):25076–87.
6. Cupp ME, Nayak SK, Adem AS, Thomsen WJ. Parathyroid hormone (PTH) and PTH-related peptide domains contributing to activation of different PTH receptor-mediated signaling pathways. *J Pharmacol Exp Ther*;**345** (3):404–18.

7. van der Lee MM, Verkaar F, Wat JW, van Offenbeek J, Timmerman M, Voorneveld, L et al. β-Arrestin-biased signaling of PTH analogs of the type 1 parathyroid hormone receptor. *Cell Signal*;**25**(2):527–38.
8. Gesty-Palmer D, Chen M, Reiter E, Ahn S, Nelson CD, Wang S, et al. Distinct beta-arrestin- and G protein-dependent pathways for parathyroid hormone receptor-stimulated ERK1/2 activation. *J Biol Chem* 2006;**281**(16):10856–64.
9. Gesty-Palmer D, Flannery P, Yuan L, Corsino L, Spurney R, Lefkowitz RJ, et al. A beta-arrestin-biased agonist of the parathyroid hormone receptor (PTH1R) promotes bone formation independent of G protein activation. *Sci Transl Med* 2009;**1**(1):1ra1.
10. Sebastian EM, Suva LJ, Friedman PA. Differential effects of intermittent PTH(1-34) and PTH(7-34) on bone microarchitecture and aortic calcification in experimental renal failure. *Bone* 2008;**43**(6):1022–30.
11. Fields S, Song O. A novel genetic system to detect protein-protein interactions. *Nature* 1989;**340**(6230):245–6.
12. Albers M, Kranz H, Kober I, Kaiser C, Klink M, Suckow J, et al. Automated yeast two-hybrid screening for nuclear receptor-interacting proteins. *Mol Cell Proteomics* 2005;**4**(2):205–13.
13. Rual JF, Venkatesan K, Hao T, Hirozane-Kishikawa T, Dricot A, Li N, et al. Towards a proteome-scale map of the human protein-protein interaction network. *Nature* 2005;**437**(7062):1173–8.
14. Daulat AM, Maurice P, Jockers R. Recent methodological advances in the discovery of GPCR-associated protein complexes. *Trends Pharmacol Sci* 2009;**30**(2):72–8.
15. Maurice P, Guillaume JL, Benleulmi-Chaachoua A, Daulat AM, Kamal M, Jockers R. GPCR-interacting proteins, major players of GPCR function. *Adv Pharmacol*;**62**:349–80.
16. Chung KY, Day PW, Velez-Ruiz G, Sunahara RK, Kobilka BK. Identification of GPCR-interacting cytosolic proteins using HDL particles and mass spectrometry-based proteomic approach. *PLoS ONE*;**8**(1):e54942.
17. Daulat A, Maurice P, Jockers R. Techniques for the discovery of GPCR-associated protein complexes. *Methods Enzymol*;**521**:329–45.
18. Daulat AM, Maurice P, Jockers R. Tandem affinity purification and identification of GPCR-associated protein complexes. *Methods Mol Biol*;**746**:399–409.
19. Burack WR, Cheng AM, Shaw AS. Scaffolds, adaptors and linkers of TCR signaling: theory and practice. *Curr Opin Immunol* 2002;**14**(3):312–6.
20. Burack WR, Shaw AS. Signal transduction: hanging on a scaffold. *Curr Opin Cell Biol* 2000;**12**(2):211–6.
21. Mahon MJ, Shimada M. Calmodulin interacts with the cytoplasmic tails of the parathyroid hormone 1 receptor and a sub-set of class b G-protein coupled receptors. *FEBS Lett* 2005;**579**(3):803–7.
22. Mahon MJ, Bonacci TM, Divieti P, Smrcka AV. A docking site for G protein betagamma subunits on the parathyroid hormone 1 receptor supports signaling through multiple pathways. *Mol Endocrinol* 2006;**20**(1):136–46.
23. Mahon MJ. The parathyroid hormone 1 receptor directly binds to the FERM domain of ezrin, an interaction that supports apical receptor localization and signaling in LLC-PK1 cells. *Mol Endocrinol* 2009;**23**(10):1691–701.
24. Romero G, Sneddon WB, Yang Y, Wheeler D, Blair HC, Friedman PA. Parathyroid hormone receptor directly interacts with dishevelled to regulate beta-Catenin signaling and osteoclastogenesis. *J Biol Chem*;**285**(19):14756–63.
25. Pierce KL, Premont RT, Lefkowitz RJ. Seven-transmembrane receptors. *Nat Rev Mol Cell Biol* 2002;**3**(9):639–50.
26. Cherezov V, Rosenbaum DM, Hanson MA, Rasmussen SG, Thian FS, Kobilka TS, et al. High-resolution crystal structure of an engineered human beta2-adrenergic G protein-coupled receptor. *Science* 2007;**318**(5854):1258–65.
27. Rosenbaum DM, Cherezov V, Hanson MA, Rasmussen SG, Thian FS, Kobilka TS, et al. GPCR engineering yields high-resolution structural insights into beta2-adrenergic receptor function. *Science* 2007;**318**(5854):1266–73.
28. Katritch V, Cherezov V, Stevens RC. Structure-function of the G protein-coupled receptor superfamily. *Annu Rev Pharmacol Toxicol*;**53**:531–56.
29. Katritch V, Cherezov V, Stevens RC. Diversity and modularity of G protein-coupled receptor structures. *Trends Pharmacol Sci*;**33**(1):17–27.
30. Wu B, Chien EY, Mol CD, Fenalti G, Liu W, Katritch V, et al. Structures of the CXCR4 chemokine GPCR with small-molecule and cyclic peptide antagonists. *Science* 2010;**330**(6007):1066–71.

31. Wu H, Wacker D, Mileni M, Katritch V, Han GW, Vardy E, et al. Structure of the human kappa-opioid receptor in complex with JDTic. *Nature* 2012;**485**(7398):327–32.
32. Xu F, Wu H, Katritch V, Han GW, Jacobson KA, Gao ZG, et al. Structure of an agonist-bound human A2A adenosine receptor. *Science* 2011;**332**(6027):322–7.
33. Abou-Samra AB, Juppner H, Force T, Freeman MW, Kong XF, Schipani E, et al. Expression cloning of a common receptor for parathyroid hormone and parathyroid hormone-related peptide from rat osteoblast-like cells: a single receptor stimulates intracellular accumulation of both cAMP and inositol trisphosphates and increases intracellular free calcium. *Proc Natl Acad Sci USA* 1992;**89**(7):2732–6.
34. Bringhurst FR, Juppner H, Guo J, Urena P, Potts Jr JT, Kronenberg HM, et al. Cloned, stably expressed parathyroid hormone (PTH)/PTH-related peptide receptors activate multiple messenger signals and biological responses in LLC-PK1 kidney cells. *Endocrinology* 1993;**132**(5):2090–8.
35. El-Hashash AH, Kimber SJ. PTHrP induces changes in cell cytoskeleton and E-cadherin and regulates Eph/Ephrin kinases and RhoGTPases in murine secondary trophoblast cells. *Dev Biol* 2006;**290**(1):13–31.
36. Maeda S, Wu S, Jüppner H, Green J, Aragay AM, Fagin JA, et al. Cell-specific signal transduction of parathyroid hormone (PTH)-related protein through stably expressed recombinant PTH/PTHrP receptors in vascular smooth muscle cells. *Endocrinology* 1996;**137**(8):3154–62.
37. Schwindinger WF, Fredericks J, Watkins L, Robinson H, Bathon JM, Pines M, et al. Coupling of the PTH/PTHrP receptor to multiple G-proteins. Direct demonstration of receptor activation of Gs, Gq/11, and Gi(1) by [alpha-32P]GTP-gamma-azidoanilide photoaffinity labeling. *Endocrine* 1998;**8**(2):201.
38. Singh AT, Gilchrist A, Voyno-Yasenetskaya T, Radeff-Huang JM, Stern PH. G alpha12/G alpha13 subunits of heterotrimeric G proteins mediate parathyroid hormone activation of phospholipase D in UMR-106 osteoblastic cells. *Endocrinology* 2005;**146**(5):2171–5.
39. Mahon MJ, Donowitz M, Yun CC, Segre GV. Na(+)/H(+) exchanger regulatory factor 2 directs parathyroid hormone 1 receptor signalling. *Nature* 2002;**417**(6891):858–61.
40. Mahon MJ, Segre GV. Stimulation by parathyroid hormone of a NHERF-1-assembled complex consisting of the parathyroid hormone I receptor, phospholipase Cbeta, and actin increases intracellular calcium in opossum kidney cells. *J Biol Chem* 2004;**279**(22):23550–8.
41. Wang B, Ardura JA, Romero G, Yang Y, Hall RA, Friedman PA. Na/H exchanger regulatory factors control parathyroid hormone receptor signaling by facilitating differential activation of G(alpha) protein subunits. *J Biol Chem*;**285**(35):26976–86.
42. Romero G, von Zastrow M, Friedman PA. Role of PDZ proteins in regulating trafficking, signaling, and function of GPCRs: means, motif, and opportunity. *Adv Pharmacol*;**62**:279–314.
43. Maurice P, Daulat AM, Broussard C, Mozo J, Clary G, Hotelier F, et al. A generic approach for the purification of signaling complexes that specifically interact with the carboxyl-terminal domain of G protein-coupled receptors. *Mol Cell Proteomics* 2008;**7**(8):1556–69.
44. Heydorn A, Søndergaard BP, Ersbøll B, Holst B, Nielsen FC, Haft CR, et al. A library of 7TM receptor C-terminal tails. Interactions with the proposed post-endocytic sorting proteins ERM-binding phosphoprotein 50 (EBP50), N-ethylmaleimide-sensitive factor (NSF), sorting nexin 1 (SNX1), and G protein-coupled receptor-associated sorting protein (GASP). *J Biol Chem* 2004;**279**(52):54291–303.
45. Bécamel C, Alonso G, Galéotti N, Demey E, Jouin P, Ullmer C, et al. Synaptic multiprotein complexes associated with 5-HT(2C) receptors: a proteomic approach. *Embo J* 2002;**21**(10):2332–42.
46. Karthikeyan S, Leung T, Birrane G, Webster G, Ladias JA. Crystal structure of the PDZ1 domain of human Na(+)/H(+) exchanger regulatory factor provides insights into the mechanism of carboxyl-terminal leucine recognition by class I PDZ domains. *J Mol Biol* 2001;**308**(5):963–73.
47. Nourry C, Grant SG, Borg JP. PDZ domain proteins: plug and play! *Sci STKE* 2003;**2003**(179):RE7.
48. Songyang Z, Fanning AS, Fu C, Xu J, Marfatia SM, Chrishti AH, et al. Recognition of unique carboxyl-terminal motifs by distinct PDZ domains. *Science* 1997;**275**(5296):73–7.
49. Wheeler DS, Barrick SR, Grubisha MJ, Brufsky AM, Friedman PA, Romero G. Direct interaction between NHERF1 and Frizzled regulates beta-catenin signaling. *Oncogene*;**30**(1):32–42.
50. Naren AP, Cobb B, Li C, Roy K, Nelson D, Heda GD, et al. A macromolecular complex of beta 2 adrenergic receptor, CFTR, and ezrin/radixin/moesin-binding phosphoprotein 50 is regulated by PKA. *Proc Natl Acad Sci USA* 2003;**100**(1):342–6.

51. Wang B, Means CK, Yang Y, Mamonova T, Bisello A, Altschuler DL, et al. Ezrin-anchored protein kinase A coordinates phosphorylation-dependent disassembly of a NHERF1 ternary complex to regulate hormone-sensitive phosphate transport. *J Biol Chem* 2012;**287**(29):24148–63.
52. Guillaume JL, Daulat AM, Maurice P, Levoye A, Migaud M, Brydon L, et al. The PDZ protein mupp1 promotes Gi coupling and signaling of the Mt1 melatonin receptor. *J Biol Chem* 2008;**283**(24):16762–71.
53. Wheeler D, Garrido JL, Bisello A, Kim YK, Friedman PA, Romero G. Regulation of parathyroid hormone type 1 receptor dynamics, traffic, and signaling by the Na + /H + exchanger regulatory factor-1 in rat osteosarcoma ROS 17/2.8 cells. *Mol Endocrinol* 2008;**22**(5):1163–70.
54. Terrillon S, Bouvier M. Roles of G-protein-coupled receptor dimerization. *EMBO Rep* 2004;**5**(1):30–4.
55. Prinster SC, Hague C, Hall RA. Heterodimerization of G protein-coupled receptors: specificity and functional significance. *Pharmacol Rev* 2005;**57**(3):289–98.
56. Gomes I, Gupta A, Filipovska J, Szeto HH, Pintar JE, Devi LA. A role for heterodimerization of mu and delta opiate receptors in enhancing morphine analgesia. *Proc Natl Acad Sci USA* 2004;**101**(14):5135–9.
57. Albizu L, Holloway T, González-Maeso J, Sealfon SC. Functional crosstalk and heteromerization of serotonin 5-HT2A and dopamine D2 receptors. *Neuropharmacology* 2011;**61**(4):770–7.
58. Ciruela F, Casadó V, Rodrigues RJ, Luján R. Burgueño, Canals M et al. Presynaptic control of striatal glutamatergic neurotransmission by adenosine A1-A2A receptor heteromers. *J Neurosci* 2006;**26**(7):2080–7.
59. AbdAlla S, Lother H, Quitterer U. AT1-receptor heterodimers show enhanced G-protein activation and altered receptor sequestration. *Nature* 2000;**407**(6800):94–8.
60. Gomes I, Ijzerman AP, Ye K, Maillet EL, Devi LA. G protein-coupled receptor heteromerization: a role in allosteric modulation of ligand binding. *Mol Pharmacol*;**79**(6):1044–52.
61. Fribourg M, Moreno JL, Holloway T, Provasi D, Baki L, Mahahan R, et al. Decoding the signaling of a GPCR heteromeric complex reveals a unifying mechanism of action of antipsychotic drugs. *Cell* 2011;**147**(5):1011–23.
62. Vilardaga JP, Nikolaev VO, Lorenz K, Ferrandon S, Zhuang Z, Lohse MJ. Conformational cross-talk between alpha2A-adrenergic and mu-opioid receptors controls cell signaling. *Nat Chem Biol* 2008;**4**(2):126–31.
63. Terrillon S, Barberis C, Bouvier M. Heterodimerization of V1a and V2 vasopressin receptors determines the interaction with beta-arrestin and their trafficking patterns. *Proc Natl Acad Sci USA* 2004;**101**(6):1548–53.
64. Rozenfeld R, Devi LA. Receptor heterodimerization leads to a switch in signaling: beta-arrestin2-mediated ERK activation by mu-delta opioid receptor heterodimers. *Faseb J* 2007;**21**(10):2455–65.
65. Jordan BA, Trapaidze N, Gomes I, Nivarthi R, Devi LA. Oligomerization of opioid receptors with beta 2-adrenergic receptors: a role in trafficking and mitogen-activated protein kinase activation. *Proc Natl Acad Sci USA* 2001;**98**(1):343–8.
66. Gurevich VV, Gurevich EV. How and why do GPCRs dimerize? *Trends Pharmacol Sci* 2008;**29**(5):234–40.
67. Gurevich VV, Gurevich EV. GPCR monomers and oligomers: it takes all kinds. *Trends Neurosci* 2008;**31**(2):74–81.
68. Gomes I, Jordan BA, Gupta A, Rios C, Trapaidze N, Devi LA. G protein coupled receptor dimerization: implications in modulating receptor function. *J Mol Med (Berl)* 2001;**79**(5–6):226–42.
69. Parnot C, Kobilka B. Toward understanding GPCR dimers. *Nat Struct Mol Biol* 2004;**11**(8):691–2.
70. Whorton MR, Jastrzebska B, Park PS, Fotiadis D, Engel A, Palczewski K, et al. Efficient coupling of transducin to monomeric rhodopsin in a phospholipid bilayer. *J Biol Chem* 2008;**283**(7):4387–94.
71. Whorton MR, Bokoch MP, Rasmussen SG, Huang B, Zare RN, Kobilka B, et al. A monomeric G protein-coupled receptor isolated in a high-density lipoprotein particle efficiently activates its G protein. *Proc Natl Acad Sci USA* 2007;**104**(18):7682–7.
72. Schmitt JM, Stork PJ. beta 2-adrenergic receptor activates extracellular signal-regulated kinases (ERKs) via the small G protein rap1 and the serine/threonine kinase B-Raf. *J Biol Chem* 2000;**275**(33):25342–50.
73. Shaw G, Morse S, Ararat M, Graham FL. Preferential transformation of human neuronal cells by human adenoviruses and the origin of HEK 293 cells. *Faseb J* 2002;**16**(8):869–71.
74. Hiller C, Kuhhorn J, Gmeiner P. Class A G-Protein-Coupled Receptor (GPCR) Dimers and Bivalent Ligands. *J Med Chem.*
75. Pin JP, Neubig R, Bouvier M, Devi L, Filizola M, Javitch JA, et al. International Union of Basic and Clinical Pharmacology. LXVII. Recommendations for the recognition and nomenclature of G protein-coupled receptor heteromultimers. *Pharmacol Rev* 2007;**59**(1):5–13.

76. Herrick-Davis K, Grinde E, Cowan A, Mazurkiewicz JE. Fluorescence correlation spectroscopy analysis of serotonin, adrenergic, muscarinic, and dopamine receptor dimerization: the oligomer number puzzle. *Mol Pharmacol.*
77. Teitler M, Klein MT. A new approach for studying GPCR dimers: drug-induced inactivation and reactivation to reveal GPCR dimer function *in vitro*, in primary culture, and *in vivo*. *Pharmacol Ther*;**133**(2):205–17.
78. Pei L, Li S, Wang M, Diwan M, Anisman H, Fletcher PJ, et al. Uncoupling the dopamine D1-D2 receptor complex exerts antidepressant-like effects. *Nat Med*;**16**(12):1393–5.
79. AbdAlla S, Lother H, el Massiery A, Quitterer U. Increased AT(1) receptor heterodimers in preeclampsia mediate enhanced angiotensin II responsiveness. *Nat Med* 2001;**7**(9):1003–9.
80. Fuxe K, Agnati LF, Jacobsen K, Hillion J, Canals M, Torvinen M, et al. Receptor heteromerization in adenosine A2A receptor signaling: relevance for striatal function and Parkinson's disease. *Neurology* 2003;**61**(11 Suppl. 6): S19–23.
81. He SQ, Zhang ZN, Guan JS, Liu HR, Zhao B, Wang HB, et al. Facilitation of mu-opioid receptor activity by preventing delta-opioid receptor-mediated codegradation. *Neuron* 2011;**69**(1):120–31.
82. Wang M, Pei L, Fletcher PJ, Kapur S, Seeman P, Liu F. Schizophrenia, amphetamine-induced sensitized state and acute amphetamine exposure all show a common alteration: increased dopamine D2 receptor dimerization. *Mol Brain* 2010;**3**:25.
83. McGraw DW, Mihlbachler KA, Schwarb MR, Rahmann FF, Small KM, Almoosa KF, et al. Airway smooth muscle prostaglandin-EP1 receptors directly modulate beta2-adrenergic receptors within a unique heterodimeric complex. *J Clin Invest* 2006;**116**(5):1400–9.
84. Waldhoer M, Fong J, Jones RM, Lunzer MM, Sharma SK, Kostenis E, et al. A heterodimer-selective agonist shows *in vivo* relevance of G protein-coupled receptor dimers. *Proc Natl Acad Sci USA* 2005;**102**(25):9050–5.
85. Yekkirala AS, Lunzer MM, McCurdy CR, Powers MD, Kalyuzhny AE, Roerig SC, et al. N-naphthoyl-beta-naltrexamine (NNTA), a highly selective and potent activator of mu/kappa-opioid heteromers. *Proc Natl Acad Sci USA*;**108**(12):5098–103.
86. Yekkirala AS, Kalyuzhny AE, Portoghese PS. Standard opioid agonists activate heteromeric opioid receptors: evidence for morphine and [d-Ala(2)-MePhe(4)-Glyol(5)]enkephalin as selective mu-delta agonists. *ACS Chem Neurosci*;**1**(2):146–54.
87. Metcalf MD, Yekkirala AS, Powers MD, Kitto KF, Fairbanks CA, Wilcox GL, et al. The delta opioid receptor agonist SNC80 selectively activates heteromeric mu-delta opioid receptors. *ACS Chem Neurosci* 2012;**3**(7):505–9.
88. Zheng Y, Akgün E, Harikumar KG, Hopson J, Powers MD, Lunzer MM, et al. Induced association of mu opioid (MOP) and type 2 cholecystokinin (CCK2) receptors by novel bivalent ligands. *J Med Chem* 2009;**52**(2):247–58.
89. Daniels DJ, Lenard NR, Etienne CL, Law PY, Roerig SC, Portoghese PS. Opioid-induced tolerance and dependence in mice is modulated by the distance between pharmacophores in a bivalent ligand series. *Proc Natl Acad Sci USA* 2005;**102**(52):19208–13.
90. Zhang Y, Gilliam A, Maitra R, Damaj MI, Tajuba JM, Seltzman HH, et al. Synthesis and biological evaluation of bivalent ligands for the cannabinoid 1 receptor. *J Med Chem*;**53**(19):7048–60.
91. Karellas P, McNaughton M, Baker SP, Scammells PJ. Synthesis of bivalent beta2-adrenergic and adenosine A1 receptor ligands. *J Med Chem* 2008;**51**(19):6128–37.
92. McRobb FM, Crosby IT, Yuriev E, Lane JR, Capuano B. Homobivalent ligands of the atypical antipsychotic clozapine: design, synthesis, and pharmacological evaluation. J Med Chem;**55**(4):1622–34.
93. McLatchie LM, Fraser NJ, Main MJ, Wise A, Brown J, Thompson N, et al. RAMPs regulate the transport and ligand specificity of the calcitonin-receptor-like receptor. *Nature* 1998;**393**(6683):333–9.
94. Kuwasako K, Cao YN, Nagoshi Y, Kitamura K, Eto T. Adrenomedullin receptors: pharmacological features and possible pathophysiological roles. *Peptides* 2004;**25**(11):2003–12.
95. Morfis M, Tilakaratne N, Furness SG, Christopoulous G, Werry TD, Christopoulous A, et al. Receptor activity-modifying proteins differentially modulate the G protein-coupling efficiency of amylin receptors. *Endocrinology* 2008;**149**(11):5423–31.
96. Christopoulos G, Perry KJ, Morfis M, Tilakaratne N, Gao Y, Fraser NJ, et al. Multiple amylin receptors arise from receptor activity-modifying protein interaction with the calcitonin receptor gene product. *Mol Pharmacol* 1999;**56**(1):235–42.

97. Poyner DR, Sexton PM, Marshall I, Smith DM, Quirion R, Born W, et al. International Union of Pharmacology. XXXII. The mammalian calcitonin gene-related peptides, adrenomedullin, amylin, and calcitonin receptors. *Pharmacol Rev* 2002;**54**(2):233–46.
98. Hilairet S, Belanger C, Bertrand J, Laperriere A, Foord SM, Bouvier M. Agonist-promoted internalization of a ternary complex between calcitonin receptor-like receptor, receptor activity-modifying protein 1 (RAMP1), and beta-arrestin. *J Biol Chem* 2001;**276**(45):42182–90.
99. Heroux M, Hogue M, Lemieux S, Bouvier M. Functional calcitonin gene-related peptide receptors are formed by the asymmetric assembly of a calcitonin receptor-like receptor homo-oligomer and a monomer of receptor activity-modifying protein-1. *J Biol Chem* 2007;**282**(43):31610–20.
100. Harikumar KG, Simms J, Christopoulos G, Sexton PM, Miller LJ. Molecular basis of association of receptor activity-modifying protein 3 with the family B G protein-coupled secretin receptor. *Biochemistry* 2009;**48**(49):11773–85.
101. Christopoulos A, Christopoulos G, Morfis M, Udawela M, Laburthe M, Couvineau A, et al. Novel receptor partners and function of receptor activity-modifying proteins. *J Biol Chem* 2003;**278**(5):3293–7.
102. Udawela M, Christopoulos G, Morfis M, Christopoulos A, Ye S, Tilakaratne N, et al. A critical role for the short intracellular C terminus in receptor activity-modifying protein function. *Mol Pharmacol* 2006;**70**(5):1750–60.
103. Evans BN, Rosenblatt MI, Mnayer LO, Oliver KR, Dickerson IM. CGRP-RCP, a novel protein required for signal transduction at calcitonin gene-related peptide and adrenomedullin receptors. *J Biol Chem* 2000;**275**(40):31438–43.
104. Egea SC, Dickerson IM. Direct interactions between calcitonin-like receptor (CLR) and CGRP-receptor component protein (RCP) regulate CGRP receptor signaling. *Endocrinology*;**153**(4):1850–60.
105. Angers S, Moon RT. Proximal events in Wnt signal transduction. *Nat Rev Mol Cell Biol* 2009;**10**(7):468–77.
106. Clevers H. Wnt/beta-catenin signaling in development and disease. *Cell* 2006;**127**(3):469–80.
107. He X, Semenov M, Tamai K, Zeng X. LDL receptor-related proteins 5 and 6 in Wnt/beta-catenin signaling: arrows point the way. *Development* 2004;**131**(8):1663–77.
108. Rao TP, Kuhl M. An updated overview on Wnt signaling pathways: a prelude for more. *Circ Res*;**106**(12):1798–806.
109. Du SJ, Purcell SM, Christian JL, McGrew LL, Moon RT. Identification of distinct classes and functional domains of Wnts through expression of wild-type and chimeric proteins in Xenopus embryos. *Mol Cell Biol* 1995;**15**(5):2625–34.
110. Wong GT, Gavin BJ, McMahon AP. Differential transformation of mammary epithelial cells by Wnt genes. *Mol Cell Biol* 1994;**14**(9):6278–86.
111. Weidinger G, Moon RT. When Wnts antagonize Wnts. *J Cell Biol* 2003;**162**(5):753–5.
112. Mikels AJ, Nusse R. Purified Wnt5a protein activates or inhibits beta-catenin-TCF signaling depending on receptor context. *PLoS Biol* 2006;**4**(4):e115.
113. Mikels A, Minami Y, Nusse R. Ror2 receptor requires tyrosine kinase activity to mediate Wnt5A signaling. *J Biol Chem* 2009;**284**(44):30167–76.
114. Glinka A, Wu W, Delius H, Monaghan AP, Blumenstock C, Niehrs C. Dickkopf-1 is a member of a new family of secreted proteins and functions in head induction. *Nature* 1998;**391**(6665):357–62.
115. Li X, Zhang Y, Kang H, Liu W, Liu P, Zhang J, et al. Sclerostin binds to LRP5/6 and antagonizes canonical Wnt signaling. *J Biol Chem* 2005;**280**(20):19883–7.
116. Cheng Z, Biechele T, Wei Z, Morrone S, Moon RT, Wang L, et al. Crystal structures of the extracellular domain of LRP6 and its complex with DKK1. *Nat Struct Mol Biol* 2011;**18**(11):1204–10.
117. Ettenberg SA, Charlat O, Daley MP, Liu S, Vincent KJ, Stuart DD, et al. Inhibition of tumorigenesis driven by different Wnt proteins requires blockade of distinct ligand-binding regions by LRP6 antibodies. *Proc Natl Acad Sci USA* 2010;**107**(35):15473–8.
118. Wu S, Pirola CJ, Green J, Yamaguchi DT, Okano K, Jueppner H, et al. Effects of N-terminal, midregion, and C-terminal parathyroid hormone-related peptides on adenosine 3′,5′-monophosphate and cytoplasmic free calcium in rat aortic smooth muscle cells and UMR-106 osteoblast-like cells. *Endocrinology* 1993;**133**(6):2437–44.
119. Ferrandon S, Feinstein TN, Castro M, Wang B, Bouley R, Potts JT, et al. Sustained cyclic AMP production by parathyroid hormone receptor endocytosis. *Nat Chem Biol* 2009;**5**(10):734–42.

120. Wehbi VL, Stevenson HP, Feinstein TN, Calero G, Romero G, Vilardaga JP. Noncanonical GPCR signaling arising from a PTH receptor-arrestin-Gbetagamma complex. *Proc Natl Acad Sci USA*;**110**(4): 1530–5.
121. Feinstein TN, Wehbi VL, Ardura JA, Wheeler DS, Ferrandon S, Gardella TJ, et al. Retromer terminates the generation of cAMP by internalized PTH receptors. *Nat Chem Biol* 2011;**7**(5):278–84.
122. Feinstein TN, Yui N, Webber MJ, Wehbi VL, Stevenson HP, King Jr JD, et al. Noncanonical control of vasopressin receptor type 2 signaling by retromer and arrestin. *J Biol Chem* 2013;**288**(39):27849–60.
123. Calebiro D, Nikolaev VO, Lohse MJ. Imaging of persistent cAMP signaling by internalized G protein-coupled receptors. *J Mol Endocrinol*;**45**(1):1–8.
124. Calebiro D, Nikolaev VO, Gagliani MC, de Filippis T, Dees C, Tacchetti C, et al. Persistent cAMP-signals triggered by internalized G-protein-coupled receptors. *PLoS Biol* 2009;**7**(8):e1000172.
125. Schwartz JH. The many dimensions of cAMP signaling. *Proc Natl Acad Sci USA* 2001;**98**(24):13482–4.
126. Buxton IL, Brunton LL. Compartments of cyclic AMP and protein kinase in mammalian cardiomyocytes. *J Biol Chem* 1983;**258**(17):10233–9.
127. Luttrell DK, Luttrell LM. Signaling in time and space: G protein-coupled receptors and mitogen-activated protein kinases. *Assay Drug Dev Technol* 2003;**1**(2):327–38.
128. Murasawa S, Mori Y, Nozawa Y, Gotoh N, Shibuya M, Masaki H, et al. Angiotensin II type 1 receptor-induced extracellular signal-regulated protein kinase activation is mediated by Ca2+/calmodulin-dependent transactivation of epidermal growth factor receptor. *Circ Res* 1998;**82**(12):1338–48.
129. Luttrell LM, Roudabush FL, Choy EW, Miller WE, Field ME, Pierce KL, et al. Activation and targeting of extracellular signal-regulated kinases by beta-arrestin scaffolds. *Proc Natl Acad Sci USA* 2001;**98**(5): 2449–54.
130. Ebner R, Derynck R. Epidermal growth factor and transforming growth factor-alpha: differential intracellular routing and processing of ligand-receptor complexes. *Cell Regul* 1991;**2**(8):599–612.
131. Roepstorff K, Grandal MV, Henriksen L, Knudsen SL, Lerdrup M, Grøvdal L, et al. Differential effects of EGFR ligands on endocytic sorting of the receptor. *Traffic* 2009;**10**(8):1115–27.
132. Vieira AV, Lamaze C, Schmid SL. Control of EGF receptor signaling by clathrin-mediated endocytosis. *Science* 1996;**274**(5295):2086–9.
133. Platta HW, Stenmark H. Endocytosis and signaling. *Curr Opin Cell Biol*;**23**(4):393–403.
134. Ahmad T, Farnie G, Bundred NJ, Anderson NG. The mitogenic action of insulin-like growth factor I in normal human mammary epithelial cells requires the epidermal growth factor receptor tyrosine kinase. *J Biol Chem* 2004;**279**(3):1713–9.
135. Balañé ME, Labriola L, Salatino M, Movsichoff F, Peters G, Charreau EH, et al. Activation of ErbB-2 via a hierarchical interaction between ErbB-2 and type I insulin-like growth factor receptor in mammary tumor cells. *Oncogene* 2001;**20**(1):34–47.
136. Jones HE, Gee JM, Hutcheson IR, Knowlden JM, Barrow D, Nicholson RI. Growth factor receptor interplay and resistance in cancer. *Endocr Relat Cancer* 2006;**13**(Suppl. 1):S45–51.
137. Reznik TE, Sang Y, Ma Y, Abounader R, Rosen EM, Xia S, et al. Transcription-dependent epidermal growth factor receptor activation by hepatocyte growth factor. *Mol Cancer Res* 2008;**6**(1):139–50.
138. Nath D, Williamson NJ, Jarvis R, Murphy G. Shedding of c-Met is regulated by crosstalk between a G-protein coupled receptor and the EGF receptor and is mediated by a TIMP-3 sensitive metalloproteinase. *J Cell Sci* 2001;**114**(Pt 6):1213–20.
139. Yarden Y, Sliwkowski MX. Untangling the ErbB signalling network. *Nat Rev Mol Cell Biol* 2001;**2**(2):127–37.
140. Harris RC, Chung E, Coffey RJ. EGF receptor ligands. *Exp Cell Res* 2003;**284**(1):2–13.
141. Guy PM, Platko JV, Cantley LC, Cerione RA, Carraway III KL. Insect cell-expressed p180erbB3 possesses an impaired tyrosine kinase activity. *Proc Natl Acad Sci USA* 1994;**91**(17):8132–6.
142. Alimandi M, Romano A, Curia MC, Muraro R, Fedi P, Aaronson SA, et al. Cooperative signaling of ErbB3 and ErbB2 in neoplastic transformation and human mammary carcinomas. *Oncogene* 1995;**10**(9): 1813–21.
143. Holbro T, Beerli RR, Maurer F, Koziczak M, Barbas III CF, Hynes NE. The ErbB2/ErbB3 heterodimer functions as an oncogenic unit: ErbB2 requires ErbB3 to drive breast tumor cell proliferation. *Proc Natl Acad Sci USA* 2003;**100**(15):8933–8.

144. Bublil EM, Yarden Y. The EGF receptor family: spearheading a merger of signaling and therapeutics. *Curr Opin Cell Biol* 2007;**19**(2):124–34.
145. Garrett TP, McKern NM, Lou M, Elleman TC, Adams TE, Lovrecz GO, et al. The crystal structure of a truncated ErbB2 ectodomain reveals an active conformation, poised to interact with other ErbB receptors. *Mol Cell* 2003;**11**(2):495–505.
146. Kharmate G, Rajput PS, Watt HL, Somvanshi RK, Chaudhari N, Qiu X, et al. Dissociation of epidermal growth factor receptor and ErbB2 heterodimers in the presence of somatostatin receptor 5 modulate signaling pathways. *Endocrinology* 2011;**152**(3):931–45.
147. Macdonald-Obermann JL, Adak S, Landgraf R, Piwnica-Worms D, Pike LJ. Dynamic analysis of the EGF receptor-ErbB2-ErbB3 network by luciferase fragment complementation imaging. *J Biol Chem*.
148. Wang YN, Yamaguchi H, Hsu JM, Hung MC. Nuclear trafficking of the epidermal growth factor receptor family membrane proteins. *Oncogene*;**29**(28):3997–4006.
149. Roudabush FL, Pierce KL, Maudsley S, Khan KD, Luttrell LM. Transactivation of the EGF receptor mediates IGF-1-stimulated shc phosphorylation and ERK1/2 activation in COS-7 cells. *J Biol Chem* 2000;**275**(29):22583–9.
150. Jo M, Stolz DB, Esplen JE, Dorko K, Michalopoulos GK, Strom SC. Cross-talk between epidermal growth factor receptor and c-Met signal pathways in transformed cells. *J Biol Chem* 2000;**275**(12):8806–11.
151. Andrae J, Gallini R, Betsholtz C. Role of platelet-derived growth factors in physiology and medicine. *Genes Dev* 2008;**22**(10):1276–312.
152. Ekman S, Thuresson ER, Heldin CH, Ronnstrand L. Increased mitogenicity of an alphabeta heterodimeric PDGF receptor complex correlates with lack of RasGAP binding. *Oncogene* 1999;**18**(15):2481–8.
153. Oates J, King G, Dixon AM. Strong oligomerization behavior of PDGF beta receptor transmembrane domain and its regulation by the juxtamembrane regions. *Biochim Biophys Acta*;**1798**(3):605–15.
154. Saito Y, Haendeler J, Hojo Y, Yamamoto K, Berk BC. Receptor heterodimerization: essential mechanism for platelet-derived growth factor-induced epidermal growth factor receptor transactivation. *Mol Cell Biol* 2001;**21**(19):6387–94.
155. Cheng C, Haasdijk RA, Tempel D, den Dekker WK, Chrifi I, Blonden LA, et al. PDGF-induced migration of vascular smooth muscle cells is inhibited by heme oxygenase-1 via VEGFR2 upregulation and subsequent assembly of inactive VEGFR2/PDGFRbeta heterodimers. *Arterioscler Thromb Vasc Biol*;**32**(5):1289–98.
156. Yeh CY, Shin SM, Yeh HH, Wu TJ, Shin JW, Chang TY, et al. Transcriptional activation of the Axl and PDGFR-alpha by c-Met through a ras- and Src-independent mechanism in human bladder cancer. *BMC Cancer* 2011;**11**:139.
157. Meyer AS, Miller MA, Gertler FB, Lauffenburger DA. The receptor AXL diversifies EGFR signaling and limits the response to EGFR-targeted inhibitors in triple-negative breast cancer cells. *Sci Signal*;**6**(287):ra66.
158. Kratchmarova I, Blagoev B, Haack-Sorensen M, Kassem M, Mann M. Mechanism of divergent growth factor effects in mesenchymal stem cell differentiation. *Science* 2005;**308**(5727):1472–7.
159. Guha U, Chaerkady R, Marimuthu A, Patterson AS, Kashyap MK, Harsha HC, et al. Comparisons of tyrosine phosphorylated proteins in cells expressing lung cancer-specific alleles of EGFR and KRAS. *Proc Natl Acad Sci USA* 2008;**105**(37):14112–7.
160. Rikova K, Guo A, Zeng Q, Possemato A, Yu J, Haack H, et al. Global survey of phosphotyrosine signaling identifies oncogenic kinases in lung cancer. *Cell* 2007;**131**(6):1190–203.
161. Huang PH, Mukasa A, Bonavia R, Flynn RA, Brewer ZE, Cavenee WK, et al. Quantitative analysis of EGFRvIII cellular signaling networks reveals a combinatorial therapeutic strategy for glioblastoma. *Proc Natl Acad Sci USA* 2007;**104**(31):12867–72.
162. Yano S, Yamada T, Takeuchi S, Minami Y, Yatable Y, Mitsudomi T, et al. Hepatocyte growth factor expression in EGFR mutant lung cancer with intrinsic and acquired resistance to tyrosine kinase inhibitors in a Japanese cohort. *J Thorac Oncol* 2011;**6**(12):2011–7.
163. Furchgott RF, Bhadrakom S. Reactions of strips of rabbit aorta to epinephrine, isopropylarterenol, sodium nitrite and other drugs. *J Pharmacol Exp Ther* 1953;**108**(2):129–43.
164. Furchgott RF, Zawadzki JV. The obligatory role of endothelial cells in the relaxation of arterial smooth muscle by acetylcholine. *Nature* 1980;**288**(5789):373–6.

165. Baron R, Kneissel M. WNT signaling in bone homeostasis and disease: from human mutations to treatments. *Nat Med*;**19**(2):179–92.
166. Sabbota AL, Kim HR, Zhe X, Fridman R, Bonfil RD, Cher ML. Shedding of RANKL by tumor-associated MT1-MMP activates Src-dependent prostate cancer cell migration. *Cancer Res*;**70**(13):5558–66.
167. Turner CH, Warden SJ, Bellido T, Plotkin LI, Kumar N, Jasiuk I, et al. Mechanobiology of the skeleton. *Sci Signal* 2009;**2**(68):pt3.
168. Hanyu R, Wehbi VL, Hayata T, Moriya S, Feinstein TN, Ezura Y, et al. Anabolic action of parathyroid hormone regulated by the beta2-adrenergic receptor. *Proc Natl Acad Sci USA* 2012;**109**(19):7433–8.

CHAPTER

3

The Application of Signaling Bias to New Therapeutic Drug Therapy for Seven Transmembrane (G Protein-coupled) Receptors: Quantifying Bias

Terry Kenakin

Department of Pharmacology, University of North Carolina School of Medicine, NC, USA

OUTLINE

INTRODUCTION

Historically, the pharmacology of agonism has been described by a parsimonious model which considers seven transmembrane receptors (7TMRs) as switches existing in an active state (that go on to activate cellular signaling pathways connected to the receptor) and an inactive state. Within this model all agonists basically produce a uniform activation with

B. Arey (Ed): Biased Signaling in Physiology, Pharmacology and Therapeutics
DOI: http://dx.doi.org/10.1016/B978-0-12-411460-9.00003-3

the only difference between agonists being their strength to do so. In terms of this model, the drug property mediating agonism, namely efficacy, has a uniform quality but variable quantity.

Over the past 20 years, data have emerged that are incompatible with this simple scheme but rather support the idea that different agonists can produce different *qualities* of agonism as well as varying quantities of agonism (for reviews see[1–3]). These different qualities of agonism are thought to emanate from different active states of the receptor stabilized by the agonists.[4] This idea was proposed on theoretical grounds some years ago but recently experimental evidence for the production of different receptor active states by different ligands has been obtained through analysis with ^{19}F-nuclear magnetic resonance.[5] Thus, instead of a single receptor active state, there are data to suggest that receptors can form numerous states and that some of these differentially interact with signaling proteins in the cell as they produce agonism. Under these circumstances, certain cellular pathways will be activated to a greater extent than others, i.e., they will produce biased signals. This chapter will review the various methods available to quantify the bias of agonists acting on seven transmembrane receptors to provide a scale where this drug property can be optimized through medicinal chemistry.

DEFINING THE PATHWAYS: β-ARRESTIN-MEDIATED SIGNALING

A first step in the discussion of signaling bias is to define the two signaling pathways involved. These could be any of the pathways controlled by a pleiotropically coupled receptor. For purposes of this discussion on how to quantify bias, the effects will be illustrated with receptor-mediated G protein signals and β-arrestin mediated signals. While G protein signaling has been known for a number of years, signaling from β-arrestin-receptor association is a relatively new finding. Initial data suggested that the interaction of receptors with β-arrestin mainly caused the termination of the G protein; however, subsequent studies have indicated a rich array of responses emanating from the β-arrestin intracellular complex[6–9] resulting in responses such as ERK1/2 signaling to the suppression of constitutive activity of receptors (i.e., μ opioid receptors,[10]). There is now a considerable body of evidence to implicate β-arrestin signaling in a host of diseases including diabetes,[11] heart failure,[12] cardiovascular disease,[12–14] central nervous system diseases involving serotonin,[15] diseases involving angiotensin[16,17] and adrenergic signaling,[18,19] and parathyroid hormone.[20] While the bulk of studies involve β-arrestin2, there are reports that β-arrestin1 signaling also can lead to selective biased signaling profiles, notably through the GLP-1 receptor for diabetes[21] and with δ-opiod receptor desensitization.[22]

One of the earliest established favorable therapeutic actions of β-arrestin activation was reported for angiotensin-1-receptor-mediated β-arrestin activation in cardiovascular disease. Selective β-arrestin activation with biased angiotensin agonists such as TRV120023 showed cardioprotection and diminished cell death in a mouse model of ischemic reperfusion injury.[23] Similarly, the biased β-arrestin angiotensin agonist SII has been shown to produce reduced size of myocardial infarction in rat ischemia-reperfusion injury.[24] Signaling bias can produce therapeutically beneficial effects by emphasizing useful

TABLE 3.1 Proposed Targets where β-Arrestin Signaling is Proposed to be Beneficial

Proposed Therapy	Target	Pharmacologic Effect	Reference
Osteoporosis	PTH	β-arrestin2 controls bone formation	Bohinc and Gesty-Palmer,[25]
Diabetes	GLP-1	β-arrestin1 mediates GLP-1 induced insulin secretion, antiapoptosis	Dalle et al.[96]/Sonoda et al.[21]
Endothelial Cytoprotection	PAR1	Activated Protein C produces β-arrestin2 mediated protective effects through PAR1R	Mosnier et al.[26]
Acute myocardial injury	AT(2)	Diminished cell death in ischemia due to β-arrestin2 mediated MAPK and Akt signaling	Kim et al.[23]
Cardiac reperfusion injury	AT(I)	β-arrestin mediated signaling reduces infarct size in myocardial ischemia	Hostrup et al.[24]
Psychosis	Dopamine D(2) R	β-arrestin2 mediated signaling mediates antipsychotic action	Allen et al.[27]/ Chen et al.[28]
Apoptosis	β-arrestin2	β-arrestin2 inhibits cell apoptosis through ERK1/2, p38 MAPK, Akt signaling	Yang et al.[29]
Cytoprotection	Activated Protein C	β-arrestin mediated signaling promotes cytoprotection	Schulte ad Shenoy,[30]/ Soh and Trejo J,[31]
Hypertension	α_2-Adrenoceptor	Hypotension with less sedation	Schmid and Bohn,[32]
Schizophrenia	Dopamine D_2 Receptor	Improved treatment of schizophrenia	Grady et al.[33]/Urban et al.[34]
Neuropsychiatric/ neurodegenerative disorders	Histamine R_2 Receptor	Reduced hyperalgesia	Galeotti et al., 2004
Thyroid hormone deficiency	Thyroid-stimulating hormone Receptor	Selective thyroid hormone synthesis	Vassart and Dumont,[35]
Congestive Heart Failure*	Angiotensin Receptors	Cardioprotection concomitant with angiotensin receptor blockade	Aplin et al.[36]/ Rajagopal et al.[37,38]/ Violin and Lefkowitz,[39]/Violin et al.[40]/Wei et al.[41]/ Zhai et al.[42]
Parkinsonism	Dopamine D_1 Receptors	Dopamine D1 receptor internalization	Ryman-Rasmussen et al.[43]/Ji et al.[44]
Treatment of Addiction	5-Hydroxytryptamine Receptors	Receptor blockade + ERK stimulation	MacKinnon et al.[45]
	Gastrin-releasing peptide/Arg vasopressin Receptors		
Small Cell Lung Cancer-Treatment of Psychosis and Depression	5-HT Receptors	5-HT receptor internalization	Willins et al.[46]

therapeutic signals and minimizing harmful secondary effects. Table 3.1 shows a partial list of targets where signaling biased toward β-arrestin signals is proposed to be a superior therapy over signaling by conventional ligands.

There are two general approaches to the application of biased signaling to therapeutic drugs. The first is to generate data to identify where a given signal is either especially beneficial or especially harmful to a defined therapeutic treatment; in this regard, the use of genetic knockout animals can be very helpful. For example, opioid receptor agonists are known to produce analgesia but also respiratory depression, an unwanted side effect. This secondary effect is greatly diminished in β-arrestin knockout mice, suggesting that a biased ligand that did not cause receptor association with β-arrestin would be a better therapy for pain.[47–50] Similarly, the activation of the nicotinic acid receptor (GPR109) in β-arrestin null mice leads to lowering of serum fatty acids without the accompanying flushing seen in normal mice,[51] indicating an agonist of GPR109 devoid of β-arrestin activating effects would be a better therapeutic approach. Finally, parathyroid hormone does not build bone or increase the number of osteoclasts in β-arrestin2 knockout mice, suggesting that this signaling pathway is the therapeutically relevant one.[52,53] This further suggests that a PTH agonist with biased signalling toward β-arrestin would be an optimal therapy for this receptor.

Another approach can be taken if it is not clear whether a bias would provide a better treatment. This strategy centers on the premise that ligands should be characterized with respect to possible bias with no preconceived idea as to which bias is preferable, and then the exemplar molecules from this process tested in more complex assays (animal models *in vivo*) to determine whether superior therapeutic phenotypes can be associated with any defined bias. Thus, the active molecules from a given screen utilizing one signaling pathway are retested in another screen for the other signaling pathway to determine differences in signaling bias for the two pathways.

WHAT IS SIGNALING BIAS?

In theory, the term 'bias' suggests simply that upon receptor activation one process is activated to a greater extent than another. This can be shown with a 'bias plot' whereby the response to one process is graphed as a function of the response produced in another. For instance, catecholamines produce myocardial inotropy (increased isometric force of contraction) and lusitropy (increased rate of relaxation) in response to elevations in intracellular cyclic AMP. It can be shown that the sensitivity of atria to isoproterenol-mediated increased relaxation is two- to three-fold greater than the inotropic response; this probably is a function of the efficiency of physiological coupling of these responses.[54] A bias plot of these two responses would thus show a curved relationship, i.e., isoproterenol is biased toward selective lusitropy as opposed to inotropy (Figure 3.1).

As shown in Figure 3.1, the same relationship holds for the β-adrenoceptor agonist pirbuterol, indicating that the bias is not associated with a unique chemical structure of the agonist but rather is a function of the system. This effect will be referred to as system bias since it involves the inherent dependence of the cell on different signaling systems customized to the specific needs of the cell. In the case described above, it would suggest that the

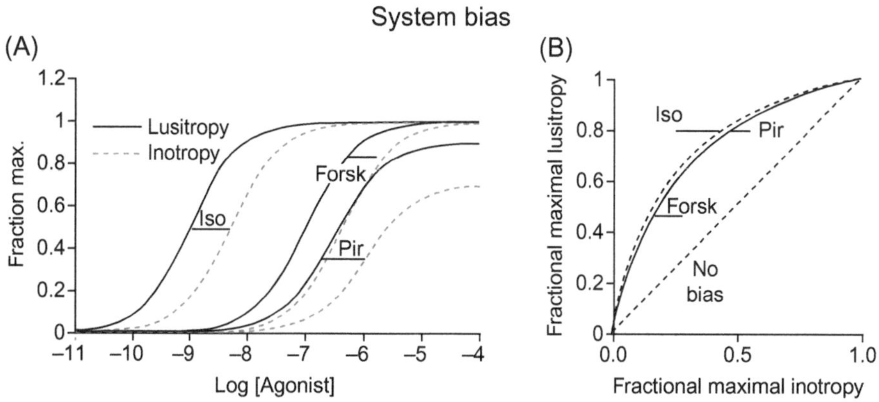

FIGURE 3.1 Comparison of lusitropic and inotropic effects of β-adrenoceptor agonists (isoproterenol, pirbuterol) and an activator of adenylate cyclase (forskolin) in rat atria. (A) shows concentration–response curves for isoproterenol (Iso), forskolin (Forsk), and pirbuterol (Pir) for increased myocardial relaxation (lusitropy: solid lines; inotropy: gray dashed lines). (B) shows a bias plot where the lusitropic response is plotted as a function of the inotropic response. The line is curved indicating a bias toward lusitropy but the effects are identical for all agonists. *Data redrawn from Kenakin et al.*[54]

relaxation mechanism in cardiac muscle is more sensitive to the elevation of the second messenger cyclic AMP than the inotropic mechanism; this is supported by data to show the same bias with exogenously added dibutryl cyclic AMP to the cardiac preparation (Figure 3.2).

Theoretically, system bias might be exploited therapeutically but the application would be very specific. Thus, if the system bias is towards the response of therapeutic interest (i.e., lusitropy) then a low efficacy agonist may be employed to selectively induce the response as in the case of prenalterol. This weak β-adrenoceptor agonist produces selective lusitropy with no concomitant inotropic effect in rat atria.[54] Such selective effects are dependent on the specific sensitivity of the tissue; in this case, the selective lusitropic effect of prenalterol would be unique only to rat atria and may not be operative in human tissue. This makes this approach of very limited value for selective therapy.

Another type of *in vitro* bias in pharmacologic experiments is called observation bias which is, namely, the effect of varying sensitivities of different assays when two assays are required to determine the signaling of the two pathways. For example, the present assays available for β-arrestin and receptor association are direct biochemical assays with no amplification mechanism (such as that seen for cyclic AMP assays). This means that β-arrestin pharmacologic assays are generally less sensitive than second messenger assays. Figure 3.3A shows the production of cyclic AMP through β-adrenergic receptor activation with β-agonists and, in a separate assay, the agonist-induced β-adrenergic receptor-mediated activation of β-arrestin.[55] It can be seen that the β-arrestin response is approximately 30-fold less sensitive in terms of the EC_{50} for agonists than the response for cyclic AMP. A bias plot for these two pathways shows a clear bias toward cyclic AMP (Figure 3.3B) but there is no distinction between the agonists. Observational bias will vary

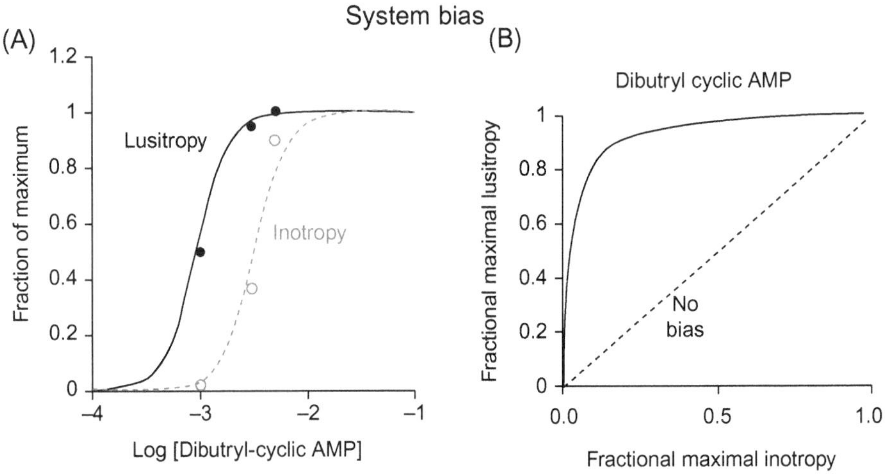

FIGURE 3.2 (A) Effects of dibutryl cyclic AMP on rat atrial inotropy and lusitropy. (B) shows a bias plot of lusitropy expressed as a function of inotropy. *Data redrawn from Kenakin et al.*[54]

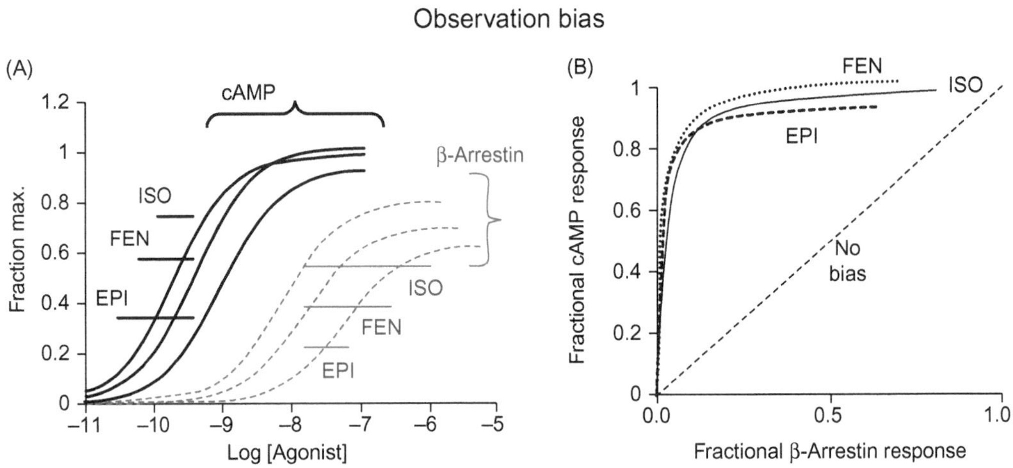

FIGURE 3.3 (A) Effects of β-adrenoceptor agonists for elevation of cyclic AMP (solid line curves) and complexation of receptor to β-arrestin (dotted gray curves). (B) shows the bias plots of cyclic AMP responses plotted as a function of β-arrestin responses. While there is a clear bias toward cyclic AMP, it is due to the difference in the sensitivities of the assays and is constant for all agonists. *Data redrawn from.*[55]

with types of assays and assay conditions and will be constant for all agonists, thus it cannot be exploited for therapeutic value.

Within system and observational bias, ligands may have the ability to stabilize different conformations of the receptor[4] coded in their chemical structure. If this happens, then there will be the possibility that some of these conformations can selectively interact with

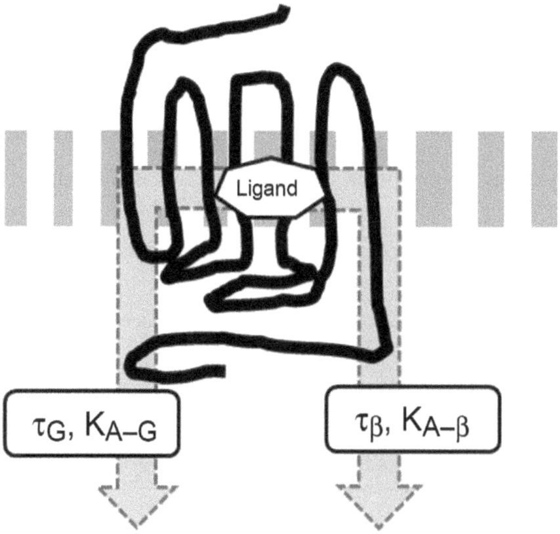

FIGURE 3.4 The mechanism of ligand-specfiic signaling bias. Ligands can stabilize different receptor conformations that then go on to interact with signaling proteins in a differential manner. Thus, an efficacy and affinity for G protein signaling (τ_G, K_{A-G}) will be distinct from the ligand-stabilized state that interacts with β-arrestin (τ_β, $K_{A-\beta}$).

signaling proteins. This mechanism is depicted in Figure 3.4, where it can be seen that different ligands principally activate G protein and β-arrestin interaction with the receptor respectively. This type of bias is intrinsically coded within the chemical structure of the molecule and thus can be used for therapeutic advantage. There are examples of where the signaling profiles of molecules is improved by the introduction of bias (see Table 3.1), but there is no *a priori* reason why this should be the case. A biased molecule simply alters the signaling characteristics of natural endogenous molecules, which may also induce a negative property. Table 3.2 shows cases where a β-arrestin activating property has been shown to be negative.

There are two settings for therapeutic bias: (1) emphasis of a given signaling pathway or (2) deletion of a given signaling pathway. This brings the discussion to the point of the active process whereby ligand bias can be optimized in the drug discovery process; a prerequisite to this process is the determination of a useful scale to gauge the bias of molecules through quantification.

THE QUANTIFICATION OF BIASED SIGNALING

Transduction Coefficients

Ligand-mediated signaling can be quantified through comparison of responses with the Black–Leff operational model of agonism[68]:

$$\text{Response} = \frac{[A]^n \tau^n E_m}{[A]^n \tau^n + ([A]+K_A)^n} \tag{3.1}$$

TABLE 3.2 Proposed Targets where β-Arrestin Signaling is Proposed to be Detrimental

Disease Area	Target	Pharmacologic Effect	Reference
Analgesia	Opioid R	β-arrestin mediates opioid respiratory depression	Xu et al.[49]/ Groer et al.[50]
Reduction of triglycerides	GPR109	β-arrestin mediates vasodilation leading to flushing	Kammermann et al.[56]
Cancer	ET1	β-arrestin1 promotes metastasis in epithelial ovarian cancer	Rosano et al.[57]
Cancer	IGF-1R	Ab for IGF-1R internalize IFG-1R to cause ERK signaling	Zheng et al.[58]
Anxiety/Schizophrenia	CB2	β-arrestin2 mediated ERK signaling through CB2R leads to upregulation of 5-HT(2A)R	Franklin et al.[59]
Alzheimers disease	β-arrestin2	β-arrestin2 over expression leads to increased amyloid-β peptide to increase amyloid pathology	Thathiah et al.[60]
Neural damage	AT(2)	β-arrestin2 mediates cerebral signaling to cause neural damage	Zhang et al.[61]
Cardiac Fibrosis	β-Adrenoceptor	metoprolol-induced cardiac fibrosis leads to cardiac dysfunction through β-arrestin2 signaling	Nakaya et al.[62]
Chronic myelogenous leukemia Thrombosis	β-arrestin2	β-arrestin2 shown to be essential for CML disease propagation	Fereshteh et al.[63]
Thrombosis	β-arrestin1	Allbb3 signaling activates β-arrestin1 promotion of thrombus formation	Schaff et al.[64]
Motor/pschoactive effects	CB1	β-arrestin2 signaling mediates tolerance, motor suppression to CB2 agonism	Nguyen et al.[65]
Prostate Cancer	β2-adrenoceptor	β-arrestin2 mediated signaling mediates cancer progression	Zhang et al.[66]
Hyperaldosteronism	AT(1)	β-arrestin1 overexpression promotes aldosterone after myocardial infarction	Lymperopoulos et al.[67]
Myeloidleukemia	β-arrestin2	β-arrestin2 essential for disease progression	Fereshteh et al. [63]

where [A] is the agonist concentration, n the slope of the concentration response curve, E_m the maximal response capability of the system, τ the efficacy of the agonist, and K_A the equilibrium dissocation constant of the agonist–receptor complex. There are three features to quantify ligand-mediated signaling that must be incorporated into a scale for biased signaling. These are: (1) the potency, namely the sensitivity to agonism in the form of the location parameter of the agonist concentration–response curve along the concentration axis, usually the pEC_{50} (which is the − Log of the molar concentration producing 50% maximal response to the agonist); (2) the maximal response to the agonist; and (3) the agonist index derived from these must culminate in a single number that can then be analyzed statistically. In addition, the agonism should be associated with the particular signaling system of interest if possible; this may not be the case for whole cell agonism.

Specifically, once the stimulus emanates into the cell cytoplasm to affect other cellular biochemical processes, the nature of these cellular processes may modify the response of the cell according to the nature of the cellular contribution to stimulus.

The most theoretically sound index of agonism to yield a single number is a ratio of τ/K_A (referred to as a transduction coefficient,[69]) utilized as $Log(\tau/K_A)$. The value τ refers to the signaling quality of the agonist–receptor complex as it interacts with the signaling proteins of the cell. The K_A value is a measure of the effective affinity of the agonist for the receptor as measured by the functional response. Log (τ/K_A) values are molecular parameters in that they describe an allosteric vector[1] of molecular quantities (agonist, receptor, signaling protein) that is constant throughout all of the tissues signaling to that agonist (Figure 3.5). Thus there will be a unique value for $Log(\tau/K_A)$ for every agonist for each biochemical signaling process associated with the receptor. It can be seen that if biased signaling is associated with a given receptor conformation formed by a unique ternary complex of agonist, receptor, and signaling protein, then there will be an efficacy (defined by τ) and affinity (defined by K_A) associated with that conformation for the agonist. While the implication of different τ values is intuitively clear, the possibility that the affinity of the agonist may differ for the receptor as it activates two separate signaling pathways requires context.

Receptors are known to exist in many inter-convertible conformational states and that efficacy is basically the stabilization of some of these (identified as active states in that they go on to initiate signaling) upon agonist binding.

$$A + R \overset{K}{\rightleftharpoons} AR \underset{\varphi}{\overset{\gamma}{\rightleftharpoons}} AR^* \tag{3.2}$$

Therefore, the functional affinity of an agonist is an amalgam of binding reactions that can be represented as[70]:

$$K_A = \frac{K}{1 + \frac{\gamma}{\varphi}} \tag{3.3}$$

where it can be seen that the actual operational affinity of the agonist for the receptor (denoted K_A in equation 3.3) will depend on the avidity with which it changes the receptor conformation. In addition, since 7TMRs are allosteric proteins that couple to signaling

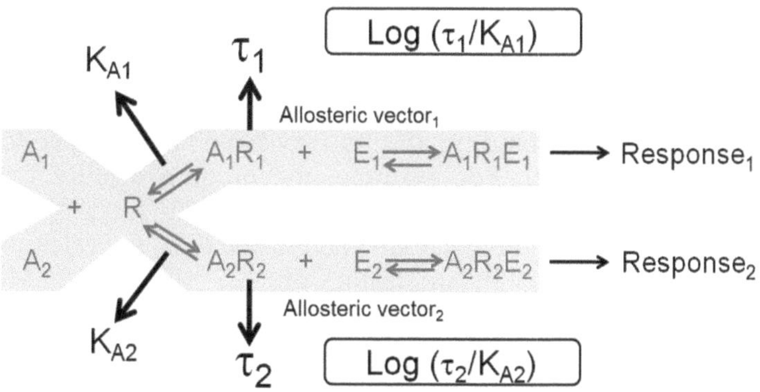

FIGURE 3.5 Application of the Black–Leff operational model to biased signaling. Values of Log (τ/K_A) can be assigned to different signaling pathways emanating from the same receptor.

protein in the cell to induce their effect, the nature and concentration of the signaling protein can affect the overall affinity of the receptor for ligands.[1,71,72] Thus, the binding of a ligand [A] in the presence of guest allosteric ligand [B] (with ligand-equilibrium dissociation constants K_A and K_B) is given by:

$$\frac{[ARB]}{[R_{tot}]} = \frac{[A]}{[A](1 + \alpha[B]/K_B) + K_A(1 + [B]/K_B)} \tag{3.4}$$

where α is the change in the affinity of the receptor for [A] in the presence of [B]. It can be seen from this equation that the observed affinity of the receptor for ligand [A] in the presence of [B] is given by:

$$K_{A(observed)} = \frac{K_A(1 + [B]/K_B)}{(1 + \alpha[B]/K_B)} \tag{3.5}$$

Equation 3.5 shows that the operational affinity of the receptor for [A] is given by the nature of (α) and the concentration of the co-binding ligand B. It can also be seen that for all values of $\alpha \neq 1$, $K_{A(observed)}$ will always be $\neq$ to K_A. There is experimental evidence to show that the affinity of agonists will be dependent upon the nature and concentration of co-binding species (e.g., G proteins, β-arrestin, other ligands) during the process of agonism. Some of this evidence comes from studies measuring guest allostery whereby the binding of an allosteric ligand alters the affinity of another ligand binding to the receptor. For example, binding studies with the allosteric ligand [^{3}H]dimethyl-W84 show that the affinity of the allosteric ligand gallamine for the free muscarinic M_2 receptor changes in the presence of the co-binding ligand N-methylscopolamine by a factor of nearly 50.[73] Similarly, the affinity of the NMDA receptor antagonist ifenprodil changes when the concentration of the co-binding ligand (NMDA) is increased by a factor of 10.[74] The same effect can be seen when the guest allosteric co-binding body is a signaling protein or a surrogate for a signaling protein.

Changes in receptor structure also have been shown with the binding of signaling proteins to receptors. For example, SCAM studies on the κ-opioid receptor show changes in conformation in transmembrane domains 6 and 7 with binding of $G_{\alpha 16}$ and/or $G_{\alpha i2}$ G protein subunits, with a resulting 18-fold change in the affinity of the ligand salvanorin.[75] Similarly, peptide fusion experiments in Sf9 membranes show a 27-fold increased affinity for isoproterenol upon the binding of nucleotide-free G_s heterotrimer protein to β_2-adrenoceptors.[76] Similarly, addition of β-arrestin and G_q protein to lipid nanodiscs containing ghrelin receptors shows clear creation of different receptor conformations through double exponential fluorescent lifetime decay analysis.[77]

In general, these data suggest that isolated binding affinity may have no relevance to the operational functional affinity in the cell. For example, the binding affinity for ^{125}I-human calcitonin in HEK 293 cells stably expressing human calcitonin receptors is 16 pM ($pK_d = 10.77$ with 95% confidence limits of 10.63–10.91), but the EC_{50} for calcium responses for human calcitonin is greater by a factor of 426 ($EC_{50} = 7.2$ nM, $pEC_{50} = 8.14 + 0.2$).[78] Presumably, the binding studies measure the total coupling of the receptor to G proteins whereas the calcium response measures a particular G protein interaction.

These data argue against the general use of binding affinity estimates in the Black–Leff equation for receptor function as a measure of K_A. Rather, the operational measure of functional affinity (K_A) through fitting the equation to functional data is a better option for the quantification of agonism. Under these circumstances, Log(τ/K_A) becomes a unique identifier of the ability of that agonist to activate that particular signaling pathway.

Figure 3.6 shows the agonist effects of four chemokines on the CCR5 chemokine receptor; Figure 3.6A shows increases in inositol phosphate IP1 while Figure 3.6B shows chemokine-induced internalization of the chemokine receptor. The bias plot, shown in Figure 3.6C, shows a bias toward the IP1 pathway but, unlike the examples of system and observation bias shown in Figures 3.1 and 3.3, the bias demonstrated by the four chemokines are quite different from each other; this is a hallmark of *conformational-specific* ligand bias. It can be seen from Figure 3.6C that CCL3L1 is uniquely the most biased towards inducing the greatest amount of CCR5 receptor internalization for a given IP1 response, when compared with CCL3, CCL4 and CCL5. In the case of CCL3L1, this may be therapeutically relevant since the gene copy number for the production of this chemokine has been associated with favorable survival after HIV-1 infection in progression to AIDS.[79]

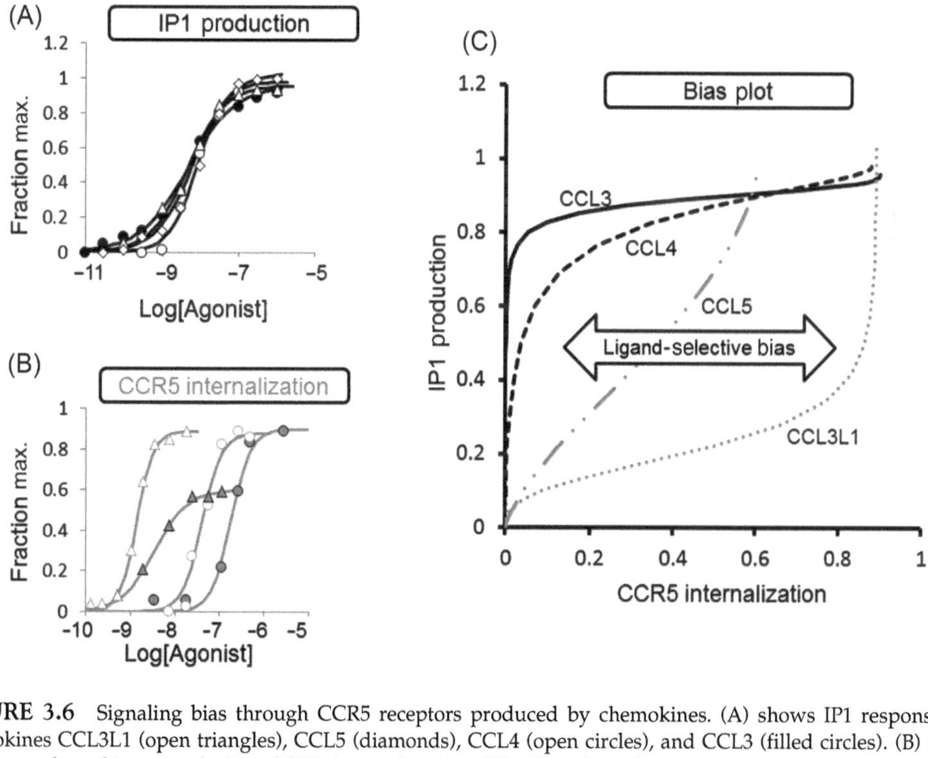

FIGURE 3.6 Signaling bias through CCR5 receptors produced by chemokines. (A) shows IP1 responses for chemokines CCL3L1 (open triangles), CCL5 (diamonds), CCL4 (open circles), and CCL3 (filled circles). (B) shows the same chemokines producing CCR5 internalization. (C) Bias plot, shows IP1 responses to each agonist expressed as a function of internalization induced by the same agonist. It can be seen that although responses are biased toward IP1 responses, the amount of bias is ligand dependent and not constant. *Data redrawn from.*[69]

TABLE 3.3 Biased Signaling for Chemokine Activation of CCR5 Receptors

Agonist	IP_1 Production		CCR5R Internalization	
	$Log(\tau/K_A)$	$\Delta log(\tau/K_A)^1$	$Log(\tau/K_A)$	$\Delta log(\tau/K_A)^1$
CCL3	7.75	0	6.58	0
CCL4	8.01	0.26	8.2	1.62
CCL5	8.27	0.52	8.53	1.95
CCL3L1	8.48	0.73	8.82	2.24

	$\Delta\Delta Log(\tau/K_A)^2$	$BIAS^3$
CCL3	0	**1**
CCL4	1.36	**23.1**
CCL5	1.43	**27**
CCL3L1	1.51	**32.4**

[1]Relative to CCL3

[2]IP1 vs Internalization

[3]$10^{\Delta\Delta Log(\tau/K_A)}$

(Data from[69])

Specifically, it has been suggested that CCL3L1-mediated internalization of the CCR5 receptor may yield protection from further HIV-1 infection by removing CCR5, the target protein used by the gp120 viral coat protein to bind to cells, from the cell surface. The 32.4-fold bias of CCL3L1 for CCR5 internalization shown in Table 3.3 and Figure 3.6 are consistent with this idea.

While the difference in ligand bias is evident from the bias plot, the actual differences between the ligands are not readily apparent without some numerical scale to grade the effects. Transduction coefficients ($\Delta Log[\tau/K_A]$ values), which indicate differences in the power of agonists to activate signaling pathways in a system-independent manner, can be used for this purpose. Table 3.3 shows $Log(\tau/K_A)$ values for the chemokines for the IP1 pathway and also for the CCR5 internalization effect. The first step is to choose an internal standard for both pathways and calculate $\Delta Log(\tau/K_A)$ values for each. For the data shown in Table 3.3, CCL3 was chosen, therefore $\Delta Log(\tau/K_A)$ values are calculated by subtracting the various $Log(\tau/K_A)$ values for each agonist from the $Log(\tau/K_A)$ value for CCL3; this step cancels the effects of system and observation bias. It is immaterial which agonist is chosen as the standard, but it is imperative that the same one be used for both pathways. For synthetic agonists, the natural endogenous agonist is often chosen since this will express the bias of the synthetic ligand in the systems compared with the natural agonist. Finally, between-pathway bias is calculated by subtracting the various $\Delta Log(\tau/K_A)$ values to yield a $\Delta\Delta Log(\tau/K_A)$ value; this is the logarithm of the bias. The bias of the agonist is then defined as:

$$BIAS = 10^{\Delta\Delta Log(\tau/K_A)} \tag{3.6}$$

The example in Table 3.3 and Figure 3.6 illustrates a potentially useful strategy for the quantification of signaling bias between two signaling pathways, namely fitting dose-response data for each pathway to the operational model for agonism (Equation 3.1), deriving $\mathrm{Log}(\tau/K_A)$ values for the agonist, relating those as a ratio of transduction coefficients to a standard agonists to derive $\Delta\mathrm{Log}(\tau/K_A)$ values, and then determining bias through ratios of the $\Delta\mathrm{Log}(\tau/K_A)$ values to obtain $\Delta\Delta\mathrm{Log}(\tau/K_A)$ values.

Equiactive Concentration Analysis

Another way to determine $\Delta\mathrm{Log}(\tau/K_A)$ values for a full vs a partial agonist is to compare equiactive concentrations of the agonists from curves in a given pathway through the method of Barlow et al.[80] This method was initially designed to estimate the relative affinity of a partial agonist by relating equiactive concentrations of a full and a partial agonist.[80,81] However, the method also can be used to calculate the relative ratios of efficacy to affinity (τ/K_A) of two agonists for a given signaling pathway for a system where the concentration–response curves have Hill coefficients that are not significantly different from unity. Expressing the response to the agonists in terms of the Black–Leff operational model (Equation 3.1), equal responses to the two agonists for a defined pathway is given by:

$$\frac{\frac{[A_1]}{[K_{A1}]}\tau_1}{\frac{[A_1]}{[K_{A1}]}(1+\tau_1)+1} = \frac{\frac{[A_2]}{[K_{A2}]}\tau_2}{\frac{[A_2]}{[K_{A2}]}(1+\tau_2)+1} \tag{3.7}$$

where $[A_1]$ and $[A_2]$ refer to concentrations producing equal responses for agonists 1 and 2, τ_1 and τ_2 refer to the efficacies of the agonists, and K_{A1} and K_{A2} refer to the equilibrium dissociation constants for the receptor–agonist complexes respectively. From this relationship, a double reciprocal plot of equiactive concentrations yields a linear plot:

$$\frac{1}{[A_1]} = \frac{1}{[A_2]}\frac{K_{A2}}{K_{A2}}\frac{\tau_1}{\tau_2} + \frac{\tau_1 - \tau_2}{K_{A1}\tau_2} \tag{3.8}$$

Thus, the slope of a line formed by a regression of $1/[A_1]$ upon $1/[A_2]$ yields the ratio of τ/K_A values for the agonist for the two agonists:

$$\mathrm{Slope} = \left[\frac{\tau_1}{K_{A1}}\right] \Big/ \left[\frac{\tau_2}{K_{A2}}\right] \tag{3.9}$$

Therefore $\Delta\mathrm{Log}(\tau/K_A)$ for the two agonists is given by Log(slope). The application of the method to dose–response curves for two theoretical agonists is shown in Figure 3.7.

σ_{lig} and β_{lig} Values

A related scale for the quantification of signaling bias has been proposed[55] that utilizes only the efficacy (τ value) of the agonists to quantify; a value termed σ_{lig} is defined for agonists 1 and 2 as:

$$\sigma_{lig} = \tau_1/\tau_2 \tag{3.10}$$

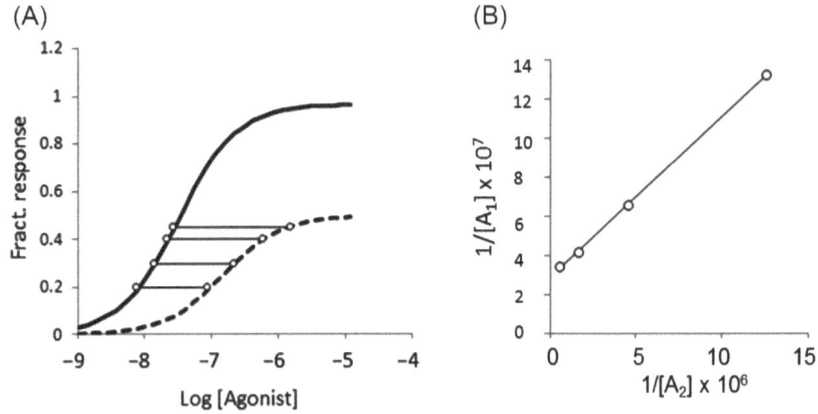

FIGURE 3.7 Method of Barlow, Scott and Stephenson[80] to assess $\Delta\mathrm{Log}(\tau/K_A)$ values for a full vs a partial agonist. (A) shows curves to a full (solid line) and partial (dashed line) agonist. In (B), equiactive agonist concentrations are identified (open circles) and used in a double reciprocal plot according to equation 3.8. The slope of this plot furnishes a ratio of τ/K_A values for the agonists.

A major difference between the σ_{lig} scale and transduction ratios is the fact that, for σ_{lig}, it is assumed that the affinity of the agonist for the receptor is identical for both signaling pathways. In fact, it is proposed that a separate estimate of ligand affinity be obtained with binding experiments. While in some cases this may be true, it is not an assumption that is supported by theoretical consideration of receptors as allosteric proteins, or by experimental data (see discussion above), and for this reason there may be errors in the estimation of bias for some ligands when using this method. The magnitude of the error will relate to the difference in the affinity of the agonist for the receptor as it interacts with two different coupling proteins mediating different signaling pathways. Thus, for two agonists 1 and 2, the logarithm of the magnitude of that error in bias, when calculated with the σ_{lig} scale, is given as:

$$\Delta\mathrm{Log\ BIAS} = \mathrm{Log}\left[\frac{K_{A-1}{}^{path2}}{K_{A-1}{}^{path1}}\right] + \mathrm{Log}\left[\frac{K_{A-2}{}^{path1}}{K_{A-2}{}^{path2}}\right] \quad (3.11)$$

where K_{A-1} and K_{A-2} refer to the equilibrium dissociation constants of the agonists for the receptor when the agonists produce activation of two pathways: path1 and path2 (denoted as superscripts in equation 3.11). It can be seen from this equation that if the affinity of the agonist does not change as the agonist activates the two pathways, then there will be no difference between the σ_{lig} and $\mathrm{Log}(\tau/K_A)$ scales.

Relative Activity (RA) Values

Another scale proposed for the quantification of signaling bias utilizes an index referred to as agonist relative activity (RA).[82–84] Specifically, RA values are the maximal response to the agonist divided by the potency expressed as an EC_{50} value (concentration producing

50% maximal response to the agonist). Interestingly, from the equations describing the maximal response to an agonist within the Black–Leff operational model[85]:

$$\mathrm{MAX} = \frac{\tau^n E_m}{(1 + \tau^n)} \tag{3.12}$$

And the potency of agonists as defined with this same model:

$$\mathrm{EC}_{50} = \frac{K_A}{((1+\tau^n)^{1/n} - 1)} \tag{3.13}$$

It can be shown that relative activity is:

$$\mathrm{RA} = \frac{\tau^n((1+\tau^n)^{1/n} - 1)E_m}{K_A(1 + \tau^n)} \tag{3.14}$$

For concentration–response curves with slopes not significantly different from unity, the RA then becomes $\tau E_m/K_A$ and relative RA values reduce to transduction ratios (ΔLog (RA) = $\Delta\mathrm{Log}(\tau/K_A)$).

PREDICTING SIGNALING BIAS *IN VIVO*

There is a great deal of emphasis on biased signaling as this is readily obtainable data *in vitro*. However, the overall effect of a biased ligand *in vivo* is comprised of such signaling effects *and* the obfuscation of endogenous signaling effects through the receptor occupancy of biased ligands. This latter effect may in fact be the more important therapeutic outcome of a biased ligand as in the case of biased antagonists. These are molecules that bind to the receptor to preclude the activation of the receptor by the endogenous agonist and then impart an added cellular signal through a selective biased efficacy. Thus, the angiotensin antagonists SII,[41] TRV120023,[23] and TRV120027[40] bind to angiotensin receptors to block the debilitating effects of angiotensin in congestive heart failure (potent vasoconstriction) but also produce cell protecting β-arrestin-based cellular signaling through a biased β-arrestin efficacy.

Specifically, studies in rats on the blockade of endogenous angiotensin with the conventional angiotensin receptor antagonists losartan or telmisartan lead to reduced mean arterial pressure and a decrease in cardiac performance. In contrast, the biased ligand TRV120027 actually increases cardiac performance and preserves cardiac stroke volume.[40] This type of effect also has been observed in canine heart failure models where cardiac unloading actions with preserved renal function have been seen with TRV120027.[86,87] Similarly, the biased histamine H4 receptor antagonist JNJ7777120 blocks agonist-induced G protein signaling of the receptor but actually promotes receptor interaction with β-arrestin.[88] In light of the deleterious effects of excess angiotensin activation of G proteins (to produce vasoconstriction) in heart failure, it may be the blockade of angiotensin signaling (the affinity of TRV 120027 for the receptor) that is the important therapeutic property of this molecule.

A biased molecule will encounter numerous tissues in the body with varying sensitivities to agonism (due to varying receptor densities and efficiencies of receptor coupling) and this can change the bias observed *in vitro*. This is because the potency of an agonist is a function of its affinity for the receptor (when it is interacting with the particular signaling protein) and its efficacy for a given pathway. However, agonism primarily based on efficacy (as opposed to affinity) is more resistant to decreases in receptor number than is agonist potency based on affinity. An example of this effect is seen by the selective diminution of the agonist effect of the high-affinity low-efficacy α-adrenoceptor agonist oxymetazoline when compared with the effect of the agonist norepinephrine (a high-efficacy low-affinity agonist), when α-adrenoceptors are diminished through alkylation with phenoxybenzamine.[89] Therefore, the relative efficacy of the biased agonist for the respective signaling pathways will determine the robustness of the bias with changing receptor densities in various tissues in the body.

Figure 3.8 shows the biased effect of an agonist on two pathways in a sensitive tissue where the agonism to pathway 2 is slightly more pronounced than that to pathway 1. However, the agonism for pathway 2 is dependent upon high affinity and low efficacy (when compared with that for pathway 2). In a less sensitive tissue (lower receptor density) the bias is reversed. Thus, the bias predicted in the sensitive tissue is not predictive of bias in all tissues *in vivo*. These effects highlight that it is the relative efficacy of the agonist as well as the bias that must be considered in predicting biased signaling *in vivo*.

Figure 3.9 shows a graph of four quadrants with the ordinate axis the relative efficacy of agonists for two pathways and the abscissae the logarithm of the bias factor (Equation 3.6). Four distinct profiles can be identified whereby the bias of the agonists will be different in a range of tissue sensitivity.

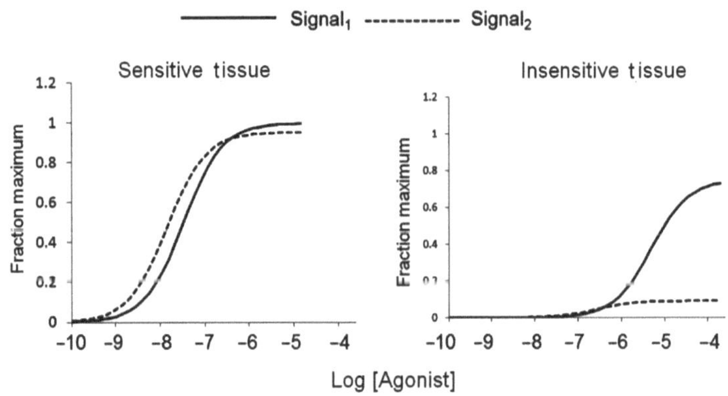

FIGURE 3.8 Effect of receptor density (tissue sensitivity) on observed agonist bias. A given agonist has a high efficacy but low potency for one signaling pathway (solid line) and a high affinity but low efficacy for another pathway (dotted line). In a sensitive tissue with high receptor number, the high potency dominates and a bias toward the dotted line pathway is observed. In a less sensitive tissue, the low efficacy for the second pathway (dashed line) causes the response for that pathway to selectively diminish over that of the solid line pathway to cause a reversal of bias in this tissue.

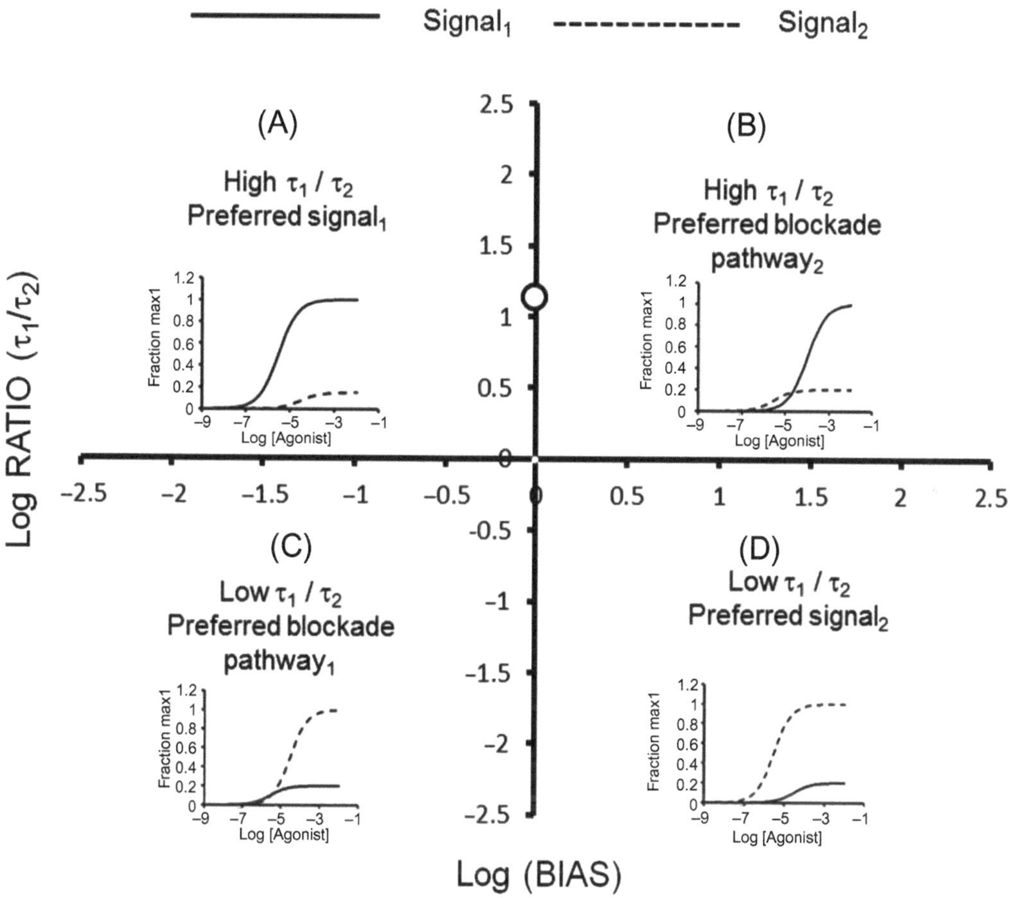

FIGURE 3.9 The ordinate axis reflects the relative intrinsic efficacy of the ligands for the two pathways while the abscissal axis reflects the bias for those pathways. It can be seen that four characteristic types of molecules can be differentiated. (A) Solid line pathway signaling preferred with low dotted line antagonism (efficacy). (B) Solid line pathway signaling preferred but high potency for antagonism of dashed line pathway. (C) Dotted line signaling preferred with low potency for antagonism of solid line pathway. (D) Dotted line signaling preferred with high potency for antagonism of solid line pathway.

PHYSIOLOGICAL RELEVANCE OF SIGNALING BIAS

Intuitively it can be seen that for pleiotropically signaling receptors, ligand bias is a practical way of fine-tuning the output of a receptor. Thus, in so-called redundant systems, such as chemokines where a number of different chemokines are the natural ligands for the same receptor, it can be seen that the different chemokines involved induce different stimulus outputs from the same receptor.[90] For example, the chemokines CCL19 and CCL21 are both natural agonists for the chemokine CCR7 receptor and both promote G protein signaling but only one (CCL19) promotes agonist-induced receptor phosphorylation and recruitment of β-arrestin.[91] Similarly, it has been shown that the activation of the CCR2 receptor by agonists produces β-arrestin response of varying stability; while CCL8 and CCL13 produce stable CCR2-β-arrestin interaction, CCL7 produced a transient complex with a half life of less than 15 min.[92] These ligands also stabilize different conformations of the CCR2 homodimer.[92]

The same type of fine-tuning may be involved in the production of receptor isoforms. For example, GPR120 has long and short splice variants and it has been shown that agonist stimulation causes β-arrestin2-receptor association and receptor internalization for both splice variants, but that the long splice variant selectively loses the ability to functionally produce G protein-dependent calcium and dynamic mass distribution effects. Thus, the short splice variant has been shown to be a naturally biased receptor.[93]

Biased signaling also can be shown to be a factor in the fine-tuning of signals through heterologous desensitization. For example, prolonged stimulation of μ-opioid receptors in brain locus ceruleus neurons has been shown to produce heterologous desensitization of α_2-adrenoceptor responses and that this effect is abolished in β-arrestin2 knockout mice. This indicates that β-arrestin2 signaling can regulate postsynaptic responsiveness in neurotransmitter release; the corollary to this finding is that biased μ-opioid agonists would therefore have differential effects on neurotransmitter release as well.[94] Finally, it should be noted that biased signaling need not be driven only through G proteins and β-arrestin but also can occur with any protein that interacts with the receptor to modify cellular response.

The introduction of the concept of signaling bias is fairly recent and with that comes a notion that it may be a rare occurrence. However, in light of ideas suggesting that ligands stabilize ensembles of receptor conformations through binding,[95] a contrary argument would suggest that bias in signaling should be a very common event. Specifically, it would be difficult to imagine that any two structurally dissimilar ligands would stabilize an identical array of conformations in the same proportions to induce identical bias in signaling. The increasing prevalence of observed signaling bias in synthetic ligands supports this view although it is a general statement yet to be validated.

CONCLUSION

The production of biased signaling by agonists is an important phenomenon that brings both promise to new therapeutic strategies and challenges to the drug discovery process. It should be stressed that bias can only be defined in terms of a reference agonist effect; a common reference point is to quantify bias to synthetic ligands within the context of effects of natural endogenous agonists. This is not to say that endogenous agonists are not biased since their effects no doubt will be biased toward the needs of the physiological system. However, synthetic ligands may produce a different bias that may, in turn, be therapeutically useful and it is this therapeutic bias that generates interest for drug discovery. In this latter regard, a quantitative scale to quantify signaling bias is essential. This chapter outlines some methods that can be used, specifically $\Delta\Delta\text{Log}(\tau/K_A)$ or $\Delta\Delta\text{Log}(RA)$ values. It will be interesting to see in the future how translational techniques correlate *in vitro*-based bias with *in vivo* therapeutic effects.

References

1. Kenakin TP, Miller LJ. Seven transmembrane receptors as shapeshifting proteins: the impact of allosteric modulation and functional selectivity on new drug discovery. *Pharmacol Rev* 2010;**62**:265–304.
2. Perez DM, Karnick SS. Multiple signaling states of G-protein coupled receptors. *Pharmacol Rev* 2005;**57**:147–61.
3. Mailman RB. GPCR functional selectivity has therapeutic impact. *Trends Pharmacol Sci* 2007;**28**:390–6.
4. Kenakin TP. Agonist-receptor efficacy II: agonist-trafficking of receptor signals. *Trends Pharmacol Sci* 1995;**16**:232–8.
5. Liu JJ, Horst R, Katritch V, Stevens RC, Wuthrich K. Biased signaling pathways in b2-adrenergic receptor characterized by 19F-NMR. *Science* 2012;**335**:1106–10.
6. Luttrell LM, Ferguson SS, Daaka Y, Miller WE, Maudsley S, Della Rocca GJ, et al. β-arrestin-dependent formation of β2 adrenergic receptor-Src protein kinase complexes. *Science* 1999;**283**:655–61.
7. DeWire SM, Ahn S, Lefkowitz RJ, Shenoy SK. Beta arrestins and cell signaling. *Annu Rev Physiol* 2007;**69**:483–510.
8. Zhan X, Kaoud TS, Dalby KN, Gurevich VV. Nonvisual arrestins function as simple scaffolds assembling the MKK4-JNK3α2 signaling complex. *Biochem* 2011;**50**:10520–9.
9. Ibrahim IA, Kurose H. β-arrestin-mediated signaling improves the efficacy of therapeutics. *J Pharmacol Sci* 2012;**118**:408–12.
10. Walwyn W, Evans CJ, Hales TG. Beta arrestin2 and c-Src regulate the constitutive activity and recycling of mu opioid receptors in dorsal root ganglion neurons. *J Neurosci* 2007;**27**:5092–104.
11. Feng X, Wang W, Liu J, Liu Y. β-arrestins: multifunctional signaling adaptors in type 2 diabetes. *Mol Bio Rep* 2011;**38**:2517–28.
12. Noor N, Patel CB, Rockman HA. B arrestin: a signaling molecule and potential therapeutic target for heart failure. *J Mol Cell Cardiol* 2011;**51**:534–41.
13. Tilley DG. G protein-dependent and G protein-independent signaling pathways and their impact on cardiac function. *Circ Res* 2011;**109**:217–30.
14. Lymperopoulos A. Beta-arrestin biased agonism/antagonism at cardiovascular seven transmembrane-spanning receptors. *Curr Pharm Des* 2012;**18**:192–8.
15. Bohn LM, Schmid CL. Serotonin receptor signaling and regulation via β-arrestins. *Crit Rev Biochem Mol Biol* 2010;**45**:555–66.
16. Tilley DG. Functional relevance of biased signaling at the angiotensin II type 1 receptor. *Endocr Metab Immune Disord Drug Targets* 2011;**11**:99–111.
17. Godin CM, Ferguson SS. Biased agonism of the angiotensin II type 1 receptor. *Mini Rev Med Chem* 2012;**12**:812–6.
18. Patel CB, Noor N, Rockman HA. Functional selectivity in adrenergic and angiotensin signaling systems. *Mol Pharmacol* 2010;**78**:983–92.
19. Shenoy SK. β-arrestin-biased signaling by the β-adrenergic receptors. *Curr Top Membr* 2011;**67**:51–78.
20. Viladarga JP, Gardella TJ, Wehbi VL, Feinstein TN. Non-canonical signaling of the PTH receptor. *Trends Pharmacol Sci* 2012;**33**:423–31.
21. Sonoda N, Imamura T, Yoshizaki T, Babendure JL, Lu J-C, Olefsky JM. β-arrestin1 mediates glucagon-like peptide-1 signaling to insulin secretion in cultured pancreatic β cells. *Proc Natl Acad Sci USA* 2008;**105**:6614–9.
22. Aquila B, Coulbault L, Davis A, Marie N, Hasbi A, Le bras F, et al. βarrestin 1-biased agonism at human δ-opioid receptor by peptidic and alkaloid ligands. *Cell Signal* 2008;**24**:699–707.
23. Kim KS, Abraham D, Williams B, Violin JD, Mao L, Rockman HA. β-arrestin-biased AT1R stimulation promotes cell survival during acute cardiac injury. *Am J Physiol Heart Circ Physiol* 2012;**303**:H1001–10.
24. Hostrup A, Christensen GL, Bentzen BH, Liang B, Aplin M, Grunnet M, et al. Functionally selective AT(1) receptor activation reduces ischemia reperfusion injury. *Cell Physiol Biochem* 2012;**30**:642–52.
25. Bohinc BN, Gesty-palmer D. β-arrestin-biased agonism at the parathyroid hormone receptor uncouples bone formation from bone resorption. *Endocr Metab Immune Disord Drug Targets* 2011;**11**:112–9.
26. Mosnier LO, Sinha RK, Burnier L, Bouwens EA, Griffin JH. Biased agonism of protease-activated receptor 1 by activated protein C caused by noncanonical cleavage of Arg46. *Blood* 2012;**120**:5237–46.
27. Allen JA, Yost JM, Setola V, Xin C, Meng C, Sassano M, et al. Discovery of β-arrestin-biased dopamine D2 ligands for probing signal transduction pathways essential for antipsychotic efficacy. *Proc Natl Acad Sci USA* 2011;**108**:18488–93.

28. Chen X, Sassano MF, Zheng L, Setola V, Chen M, Bai X, et al. Structure-functional selectivity relationship studies of β-arrestin-biased dopamine D2 receptor. *J Med Chem* 2012;**55**:7141–53.
29. Yang X, Zhou G, Ren T, Li H, Zhang Y, Yin D, et al. β-arrestin prevents cell apoptosis through pro-apoptotic ERK1/2 and p38 MAKs and anti-apoptotic Akt pathways. *Apoptosis* 2012;**17**:1019–26.
30. Schulte G, Shenoy SK. β-arrestin and dishevelled coordinate biased signaling. *Proc Natl Acad Sci USA* 2011;**108**:19839–40.
31. Soh UJK, Trejo J. Activated protein C promotes protease-activated receptor-1 cytoprotective signaling through β-arrestin and dishevelled-2 scaffolds. *Proc Natl Acad Sci USA* 2011;**108**:E1372–80.
32. Schmid CL, Bohn LM. Physiological and pharmacological implications of beta-arrestin regulation. *Pharmacol Ther* 2009;**121**:285–93.
33. Grady MA, Gasperoni TL, Kirkpatrick P. *Aripiprazole Nat Rev Drug Disc* 2003;**2**:427–8.
34. Urban JD, Vargas GA, von Zastrow M, Mailman RB. Aripirazole has functionally selective action at dopamine D2 receptor-mediated signaling pathways. *Neuropsychopharmacol* 2007;**32**:67–77.
35. Vassart G, Dumont JE. The thyrotropin receptor and the regulation of thyrocyte function and growth. *Endocr Rev* 1992;**13**:596–611.
36. Aplin M, Bonde MM, Hansen JL. Molecular determinants of angiotensin II type 1 receptor functional selectivity.. *J Mol Cell Cardiol* 2009;**46**:15–24.
37. Rajagopla K, Lefkowitz RJ, Rockman HA. When 7 transmembrane receptors are not G protein-coupled receptors. *J Clin Invest* 2005;**115**:2971–4.
38. Rajagopal K, Whalen EJ, Violin JD, Stiber JA, Rosenberg PB, Premont RT, et al. β-arrestin2 mediated inotropic effects of the angiotensin II type 1A receptor in isolated cardiac myocytes. *Proc Natl Acad Sci USA* 2006;**103**:16284–9.
39. Violin JD, Lekowitz RJ. β-arrestin-biased ligands at seven transmembrane receptors. *Trends Pharmacol Sci* 2007;**28**:416–22.
40. Violin JD, DeWire S, Yamashita D, Rominger DH, Nguyen L, Schiller L, et al. Selectively engaging β-arrestins at the angiotensin II type 1 receptor reduces blood pressure and increases cardiac performance. *J Pharmacol, Exp Ther* 2010;**335**:572–9.
41. Wei H, Ahn S, Shenoy SK, Karnik SS, Hunyady L, Luttrell LM, et al. Independent beta-arrestin 2 and G protein mediated pathways for angiotensin II activation of extracellular signal-regulated kinases 1 and 2. *Proc Natl Acad Sci USA* 2003;**100**:10782–7.
42. Zhai P, Yamamoto M, Galeotti J, Liu J, Masurekar M, Thaisz J, et al. Cardiac-specific overexpression of At1 receptor mutant lacking Gaq/Gai coupling causes hypertrophy and bradycardia in transgenic mice. *J Clin Invest* 2005;**115**:3045–56.
43. Ryman-Rasmussen JP, Griffith A, Oloff S, Vaidehi N, Brown JT, Goodard WA, et al. Functional selectivity of dopamine D1 receptor agonists in regulating the fate of internalized receptors. *Neuropharmacol* 2007;**52**:562–75.
44. Ji S-P, Zhang Y, van Cleemput J, Jiang W, Liao M, Li L, et al. Disruption of PTEN coupling with 5-HT2C receptors suppresses behavioral responses induced by drugs of abuse. *Nat Med* 2006;**12**:324–9.
45. MacKinnon AC, Tufail-Hanif U, Lucas CD, Jodrell D, Haslett C, Sethi T. Expression of V1A and GRP receptors leads to cellular transformation and increased sensitivity to substance-P analogue-induced growth inhibition. *Br J Cancer* 2005;**92**:522–31.
46. Willins DL, Berry SE, Alsayegh L, Backstrom JR, Sanders-Bush E, Friedman L, et al. Clozapine and other 5-HT2A antagonists alter the subcellular distribution of 5-hydroxytryptamine 2A receptors *in vitro* and *in vivo*. *Neuroscience* 1999;**91**:599–606.
47. Raehal KM, Walker JKL, Bohn LM. Morphine side effects in β-arrestin 2 knockout mice. *J Pharmacol Exp Ther* 2005;**314**:1195–201.
48. Bohn L, Lefkowitz RJ, Gainetdinov RR, Peppel K, Caron MG, Lin F-T. Enhanced morphine analgesia in mice lacking beta-arrestin 2. *Science* 1999;**286**:2495–8.
49. Xu H, Partilla JS, Wang X, Rutherford JM, Tidgewell K, Prisinzano TE, et al. A comparison of noninternalizing (herkinorin) and internalizing (DAMGO) mu-opioid agonists on cellular markers related to opioid tolerance and dependence. *Synapse* 2007;**61**:166–75.
50. Groer CE, Tidgewell K, Moyer RA, Harding WW, Rothman RB, Prisinzano TE, et al. An opioid agonist that does not induce μ opioid receptor—arrestin interactions or receptor internalization. *Mol Pharmacol* 2007;**71**:549–57.

51. Walters RW, Shukla A, Kovacs JJ, Violin JD, DeWire SM, Lam CM, et al. β-arrestin 1 mediates nicotinic acid-induced flushing, but not its antilipolytic effect, in mice. *J Clin Invest* 2009;**119**:1312–21.
52. Ferrari SL, Pierroz DD, Glatt V, Goddard DS, Bianchi EN, Lin FT, et al. Bone response to intermittent parathyroid hormone is altered in mice bull for (beta) arrestin 2. *Endocrinol* 2005;**146**:1854–62.
53. Gesty-Palmer D, Flannery P, Yaun L, Corsino L, Spurney R, Lefkowitz RJ, et al. A β-arrestin-biased agonist of the parathyroid hormone receptor (PTH1R) promotes bone formation independent of G protein activation. *Sci Transl Med* 2009;**1**:1ra1.
54. Kenakin TP, Ambrose JR, Irving PE. The relative efficiency of β-adrenoceptor coupling to myocardial inotropy and diastolic relaxation: organ selective treatment for diastolic dysfunction. *J Pharmacol Exp Ther* 1991;**257**:1189–97.
55. Rajagopal S, Ahn S, Rominger DH, Gowen-MacDonald W, Lam CM, DeWire SM, et al. Quantifying ligand bias at seven-transmembrane receptors. *Mol Pharmacol* 2011;**80**:367–77.
56. Kammermann M, Denelavas A, Imbach A, Grether U, Dehmlow H, Apfel CM, et al. Impedance measurement: a new method to detect ligand-biased receptor signaling. *Biochem Biophys Res Commun* 2011;**412**:419–24.
57. Rosanò L, Cianfrocca R, Tocci P, Spinella F, Di Castro V, Spadaro F, et al. β-arrestin is a nuclear transcriptional regulator of endothelin-1-induced β-catenin signaling. *Oncogene* 2013;**32**:5066–77.
58. Zheng H, Shen H, Oprea I, Worrall C, Stefanescu R, Girnita A, et al. β-Arrestin-biased agonism as the central mechanism of action for insulin-like growth factor 1 receptor-targeting antibodies in Ewing's sarcoma. *Proc Natl Acad Sci USA* 2012;**109**:20620–5.
59. Franklin JM, Vasiljevik T, Prisinzano TE, Carrasco GA. Cannabinoid agonists increase the interaction between β-arrestin and ERK1/2 and upregulate β-arrestin 2 and 5-HT(2A) receptors. *Pharmacol Res* 2012;**68**:46–58.
60. Thathiah A, Horre K, Snellinix A, Vandewyer E, Huang Y, Ciesielska M, et al. β-arrestin 2 regulates Aβ generation and Y-secretase activity in Alzheimer's disease. *Nat Med* 2013;**19**:43–9.
61. Zhang TL, Fu JL, Geng Z, Yang JJ, Sun XJ. The neuroprotective effect of losartan through inhibiting AT1/ASK1/MKK4/JNK3 pathway following cerebral I/R in rat hippocampal CA1 region. *CNS Neurosci Ther* 2012;**18**:981–7.
62. Nakaya M, Chikura S, Watari K, Mizuno N, Mochinaga K, Mangmool S, et al. Induction of cardiac fibrosis by β-blocker in G protein-independent and G protein-coupled receptor kinase 5/β-arrestin2-dependent signaling pathways. *J Biol Chem* 2012;**287**:35669–77.
63. Fereshteh M, Ito T, Kovacs JJ, Zhao C, Kwon HY, Tornini V, et al. β-arrestin 2 mediates the initiation and progression of myeloid leukemia. *Proc Natl Acad Sci USA* 2012;**109**:12532–7.
64. Schaff M, Receveur N, Boudon C, Ohlmann P, Lanza F, Gachet C, et al. β-arrestin1 participates in thrombosis and regulates integrin allbβ3 signaling without affecting P2Y receptors desensitization and function. *Thromb Haemost* 2012;**107**:735–48.
65. Nguyen PT, Schmid CL, Raehal KM, Selley DE, Bohn LM, Sim-Selley LJ. β-arrestin2 regulates cannabinoid CB1 receptor signaling and adaptation in a central nervous system region-dependent manner. *Biol Psychiatry* 2012;**71**:714–24.
66. Zhang P, He X, Tan J, Zhou X, Zou L. β-arrestin2 mediates b-2 adrenergic receptor signaling inducing prostate cancer cell progression. *Oncol Rep* 2011;**26**:1471–7.
67. Lymperopoulos A, Rengo G, Zincarelli C, Kim J, Koch WJ. Adrenal beta-arrestin 1 inhibition *in vivo* attenuates post-myocardial infarction progression to heart failure and adverse remodeling via reduction of circulating aldosterone levels. *J Am Coll Cardiol* 2011;**67**:356–65.
68. Black JW, Leff P. Operational models of pharmacological agonist. *Proc R Soc Lond [Biol]* 1983;**220**:141–62.
69. Kenakin TP, Watson C, Muniz-Medina V, Christopoulos A, Novick S. A simple method for quantifying functional selectivity and agonist bias. *ACS Chem Neurosci* 2012;**3**:193–203.
70. Colquhoun D. Imprecision in presentation of binding studies. *Trends Pharmacol Sci* 1985;**6**:197.
71. Christopoulos A, Kenakin TP. G-protein coupled receptor allosterism and Complexing. *Pharmacol Rev* 2002;**54**:323–74.
72. Christopoulos A. Allosteric binding sites on cell-surface receptors: novel targets for drug discovery. *Nature Rev Drug Disc* 2002;**1**:198–210.
73. Trankle C, Weyand A, Schroter A, Mohr K. Using a radioalloster to test predictions of the cooperativity model for gallamine binding to the allosteric site of muscarinic acetylcholine (m2) receptors. *Mol Pharmacol* 1999;**56**:962–5.
74. Kew JNC, Trube G, Kemp JA. A novel mechanism of activity-dependent NMDA receptor antagonism describes the effect of ifenprodil in rat cultured cortical neurons. *J Physiol* 1996;**497**:761–72.

75. Yan F, Mosier PD, Westkaemper RB, Roth BL. Gα-subunits differentially alter the conformation and agonist affinity of κ-opioid receptors. *Biochem* 2008;**47**:1567–78.
76. Rasmussen SGF, DeVree BT, Zou Y, Kruse AC, Chung KY, Kobilka TS, et al. Crystal structure of the β2 adrenergic receptor-Gs protein complex. *Nature* 2011;**477**:549–57.
77. Mary S, Damian M, Louet M, Floquet N, Fehrentz J-A, Marie J. Ligands and signaling proteins govern the conformational landscape explored by G protein-coupled receptor. *Proc Natl Acad Sci USA* 2012;**109**:8304–9.
78. Watson C, Chen G, Irving PE, Way J, Chen W-J, Kenakin TP. The use of stimulus-biased assay systems to detect agonist-specific receptor active states: implications for the trafficking of receptor stimulus by agonists. *Mol Pharmacol* 2000;**58**:1230–8.
79. Gonzalez E, Kulkarni H, Bolivar H, Mangano A, Sanchez R, Catano G, et al. The influence of CCL3L1 gene-containing segmental duplications on HIV-1/AIDS susceptibility. *Science* 2005;**307**:1434–40.
80. Barlow RB, Scott KA, Stephenson RP. An attempt to study the effects of chemical structure on the affinity and efficacy of compounds related to acetylcholine. *Br J Pharmacol* 1967;**21**:509–22.
81. Waud DR. On the measurement of the affinity of partial agonists for receptors. *J Pharmacol Expt Ther* 1969;**170**:117–22.
82. Ehlert FJ. Analysis of allosterism in functional assays. *J Pharmacol Exp Ther* 2005;**315**:740–54.
83. Tran JA, Chang A, Matsui M, Ehlert FJ. Estimation of relative microscopic affinity constants of agonists for the active state of the receptor in functional studies on m2 and m3 muscarinic receptors. *Mol Pharmacol* 2009;**75**:381–96.
84. Figueroa KW, Griffin MT, Ehlert FJ. Selectivity of agonists for the active state of M1 to M4 muscarinic receptor subtypes. *J Pharmacol Exp Ther* 2009;**328**:331–42.
85. Black JW, Leff P, Shankley NP, Wood Appendix J. An operational model of pharmacological agonism: the effect of E/[A] curve shape on agonist dissociation constant estimation. *Br J Pharmacol* 1985;**84**:561–71.
86. Boerrigter G, Lark MW, Whalen EJ, Soergel DG, Violin JD, Burnett JC. Cardiorenal actions of TRV120027, a novel β-arrestin-biased ligand at the angiotensin II type 1 receptor, in healthy and heart failure canines: a novel therapeutic strategy for acute heart failure. *Circ Heart Fail* 2011;**4**:770–8.
87. Boerrigter G, Soergel DG, Violin JD, Lark MW, Burnett Jr JC. TRV120027, a novel β-arrestin biased ligand at the angiotensin II type I receptor, unloads the heart and maintains renal function when added to furosemide in experimental heart failure. *Circ Heart Fail* 2012;**5**:627–34.
88. Rosethorne EM, Charlton SJ. Agonist-biased signaling at the histamine H4 receptor: JNJ7777120 recruits β-arrestin without activating G protein. *Mol Pharmacol* 2011;**79**:749–57.
89. Kenakin TP. The relative contribution of affinity and efficacy to agonist activity: Organ selectivity of noradrenaline and oxymetazoline. *Br J Pharmacol* 1984;**81**:131–41.
90. Zidar DA. Endogenous ligand bias by chemokines: implications at the front lines of infection and leukocyte trafficking. *Endocr Metab Immune Disord Drug Targets* 2011;**11**:120–31.
91. Kohout TA, Nicholas SL, Perry SJ, Reinhart G, Junger S, Struthers RS. Differential desensitization, receptor phosphorylation, α-arrestin recruitment, and ERK1/2 activation by the two endogenous ligands for the CC chemokine receptor 7. *J Biol Chem* 2004;**279**:23214–22.
92. Berchiche YA, Gravel S, Pelletier ME, St-Onge G, Heveker N. Different effects of the different natural CC chemokine receptor 2b ligands on beta-arrestin recruitment, Gαi signaling, and receptor internalization. *Mol Pharmacol* 2011;**79**:488–98.
93. Watson SJ, Brown AJ, Holliday ND. Differential signaling by splice variants of the human free fatty acid receptor GPR120. *Mol Pharmacol* 2012;**81**:631–42.
94. Dang VC, Chieng BC, Christie MJ. Prolonged stimulation of μ-opioid receptors produces β-arrestin2 mediated heterologous desensitization of α2-adrenoceptor function in locus ceruleus neurons. *Mol Pharmacol* 2012;**82**:473–80.
95. Kenakin TP. Efficacy at G protein coupled receptors. *Annu Rev Pharmacol Toxicol* 2002;**42**:349–79.
96. Dalle S, Ravier MA, Bertrand G. Emerging roles for β-arrestin1 in the control of pancreatic β-cell function and mass: new therapeutic strategies and consequences for drug screening. *Cell Signal* 2011;**23**:522–8.
97. González-Maeso J, Ang RL, Yuen T, Chan P, Weisstaub NV, López-Giménez JF, et al. Identification of a serotonin/glutamate receptor complex implicated in psychosis. *Nature* 2008;**452**:93–7.

CHAPTER

4

Biased Signaling and Conformational Dynamics in Nuclear Hormone Receptors

Colin A. Flaveny[1], *Laura A. Solt*[2], *Douglas J. Kojetin*[2] *and Thomas P. Burris*[1]

[1]Department of Pharmacological & Physiological Science, Saint Louis University School of Medicine, St. Louis, MO, USA, [2]Department of Molecular Therapeutics, The Scripps Research Institute, Jupiter, FL, USA

OUTLINE

B. Arey (Ed): *Biased Signaling in Physiology, Pharmacology and Therapeutics*
DOI: http://dx.doi.org/10.1016/B978-0-12-411460-9.00004-5

NUCLEAR RECEPTORS

Nuclear receptors generally function as ligand-regulated transcription factors, although only about half of the 48 members in the human nuclear receptor superfamily have identified physiological ligands. Almost all of the ligand-regulated receptors have been successful targets for drugs designed to treat a myriad of human diseases. As this chapter focuses on the action of ligands on NRs, only those receptors that have characterized ligands will be discussed.

NRs recognize specific DNA response elements in the promoters/enhancers of their cognate target genes, where they respond to ligands by altering their conformation thus changing the pattern of the recruitment of other transcriptional proteins that in turn alter gene expression. NRs display distinct preferences for both specific ligands and response elements. The specific ligand-mediated responses and selective binding of NRs to response elements underlie the distinct functions of each NR. Most of the NRs function as dimers, either homodimers (as in the case for the steroid receptors) or heterodimers with a particular NR, the retinoid X receptor (RXR). A limited number of NRs function as monomers, including several orphan members of the NR superfamily.

Nuclear receptors display a conserved modular domain structure (Figure 4.1), including an amino-terminal activation function-1 (AF-1) region and a highly conserved central DNA binding domain (DBD). The carboxy-terminal ligand binding domain (LBD) contains the activation function-2 (AF-2) region that serves as one of the key binding sites for co-regulator proteins. The LBD is the binding site for natural ligands of NRs. Examples of these natural ligands include 17β-estradiol for the estrogen receptor (ER), cortisol for the glucocorticoid receptor (GR), and 1α,25-dihydroxyvitamin D_3 for the vitamin D receptor (VDR). These natural ligands are key mediators of the physiological activity of their

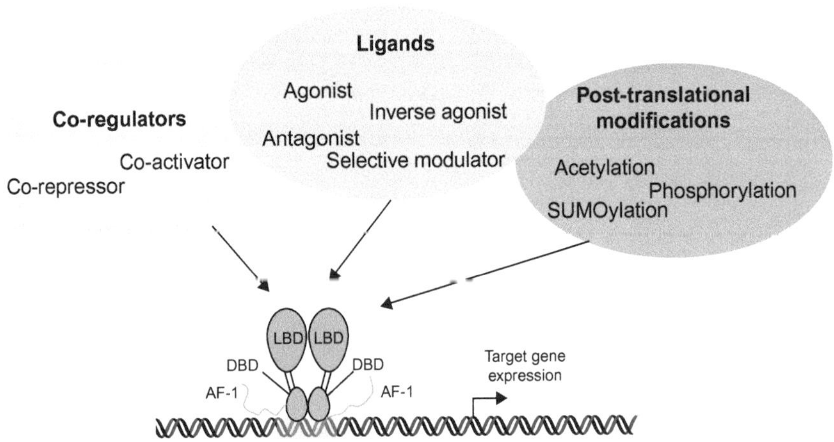

FIGURE 4.1 Modulation of nuclear receptor function. A nuclear receptor dimer is shown bound to a DNA response element on its target gene promoter. Illustrated above, within the ovals, are various mechanisms that modulate the activity of nuclear receptors. LBD = ligand binding domain, DBD = DNA binding domain, AF-1 = activation function-1.

cognate receptors and also provide a structural "blueprint" for the design of synthetic ligands. The study of natural ligand interaction with the C-terminal LBD has led to the development of a number of presently used synthetic ligands.

Nuclear receptors can be generally divided into two classes: transcriptional activators and repressors. The accepted mechanism of action for nuclear receptor transcriptional activators dictates that an agonist binds to the LBD and increases the recruitment of co-activator proteins, which in turn increases the transcription of target genes. In the classical sense, an antagonist blocks the binding of the agonist to the LBD and prevents the agonist from inducing a conformational change in the receptor. Conversely, many antagonists described can display inverse agonist activity, where binding of the ligand increases recruitment of co-repressor proteins, leading to the active repression of transcription.

The mechanism of action of nuclear receptor ligands has additional levels of complexity, as the same ligand can have different tissue-, cell- and promoter-specific action, often a function of the expression levels of co-regulator proteins, and also display graded receptor activity.[1–3] For instance, some agonists induce co-repressor recruitment to nuclear receptor transcriptional activators,[4] whereas some ligands act as agonists in certain tissues and antagonists in others. Other ligands can modulate receptor activity by altering post-translation modifications that modulate receptor activity.[5] Additionally, NRs that function primarily as transcriptional repressors, such as the Rev-erbs, are defined by their primary role as repressors whereas agonists such as the natural porphyrin heme induce co-repressor recruitment and repression.

NUCLEAR RECEPTOR ACTIVATION

A subset of NRs, including several steroid hormone receptors, undergoes translocation from the cytoplasm to the nucleus following ligand binding. In this case, these receptors are unable to exert an effect on the transcription of target genes. In the absence of ligand, parts of the receptor LBD are quite dynamic in terms of protein structural conformation. Typically, agonist binding to the LBD results in a conformational change that leads to recruitment of co-activator proteins, resulting in increased transcription of target genes. Ligand binding stabilizes the receptor LBD conformation, which facilitates interactions with co-regulator proteins that are able to modulate the rate of gene transcription. Co-regulator–receptor interaction promotes assembly of the primary transcriptional machinery and subsequent RNA polymerase binding, leading to expression of target genes (Figure 4.1). Many NRs do not undergo a ligand-dependent translocation and bind their DNA response elements both in the absence and presence of ligand, and in many cases may take an active role in target gene regulation even in the absence of ligand. One clear example of this is the thyroid hormone receptor (TR) that can actively silence target gene transcription in the absence of thyroid hormone, but transform into transcriptional activators in the presence of thyroid hormone.[6] These types of functional transformations are due to induction of conformational changes that alter the array of protein co-factors to which the NR binds. The discovery and function of NR co-factors is discussed in the following section.

NUCLEAR RECEPTOR CO-FACTORS

Nuclear receptors interact with co-factors, which include co-activators that promote gene expression or co-repressors that repress gene expression. Agonist binding induces co-activator–receptor interaction and antagonists disrupt this interaction. The role of co-activators was first elucidated when it was observed that the transactivational activity of PR could be inhibited or "squelched" by overexpression of ER. It was proposed that limiting amounts of accessory or co-activator proteins were necessary for NRs to regulate transcription.[7] Through this discovery it became clear that NRs compete for co-activator proteins that mediate activation of transcription within the cell; it also became evident that distinct transactivation domains within each NR may utilize unique co-activator proteins.[8] In contrast, receptor-mediated transcriptional silencing by the TR or retinoic acid receptor (RAR) could also be squelched, suggesting the existence of proteins that mediated suppression of target gene transcription: co-repressors.[9]

Using biochemical methods, putative intermediary proteins that interacted with ER's LBD were identified that only interacted with the ER in the presence of an agonist and a transcriptionally competent ER.[10] The first co-activator, steroid receptor co-activator-1 (SRC-1), was cloned using the LBD of PR as bait in a two-hybrid system and was shown to function as a co-activator for a range of NRs.[11] SRC-1 displayed all of the expected characteristics of an authentic co-activator, including agonist-dependent interaction with the LBD, which could be blocked by an antagonist, and the ability to rescue the activity of squelched NRs.[11] SRC-1 was the first of a family of co-activators that included SRC-2 and SRC-3, which were later characterized.[12–19] A multitude of other co-activator classes were subsequently characterized. The list of co-activator proteins has expanded to include hundreds of proteins.[20–26] Co-repressor proteins such as nuclear receptor co-repressor 1 (NCoR1) and silencing mediator for retinoid and thyroid hormone receptor (SMRT) have also been identified and characterized.[27,28] However, the list of known co-repressor proteins remains far fewer than that of co-activators. These co-activator/co-repressor proteins display an array of activities associated with regulation of transcription including histone acetyltransferase activity (HAT), histone deacetylase activity (HDAC), arginine methyltransferase activity, ubiquitin ligase activity, and ATP-dependent chromatin remodeling activity, which are only a handful of the many activities that aid in or repress transcriptional activation depending on the specific receptor and gene context.

NUCLEAR RECEPTOR MODULATORS

Classically defined ligands that bind to the LBD of nuclear receptors, whether they are endogenous or synthetic, either activate receptor activity (agonists) or block activation (antagonists). As these ligands alter receptor activity they are often called receptor "modulators." The classically defined antagonist, or neutral antagonist, simply binds to the LBD and prevents the conformational change that an agonist would cause, thus preventing co-activator recruitment and subsequent stimulation of transcription. However, this represents the most ideal scenario as the pharmacology of the NR is much more complex. Partial agonists bind to the LBD and the resulting conformational change provides only a partial activation of transcription, which is likely due to less efficient recruitment of co-activators. Partial agonists will also compete with full agonists so as to reduce the level of transactivation intrinsic to the full agonist and therefore are "antagonistic" to true agonists. Conversely, inverse agonist binding to NRs results in a conformational change that decreases the basal level of activity. In many cases, it has become clear that the conformation induced by inverse agonists may also result in recruitment of co-repressor proteins, resulting in active silencing of target gene transcription. In this case, a particular NR may not need to display any basal co-activator binding for the inverse agonist to cause a decrease in transcription of target gene transcription.

THE "HELIX 12 SWITCH" MODEL

Many advances in our understanding of nuclear receptor function have emerged via structural biology studies focused on the receptor LBD. The most common method of choice has been X-ray crystallography. Crystal structures of ligand–receptor complexes provide a "snapshot" of the molecular mechanism of action of the receptor. Hundreds of crystal structures of nuclear receptor LBDs have been reported, leading to the development of the "Helix-12 (H12) switch" model (Figure 4.2). The H12 model describes the molecular basis of ligand modulation: "on" being the transcriptionally active conformation produced by an agonist, and "off" being the transcriptionally repressed conformation produced by an antagonist. The NR-LBD adopts a three-layered α-helical sandwich fold, consisting of 12 α-helices with the ligand binding pocket.[29] The hydrophobic AF-2 surface (helix 3/4/5/12 interface) provides a binding site for co-activator proteins with a LXXLL recognition motif.[30] In the apo or unliganded form, H12 is extended away from the LBD. Agonists cause H12 to cap the ligand binding site, leaving the AF-2 surface exposed, thereby facilitating co-regulator binding.[31] Antagonists induce an unfavorable conformation for co-regulator binding. Some antagonists possess bulky portions that perturb the AF-2 surface via direct contact,[32] whereas others function in a more passive manner by failing to make the appropriate contacts in the ligand binding cavity. Passive antagonists alter H12 positioning causing an approximately 90° rotation from the agonist position and

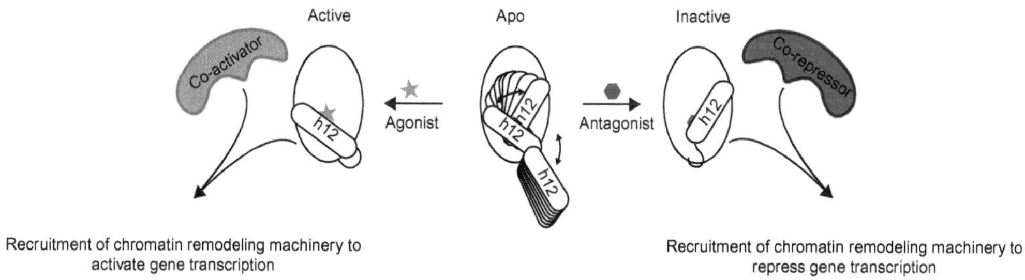

FIGURE 4.2 The Helix-12 switch model and its role in nuclear receptor activation. Illustration of the distinct H12 conformations for agonist binding leading to the active conformation of H12, and subsequent receptor activation antagonist binding leading to the inactive H12 conformation that results in decreased receptor activity.

occupying the AF-2 surface.[33] Partial agonists are thought to dynamically switch H12 between active and inactive structural conformations.[34] These structural observations have been used to understand the mechanism of action of ligands and are used as a guide for nuclear receptor virtual ligand screening and structure-based drug design efforts.

The role of positioning of H12 in response to ligand binding as the critical determinate of the function of the ligand (agonist vs antagonist) has been clearly shown in several co-crystal structures of NR–LBDs with various ligands. For example, when ER is bound to the ER agonist estradiol, H12 is positioned in a manner that completes the formation of the co-activator binding groove, allowing for recruitment of the co-activator proteins. Upon antagonist binding, the LBD assumes a conformation that does not allow co-activator NR box recognition. In contrast, the antagonist raloxifen binds to the ER causing H12 to be placed in a position that physically blocks NR box binding,[31,33] while in the case of another antagonist, ICI 182,780, binding the long aliphatic extension of this compound itself protrudes from the ligand binding pocket entryway and folds along the co-activator binding groove, thus preventing NR box interaction with co-factors.[34,35] All these investigations have helped elucidate the mechanistic differences between ligand agonist and antagonist activity.

DYNAMIC FEATURES OF GRADED RECEPTOR AGONISM

The Helix-12 switch model derived from NR–LBD crystal structures details the relationship between the two discrete (on/off) agonized or antagonized states. However, this mechanism of receptor activity being solely linked to H12 positioning does not sufficiently explain several more nuanced features of nuclear receptor function and activity. These features of receptor activity include graded transactivation, or partial agonism, and H12-independent receptor activation. Several mechanisms have been suggested for graded/partial agonist activity. Crystal structure analysis has suggested that receptor–ligand interaction can lead to an intermediate/quasi-antagonist conformation[34,36] where co-regulator interaction can switch H12 between different positions. Partial agonist binding may also promote suboptimal positioning of H12. In this scenario the partial agonist elicits a pseudo agonist conformation that does not produce a robust activation response. However, partial agonist-bound nuclear receptor LBDs have in general been difficult to crystalize,[37] and therefore the lack of partial agonist-bound LBD crystal structures has limited these interpretations.

Furthermore, these structural models describing possible mechanisms for providing a partial agonist response were primarily derived from ER crystallography studies and may not universally apply to other nuclear receptors. For example, nearly all crystal structures of PPARγ's LBD, whether in the apo form or bound to a ligand of graded agonist activity, show the same three-dimensional fold with H12 in the "active" conformation (Figure 4.2).[38–40] Therefore, H12 positioning is unlikely to be a universally applicable mechanism of graded receptor activation.

Nuclear magnetic resonance (NMR) studies were among the first to provide a hint as to the mechanism of partial agonism in nuclear receptors. The PPARγ partial agonist nTZDpa, a potent ($EC_{50} = 57$ nM) yet weak agonist, only partially stabilized the PPARγ LBD, causing a partial increase in NMR resonances when compared with NMR data for apo PPARγ

LBD.[41] Hydrogen deuterium exchange (HDX) mass spectrometry studies then produced additional mechanistic insight for a variety of PPARγ ligands with graded response profiles.[38,42] Unlike full agonists, PPARγ partial agonists do not stabilize H12 and differentially stabilize other portions of receptor ligand binding pocket, including the β-sheet surface and helix 3. Protein NMR studies have also revealed that stabilization of the receptor from intermediate conformational exchange is correlated with the graded response of ligand.[43] Full PPARγ agonists provide robust stabilization compared with apo LBD, and, conversely, weak partial PPARγ agonists provide a much lower level of conformational stabilization. Interestingly, NMR studies also revealed that PPARγ ligands bind to the LBD in more than one conformation, indicating that there is added complexity to the structure–function relationship of PPARγ-ligand interaction. Therefore, in this case receptor activity may be the cumulative result of a heterogeneous population of ligand–receptor complexes with each subpopulation contributing distinctly to overall activation.

When the PPARγ NMR and HDX studies were considered along with co-crystal structures of the PPARγ LBD complexed with various ligands of graded response profiles, a more complete picture of the mechanism of action was realized. Although the crystal structures for the PPARγ LBD complexed to MRL20 and MRL24 show H12 in the "active" conformation, only MRL20 is observed to make a hydrogen bond with Tyr473. This explained the protection from HDX on H12 observed for MRL20, as well as the slight increase in stabilization evident as assigned resonances for residues in H12. Conversely, MRL24 does not hydrogen bond to Tyr473, does not afford much protection from HDX in H12, and does not stabilize the receptor as much as MRL20, resulting in no assigned NMR resonances for residues in H12. Notably, HDX has also been used to profile full and partial vitamin-D receptor (VDR) agonists, revealing that VDR partial agonists differentially stabilize the AF-2/H12 region.[44,45] Fluorescence anisotropy studies on ERα have also suggested that estrogen receptor modulators that display weaker transactivation profiles compared with full ER agonists, increase the dynamics at the end of helix 11, which precedes the loop going into H12.[46] Thus, these techniques that report on the conformational dynamics of nuclear receptor LBDs have provided unique insight into the mechanism of action of nuclear receptor ligands, in particular those with graded function.

NUCLEAR RECEPTOR CONFORMATIONAL DYNAMICS AND DRUG DISCOVERY

Several recent studies support the use of conformational dynamics studies to help explain the mechanism of action of subclasses of ligands. Conformational dynamics data have proven especially useful for structure–activity relationship (SAR) analysis during drug discovery. HDX analyses on a variety of selective estrogen receptor modulators (SERMs) revealed dynamic profiles that correlated with the pharmacological profiles of the ligands with respect to ERα activity.[47] Analogous to the analysis on graded PPARγ agonists,[38] SERMs with various graded response profiles also displayed graded protection from HDX in several structural regions. Interestingly, clustering analysis of the ligand-induced HDX profiles allowed for the prediction of the tissue-specific activity of the ER compounds. This suggested that the HDX structural assay might be useful for designing new SERMs with enhanced tissue specificity and therefore lower risk of off target effects. In addition, as HDX is a high throughput assay in contrast to X-ray crystallography, it can be more easily integrated in current drug discovery pipelines. HDX was also utilized to show differential binding activities between two receptor subtypes, ERα and ERβ.[48] These receptors bind many of the same ligands, but have different tissue expression profiles, different affinities for the same ligand, and they are thought to oppositely regulate one another.[49,50] Interestingly, HDX studies revealed that ERβ undergoes different structural changes when compared with ERα for the same ligand, giving support to the notion that the same ligand can distinctly affect the function of ERα and ERβ.

Both of the aforementioned studies also showed that the dynamics of H12 in the ER-LBD is not altered, when compared with apo receptor, upon binding agonist or antagonist ligands. Previous crystal structures of ER-LBDs showed a clear repositioning of H12 when bound to agonists or antagonists. However, crystal structures of the apo-ER-LBD were only obtainable by introducing mutations in the ER-LBD that promoted hydrogen bond formation between both the agonist- and antagonist-bound H12 in order to stabilize the receptor conformation. This factor therefore explained the discrepancy between HDX and previous crystal structure analyses. It also suggested that ER ligands may not hydrogen bond to stabilize H12 itself, but rather these ligands interact with a number of residues within the ER-LBD pocket, possibly in concert with co-regulator proteins. This is in contrast to studies on PPARγ where crystal structures all show H12 in the agonist, or active, conformation; HDX and NMR studies show a clear stabilization of H12 for PPARγ agonist ligands.[38,43] These studies suggest that the underlying mechanism of activation of NRs can be divergent depending on the receptor in question, but can be teased out using HDX technology.

HDX studies have also been used as a guide for SAR analysis in PPARγ drug discovery. It was recently revealed that the anti-diabetic effect of synthetic ligands involves a novel mechanism—stabilization of dynamics around the PPARγ-LBD residue Ser273, which inhibits Ser273 phosphorylation by the protein kinase Cdk5.[5] Notably, blocking of Ser273 phosphorylation by PPARγ ligands was found to be independent of the transcriptional activation properties of the ligand afforded by H12 stabilization. In this study, two synthetic PPARγ compounds were studied: the full agonist, rosiglitazone, and the partial agonist, MRL24. While both ligands displayed a similar ability to block Ser273 phosphorylation, the lower transcriptional activity profile of MRL24 correlated with reduced H12 stabilization as well as a better gene expression profile when compared with rosiglitazone. Microarray analysis revealed an overlapping set of genes affected by rosiglitazone and MRL24, which included genes that are disregulated in type-2 diabetes and obesity, including adiponectin and adipsin. However, the expression of a large set of genes corresponding to the classic adipogeneic genes was increased by rosiglitazone but not by MRL24.

Subsequent to this, HDX was among several SAR assays used in the design of a proof-of-concept PPARγ non-agonist ligand, SR1664.[51] SR1664 binds directly to the PPARγ LBD and, like MRL24 but unlike rosiglitazone, SR1664 does not stabilize H12. Although SR1664 is devoid of transcriptional activation, it displays potent anti-diabetic activity with a reduced side effect profile, compared with rosiglitazone, in terms of reduced fluid retention, weight gain, and interference with bone formation. These discoveries provided evidence that the classical transcriptional activation of PPARγ could be distinguished from its anti-diabetic efficacy on the molecular level via study of HDX profiles.

NUCLEAR RECEPTOR SELECTIVE MODULATORS

The pharmacological and physiological responses produced by a given ligand are not exclusively restricted to agonist, antagonist, or inverse agonist activity. The effects of ligand–receptor interaction on a physiological level are significantly more complex and may vary depending on the tissue- or gene- specific context. This class of ligands—selective modulators—displays tissue-dependent and/or target gene-specific agonist, antagonist or inverse agonist activity. Here, we discuss the development and pharmacology of a range of selective nuclear receptor modulators.

The first examples of selective modulators were discovered through the development of synthetic ligands targeting the ER. Cancer research studies showed that compounds such as tamoxifen function as an antagonist in breast tissue, but as an agonist in bone and the uterus. It became clear through this discovery that the differential tissue-specific activity of ER can be altered. Compounds such as raloxifene were later identified that had a clinically superior tissue specificity profile of antagonist in the breast and uterus, but agonist in bone. These and other compounds that differentially modulated the tissue-specific activities of the ER were therefore dubbed selective ER modulators or SERMs. Other similar types of selective ligands were later identified for a number of other NRs. The mechanism of action that underlies the tissue/target gene specificity of most of these ligands is still not clear, but several sound theories have been proposed. It is highly likely that various selective modulators function via distinct mechanisms and that one mechanism will not be able to explain how they all display their unique pharmacological profiles.

The molecular events underlying selective NR modulator pharmacology are multitudinous, and various mechanisms have been implicated in the unique profile of ligands that target the NRs. Post-translational modifications of receptors and regulation of the modifications have been shown to lead to target-gene-specific regulation; additionally, nongenomic actions of various receptors are likely to also play a role in the action of at least some of the selective NR modulators that have been identified to date. One of the predominant theories is that distinct ligands induce conformations in the receptor that have inherent affinities for association with select co-activators/co-repressors, leading to unique higher order complexes that drive modulator pharmacology. In the following section we will review some of the key breakthroughs that have been made in the development of selective modulators for specific NRs—ER, GR and PPAR—and highlight some of the therapeutic challenges that targeting these receptors presents.

Selective Estrogen Receptor Modulators

The natural ligand for the ERs is the steroid hormone, 17β-estradiol, which binds to ERs with high affinity and activates the receptors, resulting in increased transcription of genes containing estrogen response element (ERE) promoter sequences (5′-GGTCANNNTGACCT-3′).[52] ERs form both functional homodimers (ERα-ERα; ERβ-ERβ) and heterodimers (ERα-ERβ).[53] The ER-LBD is probably the most studied NR domain in terms of NR structure, and, as discussed above, these structural studies have helped researchers understand the molecular basis of NR agonism and antagonism.

Both ERs are widely expressed, although ERα is more abundant in endometrial, breast, ovarian, and hypothalamus tissues, whereas ERβ is more abundant in bone, brain, endothelial, heart, kidney, intestine, lung, and the prostate gland. Furthermore, ERα and ERβ have both overlapping and non-overlapping functions, which combined with the tissue-specific differences in relative abundance of the receptors suggests that the specific targeting of ER isoforms could prove useful in therapeutic treatment of diseases, with perhaps limited crosstalk to other tissues or ER isoform-specific genes—essentially the primary basis of selective ligand modulation. In many cases, the actions of ERβ oppose the actions of ERα, suggesting that a single ligand may have distinct functional outcomes depending on the specific ER target.[50]

Because of their broad tissue expression patterns and important roles in development and physiology, ERs are drug targets for a variety of diseases.[54] Breast tissue development and physiology is substantially influenced by ERs. Estrogens also have a strong influence in bone, which expresses both ER isoforms, predominantly ERα in developing cortical bone and reciprocally ERβ in developing cancellous bone.[55,56] Estrogen increases HDL and triglycerides and lowers LDL levels, which is of particular importance as many post-menopausal women have an elevated risk of cardiovascular disease.[57] Estrogen therefore has a multifaceted physiological role and by extension SERMs have been used to provide multiple therapeutic benefits to breast cancer patients and post-menopausal women.[58]

SERMs have been useful for clinical treatment of a variety of disorders, including breast cancer, osteoporosis, and post-menopausal health. Treatment of ER-positive breast cancer is one of the primary clinical uses of SERMs. In breast cancer treatment, SERMs are an adjuvant therapy administered following mastectomy or lumpectomy and radiation therapy. These SERMs are used to antagonize estrogen receptor activity specifically within the breast. The first SERM used clinically, and still the most commonly prescribed, is tamoxifen (Figure 4.3B). Tamoxifen acts as an antagonist in breast tissue, but as an agonist in bone and the uterus. Tamoxifen, first discovered by ICI pharmaceuticals (AstraZeneca),[59−61] is a chemical derivative of the first non-steroidal anti-estrogen, MER25, which was discovered in the 1950s and identified as an inhibitor of estrogen actions.[62] Both of these compounds are chemical derivatives of the highly potent synthetic, non-steroidal estrogen, diethylstilbestrol.[63]

Early studies revealed that tamoxifen could compete with estrogens, in terms of preventing the accumulation of [^{3}H]-estradiol in ER target tissues (e.g., uterine, vaginal, and mammary) of mice, rats, and humans.[64−67] These studies also provided insight into the pharmacokinetic/pharmacodynamic (PK/PD) properties of tamoxifen, including its conversion into the metabolite 4-hydroxytamoxifen (Figure 4.3C), which is the primary form that binds to ERs *in vivo*.[68,69] Radioligand binding assays confirm that tamoxifen binds directly to the ER-LBD and competes with 17β-estradiol. Later, structural validation via X-ray crystallography confirmed this interaction and provided molecular details concerning the interactions between ERα with 17β-estradiol and tamoxifen.[31,33] Breast cancer patients who received tamoxifen experienced the therapeutic benefit of tamoxifen-mediated ER antagonism in the breast and increased bone density due to ER agonism in bone. Unfortunately, as it agonizes ER activity in the uterus, tamoxifen has the undesired side effect of promoting increased cell proliferation in the uterus, which enhances the risk of endometrial cancer. The complex pharmacology of tamoxifen (Figure 4.3) highlighted the need for SERMs with more refined selective activity.

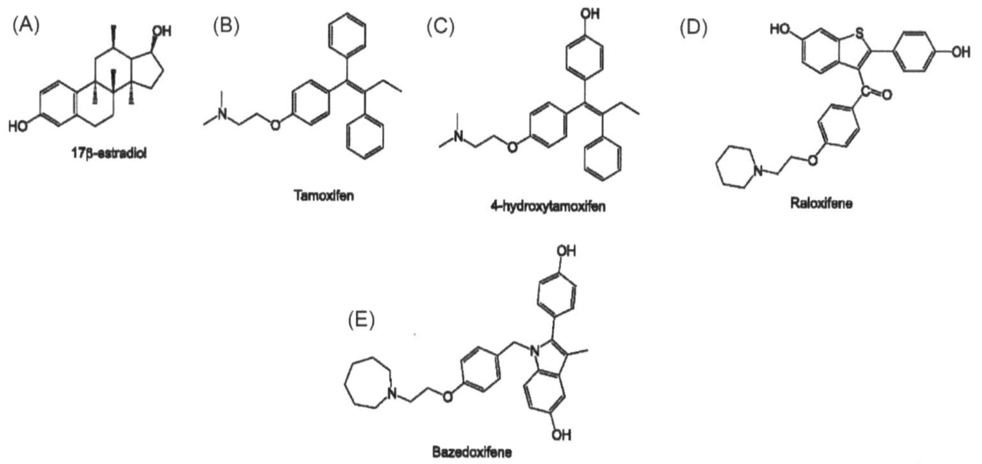

FIGURE 4.3 Structure of estrogen receptor modulators. (A) 17β-estradiol: an endogenous ER ligand. (B) The first synthetic SERM: tamoxifen. (C) The dominant ER binding metabolite of tamoxifen: 4-hydroxytamoxifen. (D) The selective ER modulators raloxifene and bazedoxifene (E) that show more desirable tissue selective ER activity profiles by inhibiting breast cancer growth, enhancing bone density, and lowering cholesterol without stimulating endometrial growth.

Breast cancer frequently affects post-menopausal women, who may also suffer from osteoporosis due to estrogen deficiency. Therefore it is desirable that clinically used SERMs display antagonistic activity in the breast and agonist activity in the bone (as with tamoxifen), thereby serving a multifaceted therapeutic role in patients. Additionally, in the cardiovascular system ER activation can also promote a healthy serum lipid profile. As post-menopausal women are more prone to cardiovascular disease, it is desirable that SERM treatments agonize ER within tissues and cell types relevant to cardiovascular health. Therefore the promotion of a healthy lipid profile is another caveat often factored into SERM design. In the search for more effective SERMs, researchers have endeavored to incorporate all these desired therapeutic qualities into newer generations of SERMs.

Raloxifene, (Figure 4.3D) discovered and marketed by Eli Lilly and Company, is another SERM used in the prevention of osteoporosis in post-menopausal women. Raloxifene has demonstrated some efficacy in reducing the risk of invasive breast cancer without increasing the risk of primary coronary events, while preserving and/or increasing bone density.[70–78] Similar to that of tamoxifen, raloxifene binds to the ER-LBD, and induces a repositioning of H12 to the AF-2 surface that blocks co-activator binding.[31] Therefore, both tamoxifen and raloxifene exert their antagonistic effect by inducing recruitment of co-repressors to ER gene promoters in breast cancer cells.[79] Raloxifene, however, does not stimulate endometrial growth as potently as tamoxifen.[80] The reduced level of agonism in uterine tissue has been shown to be due to the inability of raloxifene to stimulate SRC-1 recruitment to ER gene promoters in endometrial cells, in contrast to that of tamoxifen which potently enhances SRC-1 recruitment.[2] Therefore these studies highlighted the impact of SERM-directed differential recruitment of co-activators in tissue-specific context to ER promoters.

Bazedoxifene (WAY-140424 or Conbriza) (Figure 4.3E) was discovered as a result of collaboration between Wyeth and Ligand Pharmaceuticals and is used for the prevention and treatment of osteoporosis in post-menopausal women. This drug was developed with hopes that it would display more refined activity than the already established SERM raloxifene.[81,82] Bazedoxifene displayed general properties of a dual ERα/ERβ SERM with no cross-reactivity with other nuclear receptors. In its early discovery, it displayed the SERM gene-selective activation phenotype, where it antagonized expression on estrogen response element promoters, but agonized expression driven by a hepatic lipase promoter in the same cell line.[81] Interestingly, Bazedoxifene antagonized proliferation of both estrogen-dependent (MCF-7 and T47D) and estrogen-independent (MCF-7:5C and MCF-7:2A) cell lines[83] with limited off target effects in uterine and central nervous system tissues.

Disrupting growth of MCF-7:5C cells, in particular, is a unique feature among SERMs and occurs as a result of down-regulation of ERα protein levels via enhanced protein degradation and also suppression of cyclin D1 expression. These therapeutic effects were accomplished with concomitant increased bone density, reduced cholesterol, and even regression of endometriosis in rodent models.[81,84–86] Bazedoxifene has also demonstrated some anti-inflammatory properties as it reduces lipopolysaccharide (LPS)-induced expression of interleukin-6 (IL-6) and interferon-γ-inducible protein-10 (IP-10) through ER-dependent inhibition of NFκB p65 transactivation.[87] Collectively these preclinical studies suggested that bazedoxifene had multifactorial therapeutic benefits to breast, uterine, immune, and cardiovascular tissues.

In phase I and II clinical trials, bazedoxifene treatment preserved bone density, lowered cholesterol levels, and inhibited endometrial growth in post-menopausal women.[81,84] Other clinical trials revealed that bazedoxifene treatment prevents bone loss and reduces bone turnover, with similar efficacy to that of raloxifene, in post-menopausal women displaying normal-to-low bone mineral density, without adversely affecting breast density, incidence of hot flashes, or cardiovascular health.[88–98] Bazedoxifene also improved vasomotor symptoms and prevented endometrial hyperplasia in post-menopausal women.[99,100] Interestingly, bazedoxifene failed to increase the incidence of hot flashes, which is a common side effect shared by most SERMs. Bazedoxifene's therapeutic profile was made possible through applied knowledge of the underlying mechanism of SERM activity to the design of more refined SERMs that enhanced treatment options for post-menopausal women.

Tamoxifen, raloxifene, and bazedoxifene represent three generations of SERMs developed through iterative refinement of SERM activity based on assessment of their biological activity. The next generation SERMs will build upon our understanding of ligand-specific regulation of individual genes within a single tissue. In the future it may be possible, through protein conformational studies, to develop SERMs with not only tissue-specific agonist or antagonist activity, but also the ability to differentially regulate specific genes within a single cell type.[101]

Selective Glucocorticoid Receptor Modulators

Glucocorticoids have potent anti-inflammatory and immunosuppressive properties. The glucocorticoid receptor (GR) is expressed in a wide variety of lymphoid cells, including

T and B cell, and macrophages. Due to its expression pattern, GR is an extremely attractive target for the generation of therapeutics aimed at ameliorating inflammatory diseases caused by aberrant T cell, B cell, and macrophage function. During an inflammatory response, activation of GR leads to a decrease in pro-inflammatory cytokine expression via a mechanism that involves both activation (transactivation) and repression of gene transcription (transrepression). As is the case with most NRs, transactivation occurs when ligand-bound NRs bind their respective DNA response elements, resulting in an increased rate of gene expression. GR is also able to bind to and inhibit the transcriptional activity of the transcription factors AP-1 and NF-κB.[102,103] The anti-inflammatory effects of GR are thought to be mediated through both GR transactivation and transrepressive activities, although the role of GR transactivation and its anti-inflammatory effects are still not clearly elucidated.[104,105] Regardless, glucocorticoids were still able to inhibit inflammatory processes in a mouse model where GR was defective in its dimerization and DNA binding potential (GR^{dim}), suggesting that transrepression is a critical mechanism underlying the immunosuppressive effects of these hormones.[106,107]

Glucocorticoids also have major effects on energy balance and carbohydrate metabolism. The term "glucocorticoid' is derived from the initial observation that these hormones were involved in glucose metabolism. During fasting, cortisol stimulates several processes leading to the increase and maintenance of normal blood glucose concentrations. GR activity is involved in the expression of glucose-6-phosphatase (G6Pase) and phosphoenolpyruvate carboxykinase (PEPCK). These enzymes are involved in gluconeogenesis, which is responsible for the diabetogenic effects of glucocorticoids.

GR activation also leads to the synthesis of glucose from other non-hexose organic molecules including pyruvate, lactate, glycerol, and amino acids.[108] Prolonged GR activation is also the basis for the muscle wasting effects associated with long-term glucocorticoid use.[109] In the gluconeogenesis pathway, GR increases expression of a number of key enzymes and increases the availability of amino acids essential for this process.[110] A second mechanism involved in glucocorticoid action results in the conservation of glucose for neural tissues, such that glucose uptake is inhibited by muscle and adipose tissue.[111] Additionally, glucocorticoids stimulate lipolysis in adipose tissue. The fatty acids released from this process are used for the production of energy in tissues, including muscle. The glycerol released from lipolysis provides a substrate for gluconeogenesis. Finally, glucocorticoids inhibit leptin signaling, which is an important hormone in the maintenance of body weight and reproductive function.[112] As it has such complex and divergent physiological functions, GR is a highly attractive yet complex target for therapeutics aimed at ameliorating symptoms of immune diseases.

Various synthetic glucocorticoids are currently used as clinical treatments. The chemical structures of many of these compounds are based on the natural corticosteroids, but structural modifications have been made in order to improve efficacy in comparison to endogenous hormones, while reducing mineralocorticoid-like actions via the mineralocorticoid receptor (MR).[113] Prednisone and prednisolone were generated by introducing a C=C double bond into the first aromatic ring of cortisol, improving potency and diminishing mineralocorticoid activity.[114] Further improvement occurred when: 1) a fluoro atom was introduced at position 9α, yielding fludrocortisone; 2) a hydroxyl group was added at position 16α, yielding triamcinolone; 3) a methyl group was added at position 16α, which yielded dexamethasone and betamethasone; or 4) a methyl group was added at position 6α, which derived methylprednisolone.[115,116]

These modifications were essential in order to improve affinity for GR, minimize binding to MR, and increase the compound's half-life, thus increasing potency and efficacy. However, despite their anti-inflammatory effects, these refinements of modulator structure did not minimize the pernicious side effects experienced with classic glucocorticoid use. This fact heightened the search for novel selective GR modulators that retain only the potent anti-inflammatory effects of classic glucocorticoids. Modifications to the structure of natural glucocorticoids like hydrocortisone, prednisolone, methylprednisolone, and dexamethasone have been made to optimize the pharmacokinetics and therapeutic potential, and minimize some of the adverse side effects of GR targeting. These modifications, however, have largely failed to minimize these negative side effects. These studies have nonetheless led to the generation of compounds that can be grouped into four different classes: selective glucocorticoid receptor modulators (SGRMs), gene-selective compounds, dissociated compounds, and soft steroids.

SGRMs and SeGRAs (selective GR agonists) are compounds that retain anti-inflammatory activity but have impaired effects on bone, glucose, and lipid metabolism. Therefore, these compounds result in an improved therapeutic index *in vivo* over classic glucocorticoids.[117,118] SGRMS and SeGRAs stemmed from work where the major goal was identifying truly dissociated glucocorticoids: a compound with anti-inflammatory effects that are decoupled or dissociated from its metabolic effects. The first attempt to characterize a "dissociative" compound resulted in the identification of several steroid-based compounds, RU24858 (Figure 4.4A), RU40066, and RU24782, that were capable of separating transcriptional activation from repression *in vitro*. Vayssiere et al.[119] demonstrated that these compounds, specifically RU24858, not only bound GR with high affinity but exerted strong AP-1 inhibition with little to no agonist activity compared with dexamethasone-treated cells in a variety of *in vitro* assays. However, while RU24858 was very efficient at inhibiting both AP-1 and NF-κB-mediated gene induction *in vivo*, and was as potent an anti-inflammatory as prednisolone in a rat asthma model, it also induced side effects similar to prednisolone itself: loss of body weight and induction of osteoporosis.[120] Why RU24858 was effective and selective for the desired therapeutic effects *in vitro* but not *in vivo* remains to be determined. One possibility is that since the compound is similar in structure to classic glucocorticoids, it may have been metabolized *in vivo* and yielded a derivative, which behaved like a classic glucocorticoid.

Another novel GR ligand, A276575 (Figure 4.4B), discovered by Abbott laboratories, displayed high repression but very low transactivation activities, unlike traditional glucocorticoids. Much like dexamethasone, and with a high affinity for GR, A276575 was a potent anti-inflammatory as it inhibited IL-1β and IL-6 production in human skin fibroblasts and human lung epithelial cells, and inhibited Con A-induced proliferation of human peripheral blood mononuclear cells (PBMCs). Interestingly, A276575 was a racemic mixture, containing 7:1 (−)-Syn to anti-diastereomers, and the (+)-enantiomers were 10-fold weaker than their respective (−)-enantiomers. Additionally, the (−)-Syn enantiomer of A276575 was incapable of repressing chemokine (C-C motif) ligand 5 (CCL5/RANTES) production whereas the (−)-Anti enantiomer was highly active against CCL5/RANTES production.[115] These results highlighted that even subtle changes in a ligand structure can significantly impact receptor function. However, this compound displayed undesirable properties, including a high affinity for progesterone receptor and superagonist activity in MMTV-PR-B transfection assays, which precluded them from further use *in vivo*.[115] Further modifications of this compound failed to separate the desired transcriptional repression from undesirable transactivation characteristics.

While the compounds described above may not have been truly dissociative, they proved valuable tools for dissecting the differential effects of modulators on GR-mediated gene regulation. In fact, evaluation of these molecules revealed that the substitution pattern on the C-5 aryl group profoundly affected the compounds' functional activity. Therefore, a series of compounds were developed to investigate the effects of the C-5 aliphatic substitution, leading to the discovery of gene-selective compounds.

Gene-selective compounds act on the receptor to influence gene expression in a gene-specific or promoter-specific manner. For instance, a subset of genes might be activated, some might be repressed, but the resulting global expression profile is distinct from that of classic glucocorticoids.[121] During the search for dissociative compounds, many gene-selective compounds were identified. Screening activities identified AL-438 (Figure 4.4E), a derivative of benzoyrano[3,4f]quinoline, that appears to have partial GR agonism coupled with repressive activity. AL-438 had an almost identical binding affinity for GR as

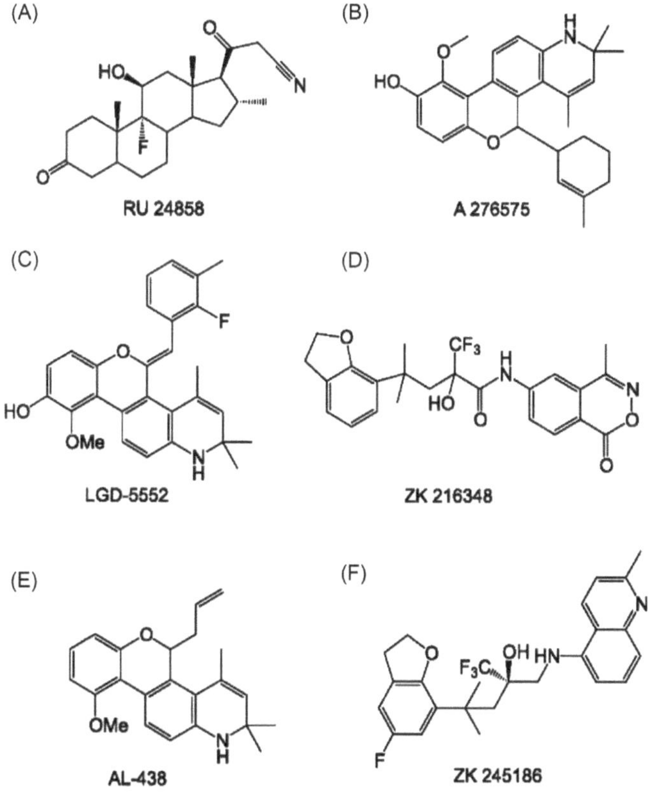

FIGURE 4.4 Chemical structures of GR modulators. (A) and (B) The "dissociative" GR selective modulators RU24858 and A276575. (C) The selective GR modulator LGD-5552. (D) The "dissociative" compound ZK216348 (E) The gene selective compound AL-438. (F) The "soft steroid" ZK245186.

prednisolone.[122] *In vitro* characterization of this compound demonstrated that AL-438 efficiently inhibited the production of IL-6 and E-selectin in transactivation assays.[121]

Due to its *in vitro* success, AL-438 was tested *in vivo* in both acute and chronic models of inflammation. In the rat carrageenan-induced paw edema assay, a model for acute inflammation, AL-438 demonstrated almost equivalent efficacy in reducing paw edema relative to prednisolone treatment. Since glucocorticoids demonstrate desirable effects on joint swelling, synovitis, and periosteal new bone formation, the rat adjuvant-induced arthritis model was used to study the effects of AL-438 in a chronic model of inflammation. AL-438 had an efficacy equal to that of prednisolone at 30 mg/kg/d, despite lower potency. Additionally, mice treated with AL-438 exhibited grooming behavior equivalent to non-adjuvant-treated controls, meaning that the behavior of AL-438-treated animals is similar to that of healthy control animals, whereas the prednisolone-treated animals exhibited signs of stress. Furthermore, AL-438 demonstrated a decreased potential to increase blood glucose, a marker for diabetes induction, and a decreased probability of inducing osteoporosis—both side effects associated with long-term glucocorticoid treatment.[121]

Molecular analysis of AL-438 elucidated the mechanism by which this improved therapeutic potential was observed; differential co-factor recruitment. AL-438 exhibited gene-specific regulation, and therefore fully regulated a subset of the genes normally regulated by GR. Using a two-hybrid assay, prednisolone was able to efficiently induce the interaction with both PCG-1 and GRIP-1, whereas AL-438 was only able to induce the interaction with GRIP-1, with an efficacy equal to prednisolone. The ability of AL-438 to recruit PGC-1 was significantly reduced to that of prednisolone.[121] Since PGC-1 is involved in the glucocorticoid regulation of hepatic glucose production, the loss of this interaction may explain why AL-438 causes less hyperglycemia *in vivo* compared with prednisolone. While these data did not characterize a fully dissociative compound, they did demonstrate that structural changes induced by AL-438 versus prednisolone are very different and that these differences are responsible for not only co-factor interaction, but altered pharmacology as well. These data not only suggested that the mechanisms of transactivation and transrepression are diverse processes, but that it is possible to achieve therapeutic benefit without complete separation between activation and repression.

After extensive high throughput screening using a GR-dependent co-transfection assay and extensive medicinal chemistry efforts, LGD-5552 (Figure 4.4C) was identified and synthesized. LGD-5552 is a non-steroidal compound similar in size to prednisolone. It exhibits selective binding to human GR, antagonizes prednisolone-induced transcriptional activation of the MMTV promoter, and displays agonistic properties in repressing IL-1β/TNF-induced activation of E-selectin and IL-6 promoters. In an adjuvant-induced arthritis model, a strong repression of serum MCP-1 and a reduced mRNA expression of joint ankle COX-2 were observed following LGD-5552 treatment. In contrast to prednisolone, the anti-inflammatory cytokine IL-10 was up-regulated by LGD-5552 treatment. Additionally, LGD-5552 binding to the LBD caused fundamental changes to the outer receptor structure that altered the ability of GR to interact with transcriptional co-factors. The conformational change resulted in effects on gene expression in a gene-specific manner leading to differential gene responses. LGD-5552 demonstrated selectivity on bone growth, blood pressure, and thymus organ weight and remains a powerful anti-inflammatory agent.[123,124]

Dissociated compounds completely dissociate transactivation from transrepression by GR. Compounds in this class fail to globally induce GR-mediated transactivation, but still significantly repress gene transcription. In an effort to identify GR agonists with dissociative profiles, several hundred compounds were screened at Bayer Schering Pharma AG. ZK216348 (Figure 4.4D) was obtained from this screen as it did not induce tyrosine amino transferase (TAT) activity but was able to repress IL-8 expression.[125] ZK216348 showed higher potency and less side effects compared with prednisolone after subcutaneous injection in mice. Similar anti-inflammatory effects were observed with this compound after systemic treatment. While ZK216348 did not increase blood glucose dose dependently in rats, no differences in ACTH suppression were observed compared were glucocorticoid treatment.[125]

The effects of ZK216348 on osteoprotegerin (OPG) and RANKL in osteoblastic cells was evaluated as both are pivotal proteins in the regulation of bone mass. RANKL stimulates bone resorption by increasing osteoclast differentiation and OPG is a decoy receptor for RANKL and inhibits bone resorption. Dexamethasone, prednisolone, deflazacort, RU24858, and other RU compounds all inhibited OPG production to a higher extent than that of AL438 and ZK216348.[126] The data suggested that these compounds induced less bone loss than traditional glucocorticoid treatment. The results also showed that there is also a difference in how these two compounds cause GR to repress OPG compared with traditional glucocorticoids. Further studies need to be performed in order to determine if ZK216348 and AL-438 recruit co-factors differently than prednisolone, to both OPG and RANKL promoters. However, because of its improved safety profile, ZK216348 presents as a promising alternative for the treatment of inflammatory disorders.

The first dissociated compound isolated from a natural source was Compound A (CpdA). This molecule is a stable analog of the hydroxyl phenyl aziridine precursor found in the Namibian shrub *Sasola tuberculatiformis*. While this compound lacks a steroidal structure, it is efficient at down-regulating NF-κB-driven genes via GR binding. What is most intriguing about CpdA is that it does not stimulate GRE-driven genes, suggesting that it is a completely dissociated compound.[127] CpdA interferes with the DNA-binding capability of NF-κB and also directly inhibits the transcriptional capacity of the NF-κB p65 subunit via activated GR. CpdA is an equally effective anti-inflammatory agent *in vivo* as dexamethasone, but presents with a significantly better side effect profile as it does not stimulate hyperglycemia.[127]

Recent evidence suggests that the transcriptional activity of GR upon agonist stimulation is correlated with an increase in the phosphorylation status of Ser-211 in the N-terminus of GR.[128] Interestingly, in contrast to dexamethasone, CpdA did not affect the phosphorylation of Ser-211, suggesting that CpdA may induce a subtly different conformational change in GR than classic glucocorticoids. Thus, the phosphorylation status of Ser-211 may reflect differences in transrepression versus transactivation and may be a valid screening method for the identification of dissociative compounds. Furthermore, it has been demonstrated that CpdA could effectively suppress experimental autoimmune neuritis—a helper T cell-mediated autoimmune demyelinating inflammatory disease of the peripheral nervous system—suggesting that CpdA could be a potent candidate for treatment of autoimmune neuropathies.[129]

While the idea of completely dissociated GR compounds is enticing, mouse models containing GR mutants suggest that complete abolishment of GR activity may not yield the anticipated results. For example, a study using the $GR^{dim/dim}$ mice demonstrated that while anti-inflammatory effects were seen in these mice, they still developed osteoporosis after a two-week systemic treatment with prednisolone.[130] These results suggest that some but not all negative side effects might be eliminated with the complete dissociation of transactivation with transrepression. Additionally, GR transactivates some genes known to negatively regulate the immune system, including thymosin β4 sulphoxide,[131] glucocorticoid-induced leucine zipper (GILZ),[132] and macrophage migration factor (MIF).[133] Complete dissociation of transactivation from transrepression would disrupt these genes' anti-inflammatory activities. Furthermore, the ability of GR to act as a "co-activator" via tethering does not appear to be affected in the $GR^{dim/dim}$ mice. For example, STAT5—a transcription factor that regulates the expression of many proinflammatory cytokines—is not affected in the dimerization and DNA binding mutant.[134]

Soft steroids refer to a class of corticosteroids that act at or near the site of administration but are rapidly inactivated by enzymes, thereby reducing systemic exposure and activity. Such compounds are used topically or inhaled for dermatologic diseases and asthma, respectively.[135,136] Glucocorticoids that follow this principle have been designed. These drugs act locally in the skin (methylprednisolone aceponate) or in the lungs (ciclesonide, butixocort 21-propionate) and show a low systemic exposure. These glucocorticoids potently inhibit pro-inflammatory cytokines and chemokines at the site of administration, while eliciting limited systemic responses.[137–139]

A selective, non-steroidal GR modulator that was identified by Bayer Schering Pharma, which is well suited for local application, is ZK-245186 (Figure 4.4F), also known as BOL-303242. This compound is expected to have a favorable therapeutic index due to its low systemic availability, because of low metabolic stability and high systemic clearance. ZK-245186 was examined in a variety of *in vitro* and *in vivo* ocular models. Primary ocular cell cultures were challenged with either lipopolysacharide or interleukin-1β and effects of ZK-245186 on NF-κB and MAPK were assessed by Luminex technology. ZK-245186 significantly reduced LPS- or IL-1β-induced inflammatory cytokine release in a dose-dependent manner. This compound also showed potency and activity similar to dexamethasone with IC_{50} values in the nanomolar range.[129] Furthermore, the anti-inflammatory actions of ZK-245186 were demonstrated in *in vitro* assays to inhibit T cell cytokine secretion and proliferation. Using a T cell-mediated allergic contact dermatitis model, ZK-245186 showed anti-inflammatory efficacy after topical application similar to the classic glucocorticoids. ZK-245186 also demonstrates a better safety profile than classic glucocorticoids as growth inhibition and induction of skin atrophy after long-term application was decreased.[140]

While our knowledge of the molecular events surrounding GR activation has greatly increased in the past decade, leading to the generation of several therapeutic selective GR modulators, the identification of a truly dissociative compound still remains at large. However, characterization of the GR^{dim} mice should give us some pause in regards to the plausibility of dissociative compounds. The idea of separating transactivation and transrepressive activities of GR are extremely enticing, but data from the GR^{dim} mouse suggest that some of the negative side effects associated with GR transactivation may not be ameliorated, as inhibition of bone formation still occurred in these mice. Additionally, a full anti-inflammatory response did not occur in the GR^{dim} mice, again suggesting that some transactivation events may be beneficial as some anti-inflammatory genes are induced by GR transactivation.[141–143]

Despite the restrictions surrounding dissociating transactivation and transrepression, there have been several successful SGRMs/dissociative compounds identified, including AL-438, ZK 216348, and ZK245186. Animal studies have revealed an improved ratio between therapeutic efficacy and side effects. However, only clinical trials will define the benefit–risk ratio in humans. With this in mind, rather than searching for more dissociative compounds, perhaps identifying differential compounds with the most favorable functional profiles appears to be a more realistic approach.

Selective PPARγ Modulators

Many genetic PPARγ manipulations yielding contradictory results have been done in mice, but, in general, they emphasize the importance of PPARγ in numerous tissues to insulin sensitivity.[144] The loss of one or both alleles of PPARγ, by dominant-negative effects, reduces adipose tissue and leads to insulin resistance in humans.[145,146] Somatic loss-of-function mutations have been found in human colon cancer.[147]

PPARγ is widely known as the target for the two thiazolidinediones (TZDs), rosiglitazone and pioglitazone, which are used clinically to increase insulin sensitivity in Type 2 diabetes mellitus, a disease with a significant worldwide health impact.[148] In addition to insulin sensitization, agonism of PPARγ with rosiglitazone or pioglitazone can lead to side effects in humans, including heart failure and edema, weight gain, increased bone fracture in women, and anti-inflammatory effects.[149–156] There are differences between the TZDs. Pioglitazone reduces non-fatal myocardial infarction and produces a healthier lipid profile, while rosiglitazone does not appear to have these properties.[149,150] Based on the observed adverse effects of rosiglitazone use, the FDA severely limited rosiglitazone use due to unwanted effects, while its European counterpart has entirely banned its use.[157] A recent report found that long-term use of pioglitazone is weakly associated with increased bladder cancer in humans, which has prompted the FDA to add new warnings to pioglitazone packaging and modify prescription guidelines.[158,159] Currently, the FDA requires two-year carcinogenicity studies in preclinical models before clinical trials lasting more than six months involving PPAR ligands.[160] A PPARγ ligand with significantly less negative side effects than the TZDs but with similar efficacy in insulin sensitization is needed.

Many studies elucidating PPARγ-dependent effects and their mechanisms have been published, which support ongoing efforts towards a safer PPARγ ligand. PPARγ plays an integral role in fat metabolism where it is both necessary and sufficient for differentiation of precursor cells into adipocytes.[161–163] This function may be mediated by endogenous PPARγ ligands, including the nitrated fatty acids linoleic acid (LNO_2) and oleic acid ($OA\text{-}NO_2$), which are potent natural PPARγ agonists that induce adipocyte differentiation.[164–166] The formation of smaller adipocytes, as found in lean subjects, along with other changes in fat cells brought on by changes in PPARγ activity may help increase insulin sensitivity.[167,168] However, PPARγ's pro-adipocyte role appears to negatively affect bone density and strength as PPARγ agonism can block osteoblast differentiation and cause osteoblast precursors to form adipocytes. This inhibits bone formation, causing increased bone marrow adipocity and apparently increasing the probability of bone fractures in women who take TZDs.[154,168,169]

In addition to negative effects in bone density, renal PPARγ appears to directly affect Na^+ excretion, which is likely a factor in edema and plasma volume expansion induced by TZDs.[170,171] It has also been shown that PPARγ in the brain, which has been agonized by TZDs, may be responsible for increased calorie intake leading to fat weight gain.[172] Interestingly, PPARγ ligands also suppress many cytokines and chemokines, with some reports showing clear PPARγ-dependent modulation of inflammation by its ligands.[173] Suppression of inflammation via PPARγ can involve direct interactions with NF-κB and NFAT, or some other indirect mechanisms.[167,173–176] Sumoylated (K395) PPARγ can block degradation of transcriptional repressors on NF-κB and AP-1 responsive genes in macrophages.[177] PPARγ's ability to suppress inflammation may be an important part of PPARγ-selective drug's efficacy in increasing insulin sensitivity in diabetes.[167,178,179]

While both clinical data and PPARγ biology point out that inducing all or most of PPARγ's effects is not desirable, inducing a subset of effects including anti-inflammatory, anti-proliferative, and insulin sensitizing, would be.[180,181] This is the goal of selective PPARγ modulator (SPPARγM) development. Many PPARγ ligands that elicit less negative effects of PPARγ agonism while still increasing insulin sensitivity have been developed.[182,183] Much of the current preclinical work suggests distinct PPARγ ligands can cause different outcomes *in vivo*. One prominent clinical example demonstrates that two PPARγ ligands can produce differing clinical results in humans: pioglitazone lowers triglyceride levels while rosiglitazone does not.[184,185] A conceptual model of PPARγ function is developing that offers insight into how selective modulators may "tune" PPARγ to produce selective effects.[186] The binding of a particular ligand (or no ligand) by PPARγ produces a unique equilibrium of conformations and dynamics that favors/disfavors interaction with a particular set of protein kinases, ubiquitin ligases, or co-regulators, thus producing ligand-specific transcriptional effects.[177,187–193] Many of the SPPARγMs display distinct co-factor binding relative to that induced by rosiglitazone or pioglitazone; however, the optimal combination of co-regulator recruitment necessary for desired physical effects remains unclear. The goal of the SPPARM concept is the design of ligands that demonstrate clear separation of weight gain, edema, and bone loss from insulin sensitization. However, no compound has yet emerged with this desired activity profile.

GW0072 (Figure 4.5) is one of the first published PPARγ partial agonists that induced less transcriptional activity from a PPRE containing reporter plasmid in a cell-based assay than the TZDs.[188,194] Oberfield and co-workers found that GW0072 did not occupy the end of the T binding pocket associated with interaction with H12, which may explain its partial agonism (15–25%), despite a binding affinity comparable to rosiglitazone. It also antagonized the activity of rosiglitazone in transcription, co-regulator recruitment, and fat cell differentiation assays. However, both rosiglitazone and GW0072 recruited far less NCoR to PPARγ compared with vehicle in a mammalian two-hybrid assay.[192] Additionally, GW0072 recruited PGC1α peptide to a similar degree as rosiglitazone, despite recruiting far less SRC1, SRC2, and SRC3 peptides.[189] Recruitment of PGC1α likely promotes insulin sensitization as it is up-regulated with exercise and likely contributes to some of the observed benefits of exercise on metabolic syndrome; therefore, a strong case for the benefits of PGC1α can be made.[195] Furthermore, skeletal muscle from $PGC1\alpha^{+/-}$ mice shows a significant increase in expression of several proinflammatory cytokines and a reduced expression of PGC1α, leading to chronic inflammation, which would be expected to

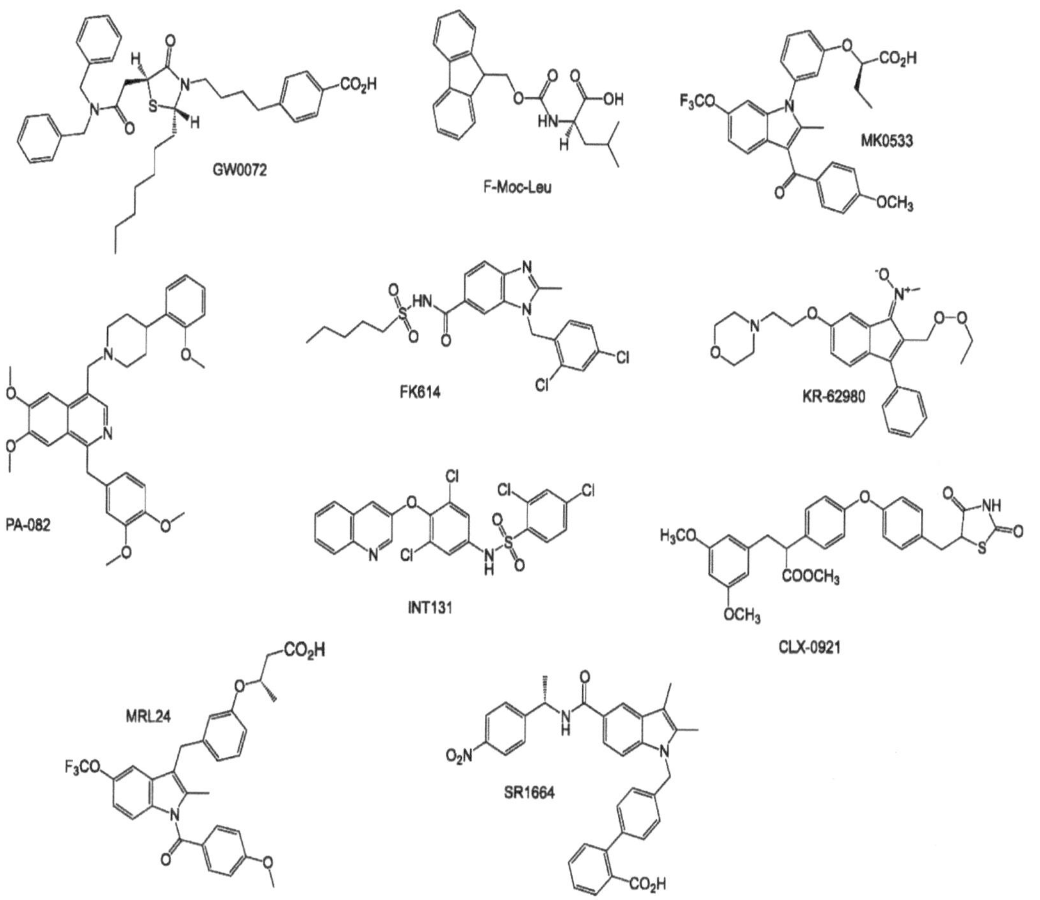

FIGURE 4.5 Structures of PPARγ modulators. Chemical structures of various PPARγ modulators are illustrated.

reduce insulin sensitivity.[179,195,196] Therefore, retaining PGC1α recruitment similar to rosiglitazone is likely beneficial.

FMoc-L-Leucine (F-L-Leu) (Figure 4.5) is a PPARγ selective partial agonist inducing transcription up to 85% that of rosiglitazone. Two F-L-Leu molecules bind to one PPARγ molecule with a K_i of 15 μM. Results from a mammalian two-hybrid demonstrated that the co-activator SRC2 (TIF-2) interacts with rosiglitazone-bound PPARγ but not with F-L-Leu-bound PPARγ, while both F-L-Leu and rosiglitazone have similar binding of SRC1.[188] These full-length data differ from peptide recruitment assay results which demonstrate that SRC 1, 2, and 3 peptides are all less recruited to PPARγ by F-L-Leu than by rosiglitazone.[188,189] The lack of SRC2 recruitment and maintained SRC1 recruitment (compared with rosiglitazone) by F-L-Leu-bound PPARγ may be at least partially responsible for the reduced weight gain in F-L-Leu treated mice. This can be seen in $SRC2^{-/-}$ mice, which are resistant to obesity, have reduced adipocyte size, increased plasma leptin concentration, and enhanced adaptive thermogenesis, while the SRC1$^{-/-}$ mice are obesity prone.[188,197] F-L-Leu induced adipocyte differentiation and was able to increase glucose tolerance in leptin receptor deficient (*db/db*) diet-induced glucose intolerant mice. However, F-L-Leu was also able to enhance glucose tolerance in normal mice while rosiglitazone did not. F-L-Leu also demonstrated anti-inflammatory efficacy in a 2,3,6-trinitrobenzene sulfonic acid-induced colitis mouse model.[188]

FK614 (Figure 4.5) is a partial agonist, inducing 75% of the transcription induced by several TZDs, while displaying different co-factor binding. FK614 recruited less CBP300, CREB-binding protein, and SRC-1 to PPARγ than the TZDs, but levels of recruitment/displacement of PGC1α, PRIP, and PPAR binding protein recruitment and NCoR and SMRT were equivalent.[191,198,199] In *db/db* mice, FK614 reduced plasma triglyceride levels and plasma glucose to levels comparable to rosiglitazone at the same dosage (3.2 mg/kg). FK614 also reduced plasma insulin and glucose in leptin deficient (*ob/ob*) mice similar to Pioglitazone at one-third the dose. Interestingly, FK614 reduced RBC counts in female rats (and not male) at doses of 320 mg/kg, while rosiglitazone affected RBC counts in both male and female rats at 32 and 100 mg/kg.[200] A reduction in RBC count can indicate an increase in blood plasma volume, a known side effect of rosiglitazone. Thus FK614 appears to have comparable effects to that of rosiglitazone and pioglitazone on plasma glucose levels in *ob/ob* mice and on insulin sensitivity in Zucker fatty rats with a lower propensity for increasing plasma volume/edema.[200–202] In white adipose tissue, FK614 also had similar effects to pioglitazone, increasing the number of small adipocytes to near lean controls and slightly decreasing the number of large adipocytes.[201] However, both rosiglitazone and FK614 trended toward increased weight in Zucker fatty rats.[202] Astellas Pharma Inc. terminated development of FK614 in 2005 due to lack of benefits over existing PPAR agonists.[203]

Another SPPARγM, INT-131 (T131, AMG131) (Figure 4.5), induces $\approx$20% of the expression of rosiglitazone in a PPRE plasmid reporter assay, while binding more tightly to PPARγ than rosiglitazone. INT-131 binds to the PPARγ LBD in a distinct manner relative to rosiglitazone with the most notable difference being that it apparently does not interact directly with H12. It also displays decreased TRAP220, SRC2, SRC3, CREB-binding protein, and p300 peptide binding when compared with rosiglitazone. Conversely, it shows similar TIF1 and RIP140 peptide binding and dramatically increased, near unbound levels of NCor and SMRT co-repressor binding compared with rosiglitazone.[190] The differences in binding and subsequent co-regulator recruitment translate into cell culture and *in vivo* differences between rosiglitazone and INT-131. Unlike rosiglitazone, INT-131 weakly promotes differentiation of preadipocytes and antagonizes the rosiglitazone-induced differentiation. INT-131 also causes similar alterations in glucose tolerance and weight gain in Zucker fatty rats, but plasma volume and heart and lung weight are less in comparison to rosiglitizone.[190]

In a phase 2b clinical trial, INT-131 was tested at 1 mg and 10 mg per day for four weeks. The 1 mg/day and 10 mg/day groups both showed significant improvement over placebo in fasting plasma glucose. The 1 mg/day group had no significant changes in hematocrit and weight, while the 10 mg/day group had significantly decreased hematocrit (indicative of increased plasma volume) and significantly increased weight along with clinical evidence of edema in 6 of 24 patients. Thus the 1 mg/day group did not display significant safety risks but did have a significant improvement in fasting plasma glucose, while 10 mg/day produced significant unwanted effects.[204] Whether INT-131 can succeed in separating edema, heart failure, weight gain, and bone marrow adipocity from adequate insulin sensitization significantly better than the TZDs in humans is still questionable.

In 2006, Kim et al.[205] described a novel non-thiazolidinedione compound, KR-62980 (Figure 4.5), which induces only 30% of the transcriptional activity compared with rosiglitazone in HIH3T3 cells using a PPARγ LBD GAL4 assay. KR-62980 also causes little differentiation of C3H10T1/2 cells and almost no induction of the aP2 transcription, and blocks differentiation and aP2 transcription induced by rosiglitazone. Additionally, KR-62980 allows for recruitment of SRC2 similar to rosiglitazone in a mammalian two-hybrid assay, but dramatically reduces TRAP220 recruitment and slightly reduces association of AIB1 and SRC1 compared with rosiglitazone. In C57BL/6J on a high fat diet, KR-62980 improved plasma glucose similar to that of rosiglitazone. It was interesting to note that KR-62980 also significantly reduced high fat diet-induced body fat and heart weight gain, while treatment with rosiglitazone did not appear to affect this.[205] KR-62980 also appears to have several anti-inflammatory effects, including IL-4 and IFNγ suppression, and improved outcomes when administered in the context of an allergic asthma model.[206]

Both KR-62980 and rosiglitazone showed neuroprotective and anti-proliferative effects in a breast cancer cell line.[205,207] Interestingly, pioglitazone significantly reduced the risk of breast cancer in the PROactive trial.[152] Further investigation into the anti-adipogenic mechanism of KR-62980 revealed that TAZ (transcriptional co-activator with PDZ-binding motif) is necessary for KR-62980 to block rosiglitazone-induced adipocyte differentiation in 3T3-L1 cells.[208] TAZ had been previously shown to bind PPARγ and functions to block adipocyte and induce osteoblast differentiation from mesenchymal stem cells.[209] KR-62980 increased transiently expressed TAZ nuclear localization in Cos7 cells. It also increased the interaction of TAZ and PPARγ in an immunoprecipitation assay using 3T3-L1 adipocytes, and decreased binding of PPRE containing DNA to PPARγ in 3T3-L1 adipocyte nuclear protein extract.[208]

Several recent reports have introduced new concepts that impact SPPARγM development, and may aid in the development of SPPARγMs with a wider separation of positive and negative effects. Choi et al.[187] offer new insight into the complex link between PPARγ and insulin sensitization. They report that phosphorylation of S273 of PPARγ by CDK5 is inhibited by rosiglitazone and that S273 phosphorylation was strongly anticorellated with *in vivo* insulin sensitization in a small human trial.[187] Rosiglitazone shares the ability to block S273 phosphorylation with a partial agonist (MRL-24), which is also capable of robust insulin sensitization. The apparent mechanism by which both these drugs block phosphorylation is through stabilization of the beta sheet region near S273, which presumably disfavors interaction with CDK5.[187] A previous report showed that stabilization of this beta sheet region is achieved by three other partial and full agonists as well, while only the two full agonists stabilized H12.[38]

These data indicate that stabilization of H12 appears necessary for full transcriptional activity but not to induce insulin sensitization. Whether S273 phosphorylation is a bystander effect or is a cause of increased insulin sensitivity is yet to be determined. However, development of drugs that specifically block S273 phosphorylation without causing transcriptional activation of PPARγ may provide a novel direction for future SPPARγM development. MRL-24 appears to be a step in this direction since it binds to PPARγ with higher affinity than rosiglitazone and blocks S273 phosphorylation more effectively, yet induces only a small fraction of the transcriptional activity of rosiglitazone. Additionally,

MRL-24 provides as good glucose correction as rosiglitazone in multiple mouse models while inducing less body weight gain.[187,210] SR1664 (Figure 4.5) goes further showing no significant induction of transcription from a PPRE-driven reporter plasmid while providing robust insulin sensitization with no indication of edema or weight gain.[211]

Two other reports by separate groups using different approaches point to the importance of brain PPARγ for weight gain and a portion of improved insulin sensitivity.[172,212] Ryan et al.[172] found that injecting a small amount of rosiglitazone or viral delivery of a constitutively active PPARγ into the brain increased feeding and weight gain. Weight gain and increased feeding induced by oral dosing of much larger amounts of rosiglitazone were blocked by ventricular injection of GW9662 (Figure 4.5), a PPARγ antagonist, which also decreased feeding and weight in rats fed a high fat diet.[172] Lu and co-workers found that both rosiglitazone-treated and untreated mice with a brain-specific knockout (BKO) of PPARγ gained less weight on a high fat diet and had increased activity compared with controls. Interestingly, the PPARγ BKO mice did show increases in insulin sensitivity from rosiglitazone treatment. However, treatment did not reduce basal and insulin-stimulated hepatic glucose production.[212] Thus, it appears that blood–brain barrier permeability of PPARγ ligands affects weight gain and possibly insulin sensitization. Perhaps modifying existing ligands to make them less brain penetrate could reduce weight gain, leaving insulin sensitization largely intact.

Preclinical research indicates that new PPARγ ligands can potentially separate edema, weight gain, and bone loss from insulin sensitization to a greater extent than TZDs, but none of these compounds has yet reached the clinic.

CONCLUSION

Analysis of nuclear protein dynamics coupled with protein crystallization of nuclear receptors bound to a variety of ligands has been fundamental to the discovery and utilization of selective modulators. These selective modulators have significantly enhanced the therapeutic benefit of pharmacological nuclear receptor targeting in a number of diseases. These techniques will no doubt continue to contribute to the development of the next generation of compounds that will more precisely target aberrant physiological pathways via modulation of nuclear receptor activity.

References

1. Kojetin DJ, Burris TP, Jensen EV, Khan SA. Implications of the binding of tamoxifen to the coactivator recognition site of the estrogen receptor. *Endocr Relat Cancer* 2008;**15**(4):851–70.
2. Shang Y, Brown M. Molecular determinants for the tissue specificity of SERMs. *Science* 2002;**295**(5564):2465–8.
3. Shang Y, Hu X, DiRenzo J, Lazar MA, Brown M. Cofactor dynamics and sufficiency in estrogen receptor-regulated transcription. *Cell* 2000;**103**(6):843–52.
4. Fernandes I, White JH. Agonist-bound nuclear receptors: not just targets of coactivators. *J Mol Endocrinol* 2003;**31**(1):1–7.
5. Choi JH, Banks AS, Estall JL, Kajimura S, Bostrom P, Laznik D, et al. Anti-diabetic drugs inhibit obesity-linked phosphorylation of PPARgamma by Cdk5. *Nature* 2010;**466**(7305):451–6.
6. Cheng SY, Leonard JL, Davis PJ. Molecular aspects of thyroid hormone actions. *Endocr Rev* 2010;**31**(2):139–70.

7. Meyer ME, Gronemeyer H, Turcotte B, Bocquel MT, Tasset D, Chambon P. Steroid hormone receptors compete for factors that mediate their enhancer function. *Cell* 1989;**57**(3):433–42.
8. Tasset D, Tora L, Fromental C, Scheer E, Chambon P. Distinct classes of transcriptional activating domains function by different mechanisms. *Cell* 1990;**62**(6):1177–87.
9. Baniahmad A, Leng XH, Burris TP, Tsai SY, Tsai MJ, Omalley BW. The tau-4 activation domain of the thyroid-hormone receptor is required for release of a putative corepressor(s) necessary for transcriptional silencing. *Mol Cell Biol* 1995;**15**(1):76–86.
10. Halachmi S, Marden E, Martin G, MacKay H, Abbondanza C, Brown M. Estrogen receptor-associated proteins: possible mediators of hormone-induced transcription. *Science* 1994;**264**(5164):1455–8.
11. Onate SA, Tsai SY, Tsai MJ, O'Malley BW. Sequence and characterization of a coactivator for the steroid hormone receptor superfamily. *Science* 1995;**270**(5240):1354–7.
12. Hong H, Kohli K, Trivedi A, Johnson DL, Stallcup MR. GRIP1, a novel mouse protein that serves as a transcriptional coactivator in yeast for the hormone binding domains of steroid receptors. *Proc Natl Acad Sci USA* 1996;**93** (10):4948–52.
13. Hong H, Kohli K, Garabedian MJ, Stallcup MR. GRIP1, a transcriptional coactivator for the AF-2 transactivation domain of steroid, thyroid, retinoid, and vitamin D receptors. *Mol Cell Biol* 1997;**17**(5):2735–44.
14. Voegel JJ, Heine MJ, Zechel C, Chambon P, Gronemeyer H. TIF2, a 160 kDa transcriptional mediator for the ligand-dependent activation function AF-2 of nuclear receptors. *EMBO J* 1996;**15**(14):3667–75.
15. Anzick SL, Kononen J, Walker RL, Azorsa DO, Tanner MM, Guan XY, et al. AIB1, a steroid receptor coactivator amplified in breast and ovarian cancer. *Science* 1997;**277**(5328):965–8.
16. Takeshita A, Cardona GR, Koibuchi N, Suen CS, Chin WW. TRAM-1, a novel 160-kDa thyroid hormone receptor activator molecule, exhibits distinct properties from steroid receptor coactivator-1. *J Biol Chem* 1997;**272**(44):27629–34.
17. Chen H, Lin RJ, Schiltz RL, Chakravarti D, Nash A, Nagy L, et al. Nuclear receptor coactivator ACTR is a novel histone acetyltransferase and forms a multimeric activation complex with P/CAF and CBP/p300. *Cell* 1997;**90**(3):569–80.
18. Li H, Gomes PJ, Chen JD. RAC3, a steroid/nuclear receptor-associated coactivator that is related to SRC-1 and TIF2. *Proc Natl Acad Sci USA* 1997;**94**(16):8479–84.
19. Torchia J, Rose DW, Inostroza J, Kamei Y, Westin S, Glass CK, et al. The transcriptional co-activator p/CIP binds CBP and mediates nuclear-receptor function. *Nature* 1997;**387**(6634):677–84.
20. McKenna NJ, O'Malley BW. SnapShot: NR coregulators. *Cell* 2010;**143**(1). 172–U6.
21. Lonard DM, O'Malley BW. Nuclear receptor coregulators: Judges, juries, and executioners of cellular regulation. *Mol Cell* 2007;**27**(5):691–700.
22. Lonard DM, Lanz RB, O'Malley BW. Nuclear receptor coregulators and human disease. *Endocr Rev* 2007;**28** (5):575–87.
23. Lonard DM, O'Malley BW. The expanding cosmos of nuclear receptor coactivators. *Cell* 2006;**125**(3):411–4.
24. McKenna NJ, O'Malley BW. Combinatorial control of gene expression by nuclear receptors and coregulators. *Cell* 2002;**108**(4):465–74.
25. Horwitz KB, Jackson TA, Bain DL, Richer JK, Takimoto GS, Tung L. Nuclear receptor coactivators and corepressors. *Mol Endocrinol* 1996;**10**(10):1167–77.
26. Xu L, Glass CK, Rosenfeld MG. Coactivator and corepressor complexes in nuclear receptor function. *Curr Opin Genet Dev* 1999;**9**(2):140–7.
27. Chen JD, Evans RM. A transcriptional co-repressor that interacts with nuclear hormone receptors. *Nature* 1995;**377**(6548):454–7.
28. Horlein AJ, Naar AM, Heinzel T, Torchia J, Gloss B, Kurokawa R, et al. Ligand-independent repression by the thyroid hormone receptor mediated by a nuclear receptor co-repressor. *Nature* 1995;**377**(6548):397–404.
29. Pike ACW. Lessons learnt from structural studies of the oestrogen receptor. *Best Pract Res* 2006;**20**(1):1–14.
30. Savkur RS, Burris TP. The coactivator LXXLL nuclear receptor recognition motif. *J Pept Res* 2004;**63**(3):207–12.
31. Brzozowski AM, Pike ACW, Dauter Z, Hubbard RE, Bonn T, Engström O, et al. Molecular basis of agonism and antagonism in the oestrogen receptor. *Nature* 1997;**389**(6652):753–8.
32. Pike ACW, Brzozowski AM, Walton J, Hubbard RE, Thorsell AG, Li Y-L, et al. Structural insights into the mode of action of a pure antiestrogen. *Structure* 2001;**9**(2):145–53.
33. Shiau AK, Barstad D, Loria PM, Cheng L, Kushner PJ, Agard DA, et al. The structural basis of estrogen receptor/coactivator recognition and the antagonism of this interaction by tamoxifen. *Cell* 1998;**95**(7):927–37.

34. Pike ACW, Brzozowski AM, Hubbard RE, Bonn T, Thorsell AG, Engström O, et al. Structure of the ligand-binding domain of oestrogen receptor beta in the presence of a partial agonist and a full antagonist. *EMBO J* 1999;**18**(17):4608–18.
35. Pike AC, Brzozowski AM, Walton J, Hubbard RE, Thorsell AG, Li YL, et al. Structural insights into the mode of action of a pure antiestrogen. *Structure* 2001;**9**(2):145–53.
36. Pike ACW, Brzozowski AM, Walton J, Hubbard RE, Bonn T, Gustafsson J-Å, et al. Structural aspects of agonism and antagonism in the oestrogen receptor. *Biochem Soc Trans* 2000;**28**(4):396–400.
37. Bruning JB, Parent AA, Gil G, Zhao M, Nowak J, Pace MC, et al. Coupling of receptor conformation and ligand orientation determine graded activity. *Nat Chem Biol* 2010;**6**(11):837–43.
38. Bruning JB, Chalmers MJ, Prasad S, Busby SA, Kamenecka TM, He Y, et al. Partial agonists activate PPARgamma using a helix 12 independent mechanism. *Structure* 2007;**15**(10):1258–71.
39. Uppenberg J, Svensson C, Jaki M, Bertilsson G, Jendeberg L, Berkenstam A. Crystal structure of the ligand binding domain of the human nuclear receptor PPARgamma. *J Biol Chem* 1998;**273**(47):31108–12.
40. Nolte RT, Wisely GB, Westin S, Cobb JE, Lambert MH, Kurokawa R, et al. Ligand binding and co-activator assembly of the peroxisome proliferator-activated receptor-gamma. *Nature* 1998;**395**(6698):137–43.
41. Berger JP, Petro AE, Macnaul KL, Kelly LJ, Zhang BB, Richards K, et al. Distinct properties and advantages of a novel peroxisome proliferator-activated protein [gamma] selective modulator. *Mol Endocrinol* 2003;**17**(4):662–76.
42. Hamuro Y, Coales SJ, Morrow JA, Molnar KS, Tuske SJ, Southern MR, et al. Hydrogen/deuterium-exchange (H/D-Ex) of PPARgamma LBD in the presence of various modulators. *Protein Sci* 2006;**15**(8):1883–92.
43. Hughes TS, Chalmers MJ, Novick S, Kuruvilla DS, Chang MR, Kamenecka TM, et al. Ligand and receptor dynamics contribute to the mechanism of graded PPARγ agonism. *Structure* 2012;**20**(1):139–50.
44. Chalmers MJ, Pascal BD, Willis S, Zhang J, Iturria SJ, Dodge JA, et al. Methods for the analysis of high precision differential hydrogen deuterium exchange data. *Int J Mass Spectrom* 2011;**302**(1–3):59–68.
45. Zhang J, Chalmers MJ, Stayrook KR, Burris LL, Garcia-Ordonez RD, Pascal BD, et al. Hydrogen/deuterium exchange reveals distinct agonist/partial agonist receptor dynamics within vitamin D receptor/retinoid X receptor heterodimer. *Structure* 2010;**18**(10):1332–41.
46. Tamrazi A, Carlson KE, Katzenellenbogen JA. Molecular sensors of estrogen receptor conformations and dynamics. *Mol Endocrinol* 2003;**17**(12):2593–602.
47. Dai SY, Chalmers MJ, Bruning J, Bramlett KS, Osborne HE, Montrose-Rafizadeh C, et al. Prediction of the tissue-specificity of selective estrogen receptor modulators by using a single biochemical method. *Proc Natl Acad Sci USA* 2008;**105**(20):7171–6.
48. Dai SY, Burris TP, Dodge JA, Montrose-Rafizadeh C, Wang Y, Pascal BD, et al. Unique ligand binding patterns between estrogen receptor alpha and beta and revealed by hydrogen/deuterium exchange. *Biochemistry* 2009;**48**(40):9668–76.
49. Zhao C, Dahlman-Wright K, Gustafsson JA. Estrogen receptor beta: an overview and update. *Nucl Recept Signal* 2008;**6**:e003.
50. Matthews J, Gustafsson J-Å. Estrogen signaling: a subtle balance between ER alpha and ER beta. *Mol Interv* 2003;**3**(5):281–92.
51. Choi J, Banks A, Kamenecka T, Busby S, Chalmers M, Kumar N, et al. Anti-diabetic actions of a non-agonist PPARγ ligand blocking Cdk5-mediated phosphorylation. *Nature* 2011;**477**(7365):477–81.
52. Maurer RA, Notides AC. Identification of an estrogen-responsive element from the 5′-flanking region of the rat prolactin gene. *Mol Cell Biol* 1987;**7**(12):4247–54.
53. Cowley SM, Hoare S, Mosselman S, Parker MG. Estrogen receptors alpha and beta form heterodimers on DNA. *J Biol Chem* 1997;**272**(32):19858–62.
54. Riggs BL, Hartmann LC. Selective estrogen-receptor modulators — mechanisms of action and application to clinical practice. *N Engl J Med* 2003;**348**(7):618–29.
55. Riggs BL, Khosla S, Melton III LJ. Sex steroids and the construction and conservation of the adult skeleton. *Endocr Rev* 2002;**23**(3):279–302.
56. Bord S, Horner A, Beavan S, Compston J. Estrogen receptors alpha and beta are differentially expressed in developing human bone. *J Clin Endocrinol Metab* 2001;**86**(5):2309–14.
57. Saarto T, Blomqvist C, Ehnholm C, Taskinen MR, Elomaa I. Antiatherogenic effects of adjuvant antiestrogens: a randomized trial comparing the effects of tamoxifen and toremifene on plasma lipid levels in postmenopausal women with node-positive breast cancer. *J Clin Oncol* 1996;**14**(2):429–33.

58. Wenger NK. Coronary heart disease: an older woman's major health risk. *BMJ* 1997;**315**(7115):1085–90.
59. Jordan VC. Tamoxifen (ICI46,474) as a targeted therapy to treat and prevent breast cancer. *Br J Pharmacol* 2006;**147**(Suppl. 1):S269–76.
60. Harper MJ, Walpole AL. A new derivative of triphenylethylene: effect on implantation and mode of action in rats. *J Reprod Fertil* 1967;**13**(1):101–19.
61. Harper MJ, Walpole AL. Mode of action of I.C.I. 46,474 in preventing implantation in rats. *J Endocrinol* 1967;**37** (1):83–92.
62. Lerner LJ, Holthaus Jr. FJ, Thompson CR. A non-steroidal estrogen antiagonist 1-(p-2-diethylaminoethoxyphenyl)-1-phenyl-2-p-methoxyphenyl ethanol. *Endocrinology* 1958;**63**(3):295–318.
63. Furr BJ, Jordan VC. The pharmacology and clinical uses of tamoxifen. *Pharmacol Ther* 1984;**25**(2):127–205.
64. Jordan VC, Dowse LJ. Tamoxifen as an anti-tumour agent: effect on oestrogen binding. *J Endocrinol* 1976;**68** (02):297–303.
65. Emmens CW. Compounds exhibiting prolonged antioestrogenic and antifertility activity in mice and rats. *J Reprod Fertil* 1971;**26**(2):175–82.
66. Jordan VC. Prolonged antioestrogenic activity of ICI 46, 474 in the ovariectomized mouse. *J Reprod Fertil* 1975;**42**(2):251–8.
67. Lunan CB, Green B. 3H-oestradiol uptake *in vivo* by human uterine endometrium: effect of tamoxifen (I.C.I. 46,474). *Clin Endocrinol (Oxf)* 1974;**3**(4):465–80.
68. Rochefort H, Garcia M, Borgna JL. Absence of correlation between antiestrogenic activity and binding affinity for the estrogen receptor. *Biochem Biophys Res Commun* 1979;**88**(2):351–7.
69. Rochefort H, Borgna JL. Differences between oestrogen receptor activation by oestrogen and antioestrogen. *Nature* 1981;**292**(5820):257–9.
70. Jordan VC, Phelps E, Lindgren JU. Effects of anti-estrogens on bone in castrated and intact female rats. *Breast Cancer Res Treat* 1987;**10**(1):31–5.
71. Black LJ, Sato M, Rowley ER, Magee DE, Bekele A, Williams DC, et al. Raloxifene (LY139481 HCl) prevents bone loss and reduces serum cholesterol without causing uterine hypertrophy in ovariectomized rats. *J Clin Invest* 1994;**93**(1):63–9.
72. Sato M, McClintock C, Kim J, Turner CH, Bryant HU, Magee D, et al. Dual-energy x-ray absorptiometry of raloxifene effects on the lumbar vertebrae and femora of ovariectomized rats. *J Bone Miner Res* 1994;**9**(5):715–24.
73. Sato M, Kim J, Short LL, Slemenda CW, Bryant HU. Longitudinal and cross-sectional analysis of raloxifene effects on tibiae from ovariectomized aged rats. *J Pharmacol Exp Ther* 1995;**272**(3):1252–9.
74. Turner CH, Sato M, Bryant HU. Raloxifene preserves bone strength and bone mass in ovariectomized rats. *Endocrinology* 1994;**135**(5):2001–5.
75. Gottardis MM, Jordan VC. Antitumor actions of keoxifene and tamoxifen in the N-nitrosomethylurea-induced rat mammary carcinoma model. *Cancer Res* 1987;**47**(15):4020–4.
76. Clemens JA, Bennett DR, Black LJ, Jones CD. Effects of a new antiestrogen, keoxifene (LY156758), on growth of carcinogen-induced mammary tumors and on LH and prolactin levels. *Life Sci* 1983;**32**(25):2869–75.
77. Anzano MA, Peer CW, Smith JM, Mullen LT, Shrader MW, Logsdon DL, et al. Chemoprevention of mammary carcinogenesis in the rat: combined use of raloxifene and 9-cis-retinoic acid. *J Natl Cancer Inst* 1996;**88** (2):123–5.
78. Barrett-Connor E, Mosca L, Collins P, Geiger MJ, Grady D, Kornitzer M, et al. Effects of raloxifene on cardiovascular events and breast cancer in postmenopausal women. *N Engl J Med* 2006;**355**(2):125–37.
79. Shang YF, Brown M. Molecular determinants for the tissue specificity of SERMs. *Science* 2002;**295** (5564):2465–8.
80. Gottardis MM, Ricchio ME, Satyaswaroop PG, Jordan VC. Effect of steroidal and nonsteroidal antiestrogens on the growth of a tamoxifen-stimulated human endometrial carcinoma (EnCa101) in athymic mice. *Cancer Res* 1990;**50**(11):3189–92.
81. Komm BS, Lyttle CR. Developing a SERM: stringent preclinical selection criteria leading to an acceptable candidate (WAY-140424) for clinical evaluation. *Ann N Y Acad Sci* 2001;**949**:317–26.
82. Bazedoxifene: bazedoxifene acetate, TSE 424, TSE-424, WAY 140424. *Drugs R D* 2008;**9**(3):191–6.
83. Lewis-Wambi JS, Kim H, et al. The selective estrogen receptor modulator bazedoxifene inhibits hormone-independent breast cancer cell growth and down-regulates estrogen receptor alpha and cyclin D1. *Mol Pharmacol* 2011;**80**(4):610–20.

84. Ronkin S, Northington R, Baracat E, Nunes MG, Archer DF, Constantine G, et al. Endometrial effects of bazedoxifene acetate, a novel selective estrogen receptor modulator, in postmenopausal women. *Obstet Gynecol* 2005;**105**(6):1397–404.
85. Komm BS, Kharode YP, Bodine PV, Harris HA, Miller CP, Lyttle CR. Bazedoxifene acetate: a selective estrogen receptor modulator with improved selectivity. *Endocrinology* 2005;**146**(9):3999–4008.
86. Kulak Jr. J, Fischer C, Komm B, Taylor HS. Treatment with bazedoxifene, a selective estrogen receptor modulator, causes regression of endometriosis in a mouse model. *Endocrinology* 2011;**152**(8):3226–32.
87. Cerciat M, Unkila M, Garcia-Segura LM, Arevalo MA. Selective estrogen receptor modulators decrease the production of interleukin-6 and interferon-gamma-inducible protein-10 by astrocytes exposed to inflammatory challenge *in vitro*. *Glia* 2010;**58**(1):93–102.
88. Miller PD, Chines AA, Christiansen C, Hoeck HC, Kendler DL, Lewiecki EM, et al. Effects of bazedoxifene on BMD and bone turnover in postmenopausal women: 2-yr results of a randomized, double-blind, placebo-, and active-controlled study. *J Bone Miner Res* 2008;**23**(4):525–35.
89. Silverman SL, Christiansen C, Genant HK, Vukicevic S, Zanchetta JR, de Villiers TJ, et al. Efficacy of bazedoxifene in reducing new vertebral fracture risk in postmenopausal women with osteoporosis: results from a 3-year, randomized, placebo-, and active-controlled clinical trial. *J Bone Miner Res* 2008;**23**(12):1923–34.
90. Kanis JA, Johansson H, Oden A, McCloskey EV. Bazedoxifene reduces vertebral and clinical fractures in postmenopausal women at high risk assessed with FRAX. *Bone* 2009;**44**(6):1049–54.
91. de Villiers TJ, Chines AA, Palacios S, Lips P, Sawicki AZ, Levine AB, et al. Safety and tolerability of bazedoxifene in postmenopausal women with osteoporosis: results of a 5-year, randomized, placebo-controlled phase 3 trial. *Osteoporos Int* 2011;**22**(2):567–76.
92. Xu L, Tsai KS, Kim GS, Wu Y, Vincendon P, Chines AA, et al. Efficacy and safety of bazedoxifene in postmenopausal Asian women. *Osteoporos Int* 2011;**22**(2):559–65.
93. Christiansen C, Chesnut III CH, Adachi JD, Brown JP, Fernandes CE, Kung AW, et al. Safety of bazedoxifene in a randomized, double-blind, placebo- and active-controlled Phase 3 study of postmenopausal women with osteoporosis. *BMC Musculoskelet Disord* 2010;**11**:130.
94. Silverman SL, Chines AA, Kendler DL, Kung AW, Teglbjaerg CS, Felsenberg D, et al. Sustained efficacy and safety of bazedoxifene in preventing fractures in postmenopausal women with osteoporosis: results of a 5-year, randomized, placebo-controlled study. *Osteoporos Int* 2011.
95. Harvey JA, Holm MK, Ranganath R, Guse PA, Trott EA, Helzner E. The effects of bazedoxifene on mammographic breast density in postmenopausal women with osteoporosis. *Menopause* 2009;**16**(6):1193–6.
96. Bachmann G, Crosby U, Feldman RA, Ronkin S, Constantine GD. Effects of bazedoxifene in nonflushing postmenopausal women: a randomized phase 2 trial. *Menopause* 2011;**18**(5):508–14.
97. Archer DF, Pinkerton JV, Utian WH, Menegoci JC, de Villiers TJ, Yuen CK, et al. Bazedoxifene, a selective estrogen receptor modulator: effects on the endometrium, ovaries, and breast from a randomized controlled trial in osteoporotic postmenopausal women. *Menopause* 2009;**16**(6):1109–15.
98. Pinkerton JV, Archer DF, Utian WH, Menegoci JC, Levine AB, Chines AA, et al. Bazedoxifene effects on the reproductive tract in postmenopausal women at risk for osteoporosis. *Menopause* 2009;**16**(6):1102–8.
99. Pinkerton JV, Utian WH, Constantine GD, Olivier S, Pickar JH. Relief of vasomotor symptoms with the tissue-selective estrogen complex containing bazedoxifene/conjugated estrogens: a randomized, controlled trial. *Menopause* 2009;**16**(6):1116–24.
100. Pickar JH, Yeh IT, Bachmann G, Speroff L. Endometrial effects of a tissue selective estrogen complex containing bazedoxifene/conjugated estrogens as a menopausal therapy. *Fertil Steril* 2009;**92**(3):1018–24.
101. Bramlett KS, Burris TP. Target specificity of selective estrogen receptor modulators within human endometrial cancer cells. *J Steroid Biochem Mol Biol* 2003;**86**(1):27–34.
102. Herrlich P, Ponta H. Mutual cross-modulation of steroid/retinoic acid receptor and AP-1 transcription factor activities: a novel property with practical implications. *Trends Endocrinol Metab* 1994;**5**(8):341–6.
103. Lucibello FC, Slater EP, Jooss KU, Beato M, Muller R. Mutual transrepression of Fos and the glucocorticoid receptor: involvement of a functional domain in Fos which is absent in FosB. *EMBO J* 1990;**9**(9):2827–34.
104. Newton R, Holden NS. Separating transrepression and transactivation: a distressing divorce for the glucocorticoid receptor? *Mol Pharmacol* 2007;**72**(4):799–809.
105. Clark AR. Anti-inflammatory functions of glucocorticoid-induced genes. *Mol Cell Endocrinol* 2007;**275**(1–2):79–97.

106. Tuckermann JP, Reichardt HM, Arribas R, Richter KH, Schutz G, Angel P. The DNA binding-independent function of the glucocorticoid receptor mediates repression of AP-1-dependent genes in skin. *J Cell Biol* 1999;**147**(7):1365–70.
107. Reichardt HM, Kaestner KH, Tuckermann J, Kretz O, Wessely O, Bock R, et al. DNA binding of the glucocorticoid receptor is not essential for survival. *Cell* 1998;**93**(4):531–41.
108. van Raalte DH, Ouwens DM, Diamant M. Novel insights into glucocorticoid-mediated diabetogenic effects: towards expansion of therapeutic options? *Eur J Clin Invest* 2009;**39**(2):81–93.
109. Carballo-Jane E, Pandit S, Santoro JC, Freund C, Luell S, Harris G, et al. Skeletal muscle: a dual system to measure glucocorticoid-dependent transactivation and transrepression of gene regulation. *J Steroid Biochem Mol Biol* 2004;**88**(2):191–201.
110. Pilkis SJ, Granner DK. Molecular physiology of the regulation of hepatic gluconeogenesis and glycolysis. *Annu Rev Physiol* 1992;**54**:885–909.
111. McMahon M, Gerich J, Rizza R. Effects of glucocorticoids on carbohydrate-metabolism. *Diabetes Metab Rev* 1988;**4**(1):17–30.
112. Zakrzewska KE, Cusin I, Sainsbury A, Rohner-Jeanrenaud F, Jeanrenaud B. Glucocorticoids as counterregulatory hormones of leptin—Toward an understanding of leptin resistance. *Diabetes* 1997;**46**(4):717–9.
113. Schaecke H, Berger M, Rehwinkel H, Asadullah K. Selective glucocorticoid receptor agonists (SEGRAs): novel ligands with an improved therapeutic index. *Mol Cell Endocrinol* 2007;**275**(1–2):109–17.
114. Lutsky BN, Berkenkopf J, Fernandez X, Monahan M, Shue HJ, Tiberi RL, et al. Novel class of potent topical anti-inflammatory agents—17alpha benzoylated, 7 alpha-halogeno substituted corticosteroids. *Arzneimittelforschung/Drug Res* 1979;**29–2**(11):1662–7.
115. Gessi S, Merighi S, Borea PA. Glucocorticoids' pharmacology: past, present and future. *Curr Pharm Des* 2010;**16**(32):3540–53.
116. Brattsand R, Thalen A, Roempke K, Kallstrom L, Gruvstad E. Influence of 16-alpha, 17-alpha acetal substitution and steroid nucleus fluorination on the topical to systemic activity ration of glucocorticoids. *J Steroid Biochem Mol Biol* 1982;**16**(6):779–86.
117. Miner JN. Designer glucocorticoids. *Biochem Pharmacol* 2002;**64**(3):355–61.
118. Schacke H, Hennekes H, Schottelius A, Joroch S, Lehmann M, Schmees N, et al. SEGRAs: a novel class of anti-inflammatory compounds. *Ernst Schering Res Found Workshop* 2002;**40**:357–71.
119. Vayssière BM, Dupont S, Choquart A, Petit F, Garcia T, Marchandeau C, et al. Synthetic glucocorticoids that dissociate transactivation and AP-1 transrepression exhibit antiinflammatory activity *in vivo*. *Mol Endocrinol* 1997;**11**:1245–55.
120. Belvisi MG, Wicks SL, Battram CH, Bottoms SEW, Redford JE, Woodman P, et al. Therapeutic benefit of a dissociated glucocorticoid and the relevance of *in vitro* separation of transrepression from transactivation activity. *J Immunol* 2001;**166**(3):1975–82.
121. Coghlan MJ, Jacobson PB, Lane B, Nakane M, Lin CW, Elmore SW, et al. A novel antiinflammatory maintains glucocorticoid efficacy with reduced side effects. *Mol Endocrinol* 2003;**17**(5):860–9.
122. Kassel O, Schneider S, Heilbock C, Litfin M, Gottlicher M, Herrlich P. A nuclear isoform of the focal adhesion LIM-domain protein Trip6 integrates activating and repressing signals at AP-1- and NF-kappa B-regulated promotes. *Genes Dev* 2004;**18**(20):2518–28.
123. López FJ, Ardecky RJ, Bebo B, Benbatoul K, De Grandpre L, Liu S, et al. LGD-5552, an antiinflammatory glucocorticoid receptor ligand with reduced side effects, *in vivo*. *Endocrinology* 2008;**149**(5):2080–9.
124. Miner JN, Ardecky B, Benbatoul K, Griffiths K, Larson CJ, Mais DE, et al. Anti-inflammatory glucocorticoid receptor ligand with reduced side effects exhibits an altered protein-protein interaction profile. *Proc Natl Acad Sci USA* 2007;**104**(49):19244–9.
125. Schacke H, Schottelius A, Docke WD, Strehlke P, Jaroch S, Schmees N, et al. Dissociation of transactivation from transrepression by a selective glucocorticoid receptor agonist leads to separation of therapeutic effects from side effects. *Proc Natl Acad Sci USA* 2004;**101**(1):227–32.
126. Humphrey EL, Williams JHH, Davie MWJ, Marshall MJ. Effects of dissociated glucocorticoids on OPG and RANKL in osteoblastic cells. *Bone* 2006;**38**(5):652–61.
127. De Bosscher K, Vanden Berghe W, Beck IME, Van Molle W, Hennuyer N, Hapgood J, et al. A fully dissociated compound of plant origin for inflammatory gene repression. *Proc Natl Acad Sci USA* 2005;**102**(44):15827–32.

128. Wang Z, Frederick J, Garabedian MJ. Deciphering the phosphorylation "code" of the glucocorticoid receptor *in vivo*. *J Biol Chem* 2002;**277**(29):26573–80.
129. Zhang ZR, Zhang ZY, Schluesener HJ. Compound A, a plant origin ligand of glucocorticoid receptors, increases regulatory T cells and M2 macrophages to attenuate experimental autoimmune neuritis with reduced side effects. *J Immunol* 2009;**183**(5):3081–91.
130. Rauch A, Seitz S, Baschant U, Schilling AF, Illing A, Stride B, et al. Glucocorticoids suppress bone formation by attenuating osteoblast differentiation via the monomeric glucocorticoid receptor. *Cell Metab* 2010;**11**(6):517–31.
131. Young JD, Lawrence AJ, MacLean AG, Leung BP, McInnes IB, Canas B, et al. Thymosin beta 4 sulfoxide is an anti-inflammatory agent generated by monocytes in the presence of glucocorticoids. *Nat Med* 1999;**5**(12):1424–7.
132. Berrebi D, Bruscoli S, Cohen N, Foussat A, Migliorati G, Bouchet-Delbos L, et al. Synthesis of glucocorticoid-induced leucine zipper (GILZ) by macrophages: an anti-inflammatory and immunosuppressive mechanism shared by glucocorticoids and IL-10. *Blood* 2003;**101**(2):729–38.
133. Calandra T, Bernhagen J, Metz CN, Spiegel LA, Bacher M, Donnelly T, et al. MIF as a glucocorticoid-induced modulator of cytokine production. *Nature* 1995;**377**(6544):68–71.
134. Stocklin E, Wissler M, Gouilleux F, Groner B. Functional interactions between Stat5 and the glucocorticoid receptor. *Nature* 1996;**383**(6602):726–8.
135. Lee HJ, Ko DH. A novel approach to the discovery of non-systemic anti-inflammatory steroids; Antedrug. *Arch Pharm Res* 1999;**22**(3):279–87.
136. Belvisi MG, Hele DJ. Soft steroids: a new approach to the treatment of inflammatory airways diseases. *Pulm Pharmacol Ther* 2003;**16**(6):321–5.
137. Welker P, Lippert U, Nurnberg W, KrugerKrasagakes S, Moller A, Czarnetzki BM. Glucocorticoid-induced modulation of cytokine secretion from normal and leukemic human myelomonocytic cells. *Int Arch Allergy Immunol* 1996;**109**(2):110–5.
138. Gunther C, Kecskes A, Staks T, Tauber U. Percutaneous absorption of methylprednisolone aceponate following topical application of Advantan (R) lotion on intact, inflamed and stripped skin of male volunteers. *Skin Pharmacol Appl Skin Physiol* 1998;**11**(1):35–42.
139. O'Connell EJ. Review of the unique properties of budesonide. *Clin Ther* 2003;**25**:C42–60.
140. Schacke H, Zollner TM, Docke WD, Rehwinkel H, Jaroch S, Skuballa W, et al. Characterization of ZK 245186, a novel, selective glucocorticoid receptor agonist for the topical treatment of inflammatory skin diseases. *Br J Pharmacol* 2009;**158**(4):1088–103.
141. Grose R, Werner S, Kessler D, Tuckermann J, Huggel K, Durka S, et al. A role for endogenous glucocorticoids in wound repair. *EMBO Rep* 2002;**3**(6):575–82.
142. Kleiman A, Tuckermann JP. Glucocorticoid receptor action in beneficial and side effects of steroid therapy: lessons from conditional knockout mice. *Mol Cell Endocrinol* 2007;**275**(1–2):98–108.
143. Tuckermann JP, Kleiman A, Moriggl R, Spanbroek R, Neumann A, Illing A, et al. Macrophages and neutrophils are the targets for immune suppression by glucocorticoids in contact allergy. *J Clin Invest* 2007;**117**(5):1381–90.
144. Barak Y, Kim S. Genetic manipulations of PPARs: effects on obesity and metabolic Disease. *PPAR Res* 2007;**2007**:12781.
145. Francis GA, Li G, Casey R, Wang J, Cao HN, Leff T, et al. Peroxisomal proliferator activated receptor-gamma deficiency in a Canadian kindred with familial partial lipodystrophy type 3 (FPLD3). *BMC Med Genet* 2006;7.
146. Barroso I, Gurnell M, Crowley VEF, Agostini M, Schwabe JW, Soos MA, et al. Dominant negative mutations in human PPAR gamma associated with severe insulin resistance, diabetes mellitus and hypertension. *Nature* 1999;**402**(6764):880–3.
147. Sarraf P, Mueller E, Smith WM, Wright HM, Kum JB, Aaltonen LA, et al. Loss-of-function mutations in PPAR gamma associated with human colon cancer. *Mol Cell* 1999;**3**(6):799–804.
148. Wild S, Roglic G, Green A, Sicree R, King H. Global prevalence of diabetes—Estimates for the year 2000 and projections for 2030. *Diabetes Care* 2004;**27**(5):1047–53.
149. Home PD, Pocock SJ, Beck-Nielsen H, Curtis PS, Gomis R, Hanefeld M, et al. Rosiglitazone evaluated for cardiovascular outcomes in oral agent combination therapy for type 2 diabetes (RECORD): a multicentre, randomised, open-label trial. *Lancet* 2009;**373**(9681):2125–35.

150. Lago RM, Singh PP, Nesto RW. Congestive heart failure and cardiovascular death in patients with prediabetes and type 2 diabetes given thiazolidinediones: a meta-analysis of randomised clinical trials. *Lancet* 2007;**370**(9593):1129–36.
151. Kahn SE, Haffner SM, Heise MA, Herman WH, Holman RR, Jones NP, et al. Glycemic durability of rosiglitazone, metformin, or glyburide monotherapy. *N Engl J Med* 2006;**355**(23):2427–43.
152. Dormandy JA, Charbonnel B, Eckland DJA, Erdmann E, Massi-Benedetti M, Kmoules IK, et al. Secondary prevention of macrovascular events in patients with type 2 diabetes in the PROactive Study (PROspective pioglitAzone Clinical Trial In macroVascular Events): a randomised controlled trial. *Lancet* 2005;**366**(9493):1279–89.
153. Erdmann E, Dormandy J, Wilcox R, Massi-Benedetti M, Charbonnel B. PROactive 07: pioglitazone in the treatment of type 2 diabetes: results of the PROactive study. *Vasc Health Risk Manag* 2007;**3**:355–70.
154. Loke YK, Singh S, Furberg CD. Long-term use of thiazolidinediones and fractures in type 2 diabetes: a meta-analysis. *Can Med Assoc J* 2009;**180**(1):32–9.
155. Bongartz T, Coras B, Vogt T, Scholmerich J, Muller-Ladner U. Treatment of active psoriatic arthritis with the PPAR gamma ligand pioglitazone: an open-label pilot study. *Rheumatology* 2005;**44**(1):126–9.
156. Haffner SM, Greenberg AS, Weston WM, Chen HZ, Williams K, Freed MI. Effect of rosiglitazone treatment on nontraditional markers of cardiovascular disease in patients with type 2 diabetes mellitus. *Circulation* 2002;**106**(6):679–84.
157. Grether U, Klaus W, Kuhn B, Maerki HP, Mohr P, Wright MB. New insights on the mechanism of PPAR-targeted drugs. *ChemMedChem* 2010;**5**(12):1973–6.
158. Lewis JD, Ferrara A, Peng T, Hedderson M, Bilker WB, Quesenberry CP, et al. Risk of Bladder cancer among diabetic patients treated with pioglitazone: interim report of a longitudinal cohort study. *Diabetes Care* 2011;**34**(4):916–22.
159. FDA Drug Safety Communication: Update to ongoing safety review of Actos (pioglitazone) and increased risk of bladder cancer. Available from: <http://wwwfedgov/Drugs/DrugSafety/ucm259150htm>2011.
160. Aoki T. Current status of carcinogenicity assessment of peroxisome proliferator-activated receptor agonists by the US FDA and mode-of-action approach to the carcinogenic potential. *J Toxicol Pathol* 2007;**20**(4):611–7.
161. Rosen ED, Sarraf P, Troy AE, Bradwin G, Moore K, Milstone DS, et al. PPAR gamma is required for the differentiation of adipose tissue *in vivo* and *in vitro*. *Mol Cell* 1999;**4**(4):611–7.
162. Barak Y, Nelson MC, Ong ES, Jones YZ, Ruiz-Lozano P, Chien KR, et al. PPAR gamma is required for placental, cardiac, and adipose tissue development. *Mol Cell* 1999;**4**(4):585–95.
163. Tontonoz P, Hu ED, Spiegelman BM. Stimulation of adipogenesis in fibroblasts by PPARgamma2, a lipid activated transcription factor. *Cell* 1994;**79**(7):1147–56.
164. Baker PR, Lin Y, Schopfer FJ, Woodcock SR, Groeger AL, Batthyany C, et al. Fatty acid transduction of nitric oxide signaling: multiple nitrated unsaturated fatty acid derivatives exist in human blood and urine and serve as endogenous peroxisome proliferator-activated receptor ligands. *J Biol Chem* 2005;**280**(51):42464–75.
165. Schopfer FJ, Lin Y, Baker PR, Cui T, Garcia-Barrio M, Zhang J, et al. Nitrolinoleic acid: an endogenous peroxisome proliferator-activated receptor gamma ligand. *Proc Natl Acad Sci USA* 2005;**102**(7):2340–5.
166. Li Y, Zhang J, Schopfer FJ, Martynowski D, Garcia-Barrio MT, Kovach A, et al. Molecular recognition of nitrated fatty acids by PPAR gamma. *Nat Struct Mol Biol* 2008;**15**(8):865–7.
167. Tontonoz P, Spiegelman BM. Fat and beyond: the diverse biology of PPAR gamma. *Annu Rev Biochem* 2008;**77**:289–312.
168. Kawai M, Rosen CJ. PPAR gamma: a circadian transcription factor in adipogenesis and osteogenesis. *Nat Rev Endocrinol* 2010;**6**(11):629–36.
169. Shockley KR, Lazarenko OP, Czernik PJ, Rosen CJ, Churchill GA, Lecka-Czemik B. PPAR gamma 2 nuclear receptor controls multiple regulatory pathways of osteoblast differentiation from marrow mesenchymal stem cells. *J Cell Biochem* 2009;**106**(2):232–46.
170. Guan Y, Hao C, Cha DR, Rao R, Lu W, Kohan DE, et al. Thiazolidinediones expand body fluid volume through PPARgamma stimulation of ENaC-mediated renal salt absorption. *Nat Med* 2005;**11**(8):861–6.
171. Savkur RS, Miller AR. Investigational PPAR-gamma agonists for the treatment of Type 2 diabetes. *Expert Opin Investig Drugs* 2006;**15**:763–78.
172. Ryan KK, Li B, Grayson BE, Matter EK, Woods SC, Seeley RJ. A role for central nervous system PPAR-gamma in the regulation of energy balance. *Nat Med* 2011;**17**(5). 623–U145

173. Ricote M, Glass CK. PPARs and molecular mechanisms of transrepression. *Biochim Biophys Acta (BBA) – Mol Cell Biol Lipids* 2007;**1771**(8):926–35.
174. Chung SW, Kang BY, Kim SH, Pak YK, Cho D, Trinchieri G, et al. Oxidized low density lipoprotein inhibits interleukin-12 production in lipopolysaccharide-activated mouse macrophages via direct interactions between peroxisome proliferator-activated receptor-gamma and nuclear factor-kappa B. *J Biol Chem* 2000;**275**(42):32681–7.
175. Yang XY, Wang LH, Chen TS, Hodge DR, Resau JH, DaSilva L, et al. Activation of human T lymphocytes is inhibited by peroxisome proliferator-activated receptor gamma (PPAR gamma) agonists—PPAR gamma co-association with transcription factor NFAT. *J Biol Chem* 2000;**275**(7):4541–4.
176. Zhao W, Wang LJ, Zhang M, Wang P, Zhang L, Yuan C, et al. Peroxisome proliferator-activated receptor gamma negatively regulates IFN-beta production in toll-like receptor (TLR) 3- and TLR4-stimulated macrophages by preventing interferon regulatory factor 3 binding to the IFN-beta promoter. *J Biol Chem* 2011;**286**(7):5519–28.
177. Pascual G, Fong AL, Ogawa S, Gamliel A, Li AC, Perissi V, et al. A SUMOylation-dependent pathway mediates transrepression of inflammatory response genes by PPAR-gamma. *Nature* 2005;**437**(7059):759–63.
178. Kallenberger BC, Love JD, Chatterjee VKK, Schwabe JWR. A dynamic mechanism of nuclear receptor activation and its perturbation in a human disease. *Nat Struct Biol* 2003;**10**(2):136–40.
179. Hotamisligil GS. Inflammation and metabolic disorders. *Nature* 2006;**444**(7121):860–7.
180. Straus DS, Glass CK. Anti-inflammatory actions of PPAR ligands: new insights on cellular and molecular mechanisms. *Trends Immunol* 2007;**28**(12):551–8.
181. Han S, Roman J. Peroxisome proliferator-activated receptor gamma: a novel target for cancer therapeutics? *Anticancer Drugs* 2007;**18**(3):237–44.
182. Rangwala SM, Lazar MA. The dawn of the SPPARMs. *Sci STKE* 2002;**2002**(121):pe9.
183. Higgins LS, DePaoli AM. Selective peroxisome proliferator-activated receptor gamma (PPAR gamma) modulation as a strategy for safer therapeutic PPAR gamma activation. *Am J Clin Nutr* 2010;**91**(1):267S–272SS.
184. Goldberg RB, Kendall DM, Deeg MA, Buse JB, Zagar AJ, Pinaire JA, et al. A comparison of lipid and glycemic effects of pioglitazone and rosiglitazone in patients with type 2 diabetes and dyslipidemia. *Diabetes Care* 2005;**28**(7):1547–54.
185. Chiquette E, Ramirez G, DeFronzo R. A meta-analysis comparing the effect of thiazolidinediones on cardiovascular risk factors. *Arch Intern Med* 2004;**164**(19):2097–104.
186. Wu Y, Chin WW, Wang Y, Burris TP. Ligand and co-activator identity determines the requirement of the charge clamp for co-activation of the peroxisome proliferator-activated receptor gamma. *J Biol Chem* 2003;**278**(10):8637–44.
187. Choi JH, Banks AS, Estall JL, Kajimura S, Bostrom P, Laznik D, et al. Anti-diabetic drugs inhibit obesity-linked phosphorylation of PPAR gamma by Cdk5. *Nature* 2010;**466**(7305). 451–U1
188. Rocchi S, Picard F, Vamecq J, Gelman L, Potier N, Zeyer D, et al. A unique PPAR gamma ligand with potent insulin-sensitizing yet weak adipogenic activity. *Mol Cell* 2001;**8**(4):737–47.
189. Burgermeister E, Schnoebelen A, Flament A, Benz J, Stihle M, Gsell B, et al. A novel partial agonist of peroxisome proliferator-activated receptor-gamma (PPAR gamma) recruits PPAR gamma-coactivator-1 alpha, prevents triglyceride accumulation, and potentiates insulin signaling *in vitro*. *Mol Endocrinol* 2006;**20**(4):809–30.
190. Motani A, Wang ZL, Weiszmann J, McGee LR, Lee G, Liu QX, et al. INT131: A selective modulator of PPAR gamma. *J Mol Biol* 2009;**386**(5):1301–11.
191. Fujimura T, Sakuma H, Konishi S, Oe T, Hosogai N, Kimura C, et al. FK614, a novel peroxisome proliferator-activated receptor gamma modulator, induces differential transactivation through a unique ligand-specific interaction with transcriptional coactivators. *J Pharmacol Sci* 2005;**99**(4):342–52.
192. Oberfield JL, Collins JL, Holmes CP, Goreham DM, Cooper JP, Cobb JE, et al. A peroxisome proliferator-activated receptor gamma ligand inhibits adipocyte differentiation. *Proc Natl Acad Sci USA* 1999;**96**(11):6102–6.
193. Kim KY, Kim SS, Cheon HG. Differential anti-proliferative actions of peroxisome proliferator-activated receptor-gamma agonists in MCF-7 breast cancer cells. *Biochem Pharmacol* 2006;**72**(5):530–40.
194. Lehmann JM, Moore LB, Smith-Oliver TA, Wilkison WO, Willson TM, Kliewer SA. An antidiabetic thiazolidinedione is a high affinity ligand for peroxisome proliferator-activated receptor gamma. *J Biol Chem* 1995;**270**(22):12953–6.

195. Handschin C, Spiegelman BM. The role of exercise and PGC1 alpha in inflammation and chronic disease. *Nature* 2008;**454**(7203):463–9.
196. Handschin C, Choi CS, Chin S, Kim S, Kawamori D, Kurpad AJ, et al. Abnormal glucose homeostasis in skeletal muscle-specific PGC-1 alpha knockout mice reveals skeletal muscle-pancreatic beta cell crosstalk. *J Clin Invest* 2007;**117**(11):3463–74.
197. Picard F, Gehin M, Annicotte JS, Rocchi S, Champy MF, O'Malley BW, et al. SRC-1 and TIF2 control energy balance between white and brown adipose tissues. *Cell* 2002;**111**(7):931–41.
198. Fujimura T, Kimura C, Oe T, Takata Y, Sakuma H, Aramori I, et al. A selective peroxisome proliferator-activated receptor gamma modulator with distinct fat cell regulation properties. *J Pharmacol Exp Ther* 2006;**318**(2):863–71.
199. Fujimura T, Sakuma H, Ohkubo-Suzuki A, Aramori I, Mutoh S. Unique properties of coactivator recruitment caused by differential binding of FK613, an anti-diabetic agent, to peroxisome proliferator-activated receptor gamma. *Biol Pharm Bull* 2006;**29**:423–9.
200. Minoura H, Takeshita S, Ita M, Hirosumi J, Mabuchi M, Kawamura I, et al. Pharmacological characteristics of a novel nonthiazolidinedione insulin sensitizer, FK614. *Eur J Pharmacol* 2004;**494**(2–3):273–81.
201. Minoura H, Takeshita S, Kimura C, Hirosumi J, Takakura S, Kawamura I, et al. Mechanism by which a novel non-thiazolidinedione peroxisome proliferator-activated receptor gamma agonist, FK614, ameliorates insulin resistance in Zucker fatty rats. *Diabetes Obes Metab* 2007;**9**(3):369–78.
202. Minoura H, Takeshita S, Yamamoto T, Mabuchi M, Hirosumi J, Takakura S, et al. Ameliorating effect of FK614, a novel nonthiazolidinedione peroxisome proliferator-activated receptor gamma agonist, on insulin resistance in Zucker fatty rat. *Eur J Pharmacol* 2005;**519**(1–2):182–90.
203. Shimizu M. Overview of R&D pipeline. Available at: <http://www.astellas.com/en/ir/library/pdf/1q2013_rd_en.pdf>. 2005.
204. Dunn FL, Higgins LS, Fredrickson J, DePaoli AM, Grp INT. Selective modulation of PPAR gamma activity can lower plasma glucose without typical thiazolidinedione side effects in patients with Type 2 diabetes. *J Diabetes Complications* 2011;**25**(3):151–8.
205. Kim KR, Lee JH, Kim SJ, Dal Rhee S, Jung WH, Yang SD, et al. KR-62980: a novel peroxisome proliferator-activated receptor gamma agonist with weak adipogenic effects. *Biochem Pharmacol* 2006;**72**(4):446–54.
206. Won HY, Min HJ, Ahn JH, Yoo SE, Bae MA, Hong JH, et al. Anti-allergic function and regulatory mechanisms of KR62980 in allergen-induced airway inflammation. *Biochem Pharmacol* 2010;**79**(6):888–96.
207. Kim KY, Cho HS, Lee SH, Ahn JH, Cheon HG. Neuroprotective effects of KR-62980, a new PPAR gamma agonist, against chemical ischemia-reperfusion in SK-N-SH cells. *Brain Res* 2011;**1372**:103–14.
208. Jung H, Lee MS, Jang EJ, Ahn JH, Kang NS, Yoo SE, et al. Augmentation of PPAR gamma-TAZ interaction contributes to the anti-adipogenic activity of KR62980. *Biochem Pharmacol* 2009;**78**(10):1323–9.
209. Hong JH, Hwang ES, McManus MT, Amsterdam A, Tian Y, Kalmukova R, et al. TAZ, a transcriptional modulator of mesenchymal stem cell differentiation. *Science* 2005;**309**(5737):1074–8.
210. Acton JJ, Black RM, Jones AB, Moller DE, Colwell L, Doebber TW, et al. Benzoyl 2-methyl indoles as selective PPAR gamma modulators. *Bioorg Med Chem Lett* 2005;**15**(2):357–62.
211. Choi JH, Banks AS, Kamenecka TM, Busby SA, Chalmers MJ, Kumar N, et al. Antidiabetic actions of a non-agonist PPAR-gamma ligand blocking Cdk5-mediated phosphorylation. *Nature* 2011;**477**(7365):477–81.
212. Lu M, Sarruf DA, Talukdar S, Sharma S, Li PP, Bandyopadhyay G, et al. Brain PPAR-gamma promotes obesity and is required for the insulin-sensitizing effect of thiazolidinediones. *Nat Med* 2011;**17**(5). 618–U140

CHAPTER

5

Biasing Receptor Tyrosine Kinase Signaling Pathways

John Watson[1], Brian J. Arey[2] and Andrew Alt[1]

[1]Lead Discovery and Optimization, Research and Development, Bristol-Myers Squibb Co., Wallingford, CT, USA, [2]Cardiovascular Drug Discovery Biology, Research and Development, Bristol-Myers Squibb Co., Hopewell, NJ, USA

OUTLINE

INTRODUCTION

Post-translational modification of proteins by phosphorylation and dephosphorylation is an evolutionarily ancient way of controlling the activity and localization of cellular proteins. Protein kinases, by virtue of their ability to add phosphate residues to target proteins, can be considered to be the master orchestrators of virtually all cellular processes. The majority of these protein kinases phosphorylate serine and threonine (ser/thr)

B. Arey (Ed): *Biased Signaling in Physiology, Pharmacology and Therapeutics*
DOI: http://dx.doi.org/10.1016/B978-0-12-411460-9.00005-7

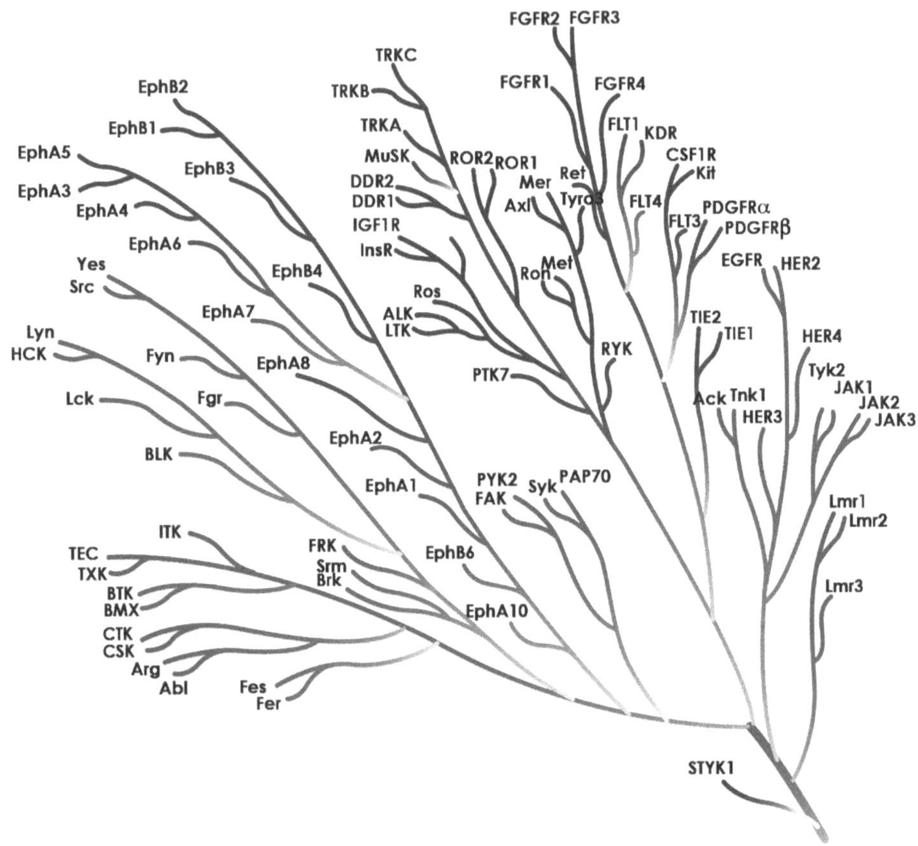

FIGURE 5.1 Phylogenetic relationships of the human tyrosine kinases. Both receptor tyrosine kinases and non-receptor tyrosine kinases are shown. *Reproduced with permission from Cell Signaling Technology (www.cellsignal.com).*

residues on target proteins and indeed the first two protein kinases discovered, protein phosphorylase and protein kinase A (PKA), are ser/thr kinases.[1] The chance discovery 30 years ago that the protein kinase activities associated with the transforming polyoma and Rous sarcoma (Src) viruses phosphorylated target proteins on tyrosine residues,[2,3] combined with the observation three years later that the catalytic domains of the v-Src tyrosine kinase and the PKA ser/thr kinase were structurally related, served to define the boundaries of the protein kinase portion of the human proteome, known as the kinome.

Whole human genome sequencing has revealed the existence of 518 protein kinases.[4] There are 90 tyrosine kinases, 58 of which are receptor tyrosine kinases (RTKs), representing 11% of the human kinome (Figure 5.1). All RTKs share a common architecture consisting of a ligand binding extracellular domain, a single transmembrane domain, and a cytoplasmic domain that harbors the tyrosine kinase as well as additional sequences important for kinase regulation and activation of downstream signaling pathways (Figure 5.2). The human RTK kinome has been subdivided, on the basis of primary protein

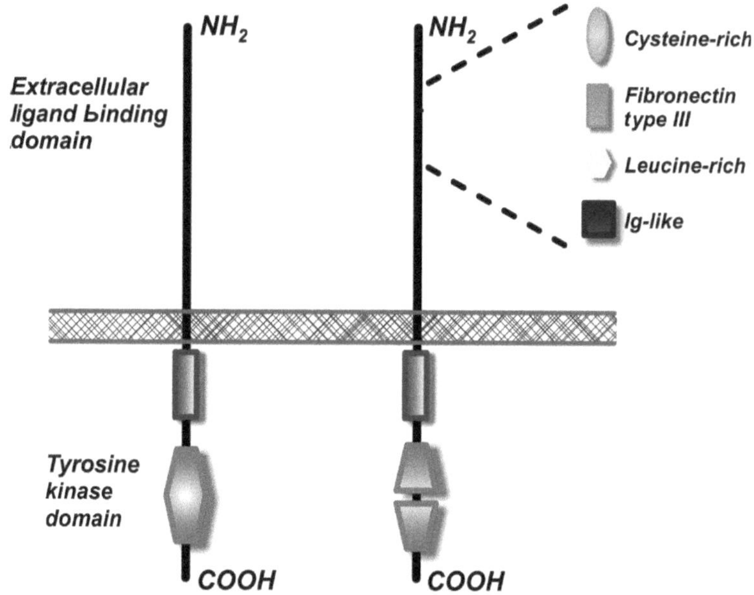

FIGURE 5.2 Schematic representation of the modular architecture of an RTK. The extracellular ligand binding domain is composed of structural motifs (typically repeated) that mediate ligand binding and, in some cases, participate in dimerization. Some of the more common motifs are indicated on the right-hand side, although other specific motifs are found in certain RTKs. The cytoplasmic tyrosine kinase domain is indicated by the box and the juxtamembrane regions by the cylinders. The carboxyl tails and the cytoplasmic juxtamembrane regions form important docking sites for RTK signal transducers and in some cases participate in regulation of the kinase domain itself. Some RTK tyrosine kinase domains are split with the two portions connected by a linker; the split kinase linker region can serve regulatory and protein binding functions.

sequence, into 20 families[5,6] (Table 5.1). Overall RTK structure and signaling properties have been evolutionarily conserved throughout metazoan evolution, from *C. elegans* and *Drosophila* to humans,[7] a compelling indication of the essential cellular functions they perform. Since their discovery, RTKs have come to be appreciated as fundamentally important signaling molecules and regulators of a multitude of biological processes, including embryonic development, differentiation, proliferation, migration, survival, cell cycle control, and metabolic activity.[6,8–10] Moreover, mutations in RTK genes and dysregulation of their intracellular signaling pathways have been causally linked to a variety of chronic human diseases such as cancer, diabetes, inflammation, neurological disease, and pulmonary disease.[11–14]

In the 35 years since their discovery, RTKs have been studied in detail, including studies of structural determinations of complexes of RTK extracellular domains, with and without their peptide ligands, and of active and inactive tyrosine kinase domains. The binding of activating ligands (peptides) to the extracellular portion of an RTK results in the formation of a stable dimer consisting of two receptor monomers that is able to activate the cytosolic tyrosine kinase domain. The peptide ligands of RTKs can be, but are not necessarily, dimers themselves. Exceptions to this general rule are the insulin receptor (IR)

TABLE 5.1 The Receptor Tyrosine Kinase Families

Family	Family Members	Family	Family Members
ErbB	ErbB1 (EGFR)	MuSL	MuSK
	ErbB2		
	ErbB3	MET	Met
	ErbB4		Ron
INS	InsR	AXL	Axl
	IGF1R		Mer
	InsRR		Tyro3
PDGF	PDGFRα	TIE	Tie1
	PDGFRβ		Tie2
	CSFR1		
	Kit	EPH	EphA1–EphA8
	Flt3		EphA10
			EphB1–EphB6
VEGF	VEGFR1		
	VEGFR2	RET	Ret
	VEGFR3		
		RYK	Ryk
FGF	FGFR1		
	FGFR2	DDR	DDR1
	FGFR3		DDR2
	FGFR4		
	FGFRL1	ROS	Ros
PTK7	PTK7	LMR	LMR1
			LMR2
TRK	TrkA (NGFR)		LMR3
	TrkB		
	TrkC	ALK	ALK
			LTK
RER	Rer1		
	Rer2	STYK1	STYK1

and the insulin-like growth factor 1 (IGF-1) receptor which exist in the membrane as stable disulfide-linked homo- and heterodimers. The IR and IGF-1 receptors are discussed in more detail, below.

Although conformational changes in the tyrosine kinase domain that accompany kinase activation have been elucidated in detail for some RTKs, it remains to be determined exactly how the conformational change that results from ligand binding to the extracellular portion of an RTK is propagated to the cytoplasmic portion of the receptor to produce an increase in enzyme activity. Available crystal structures of RTK tyrosine kinase domains make it clear, however, that all activated RTK tyrosine kinase domains (indeed all kinase domains in general) are quite similar due, at least in part, to the geometric requirements needed to support phosphotransfer from ATP to the bound substrate.[15,16] For example, all kinase domains contain a conserved core structure consisting of a small N-terminal lobe and a larger C-terminal lobe (N-lobe and C-lobe), with residues essential for catalytic activity occupying the cleft between the two lobes.[15,17,18] Kinase activation is typically characterized by conserved rearrangements of structures, such as the N-lobe αC helix, and the activation loop in the C-lobe, to produce and stabilize the conformations necessary to support phosphorylation of peptide substrates.[18,19]

Interestingly, the structures of inactive kinase domains are surprisingly different, a reflection that all kinases, including RTKs, have evolved a diversity of regulatory mechanisms to control catalytic activity. Different RTKs have adopted different strategies to inhibit kinase activity in the absence of an activating ligand. These inhibitory mechanisms are termed cis-inhibitions since tyrosine kinase activity is inhibited by contacts with residues in different tyrosine-containing portions of the RTK cytosolic regions. These cis-inhibitory regions can be located in the tyrosine kinase domain itself (i.e., the activation loop), in the cytoplasmic juxtamembrane (JM) region, or the cytoplasmic tail. Ligand binding and dimerization results in trans-phosphorylation by one monomeric tyrosine kinase domain of tyrosine residues in the cis-inhibitory region(s) of the other, resulting in loss of inhibition and dramatic increases in catalytic activity.

An example of the release of cis-inhibitory contacts in the activation loop of the fibroblast growth factor receptor 1 (FGFR1) is shown in Figure 5.3. The activation loop of FGFR1 (and the closely related FGFR2 RTK) does not sterically inhibit access to the catalytic site. Instead, contacts between the activation loop and N-lobe stabilize an inactive conformation of the kinase.[20,21] Phosphorylation of a tyrosine (Y653) in the FGFR1 activation loop relieves the cis-inhibition such that there is a global rotation of the N-lobe and both the activation loop and the αC helix adopt conformations relative to the catalytic loop necessary for substrate phosphorylation.[15,19]

While all RTKs are cis-inhibited, they do not universally follow the mechanism adopted by the FGFR1. Some receptors, such as the angiopoietin receptor Tie2, use sequences found in the carboxy terminus[22] to regulate their catalytic activity. Still other RTKs have been shown to be cis-inhibited by their cytosolic juxtamembrane regions.[23] As with inhibition by activation loop sequences, auto-phosphorylation of key tyrosine residues in the juxtamembrane or carboxy terminal regions disrupts the inhibitory interactions and allows the kinase domain to assume an active configuration. Finally, the epidermal growth factor family of RTKs (EGFR and ErbB2-4) does not rely on auto-phosphorylation of any critical tyrosines to induce the active state. Instead, these kinases use an allosteric method of

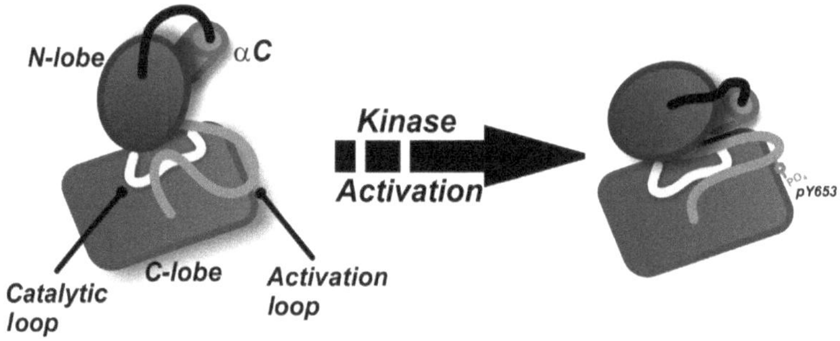

FIGURE 5.3 Simplified cartoon of the structure of the inactive (left) and active (right) tyrosine kinase domain of the fibroblast growth factor receptor 1 (FGFR1). In the inactive FGFR1 kinase domain, catalytic activity is inhibited by contacts of the activation loop that stabilize an inactive kinase conformation. The catalytic loop is also shown. Upon phosphorylation of tyrosine 653 (Y653), the activation loop rearranges, thereby allowing further rearrangements of the N-lobe and the αC helix (αC), resulting in substrate binding and kinase activation. The relative rearrangements of the N-lobe, αC helix, and activation loop are exaggerated to illustrate the conformational changes that occur during kinase activation. *Adapted from Ref. 20 and used with permission of PNAS.*

catalytic domain activation which involves intra-kinase protein–protein interactions to disrupt the auto-inhibitory contacts provided by the carboxy terminal region.[24]

For most RTKs, the most important initial phosphorylation reactions that take place following dimerization and activation usually target regulatory portions of the kinase domain itself. Trans-auto-phosphorylation of activation loop tyrosines in the FGFR1, for example, can increase kinase activity 50-fold,[25] thereby setting the stage for additional auto-phosphorylation reactions on RTK cytoplasmic tyrosines that serve as recruitment and binding sites for RTK substrates and adapter proteins, which propagate the activated RTK signal to downstream signaling cascades. The recruited proteins typically contain phosphotyrosine binding motifs such as Src-homology (SH2/SH3) or phosphotyrosine binding (PTB) domains[26,27] and may themselves contain multiple tyrosine residues that when phosphorylated by the activated RTK kinase domain serve as scaffolding or docking sites for the assembly of large multi-protein signaling complexes.[28] Some of these docking proteins are the insulin receptor substrates, IRS1/IRS2, and the FGFR substrate, FRS2α, and the scaffolding protein, Shc.

Other proteins, so-called adapter proteins, are recruited to the activated RTKs or to associated docking proteins and are phosphorylated. Some of the more common adapters are cytosolic kinases like protein kinase D (PKD), Akt, phosphatidyl-3-kinases (PI3K), protein tyrosine phosphatases such as Shp2, or enzymes such as phospholipase C gamma (PLCγ). These adapters in turn regulate the activity of downstream signaling cascades such as the Ras/Raf/MAPK (ERK1/2) mitogenic pathway, the Jak-STAT transcriptional activation pathway, and the inositol phosphate/Ca^{++}/PKC pathway. While certain of the docking and adapter proteins are unique for a given RTK, many are shared among many receptors. Because of the overlapping nature of the RTK signaling networks, the specificity and integrity of a given RTK signal can be thought of as the spatial and temporal complement of available interacting proteins within the cell.[29]

The networks regulated by RTKs are themselves extraordinarily complex. For example, a systems biology analysis of the epidermal growth factor receptor (EGFR) signaling

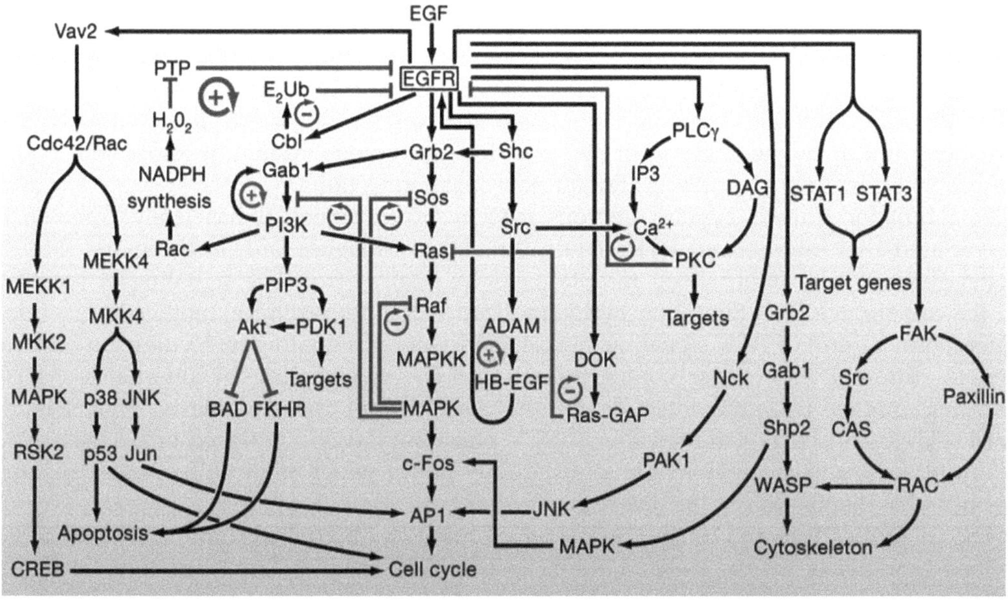

FIGURE 5.4 Complexity of an RTK signaling network. A portion of the epidermal growth factor receptor (EGFR) signaling network is shown. Positive feedback loops are indicated by the circles or (+). Negative regulatory circuits and feedback loops are shown as (−). The striking complexity of the network creates both opportunities and challenges for the identification of pathway-biased ligands which potentially might activate (or inhibit) specific subsets of the reactions. *Figure adapted from Ref. 5 and used with permission of Elsevier.*

network estimated that the network consists of 211 interactions involving 202 proteins, including 10 different receptors, 61 different enzymes (32 of them kinases), 22 adapter (docking) proteins, and 3 ion channels.[30] A portion of the EGFR interaction network is shown in Figure 5.4. However, the complexity of an RTK signaling network creates multiple opportunities to identify pathway-biased modulators of RTK signaling, i.e., synthetic peptides or small molecule ligands that can selectively activate or inhibit a subset of an RTK signaling network. Such pathway-biased ligands might be expected to display greater specificity of response and control of signaling than unbiased ligands that activate or inhibit the entire signaling cascade. The enormous complexity of an RTK signaling network creates concomitant challenges as well, that stem from the sheer number of potential interactions and the difficulties in their quantitative description. Nonetheless, the implications for control of RTK-mediated signaling by pathway-biased ligands are significant; depending on the signaling outputs considered, pathway bias can result in dramatic shifts in, or even reversals of, ligand efficacy.[31]

Subsequent to dimerization and activation induced by ligand binding, activated RTKs are subject to clustering on the cell surface followed by clathrin- or caveolin-dependent internalization, degradation, and/or recycling.[32] This behavior is highly reminiscent of ligand-induced receptor internalization in G protein-coupled receptor (GPCR) signaling,[33] and similarly is important for cellular regulation (desensitization) of RTK signaling intensity and duration.[34] Historically, however, it has long been appreciated that there was potentially more to the RTK internalization than simple signal termination. For example, internalization and retrograde transport of nerve growth factor (NGF)-TrkA receptor (and other neurotrophin-Trk receptor) complexes at axon terminals has been known to be intimately linked to the survival of neurons during nervous system development, and to the proper formation and preservation of neuronal connections.[35]

In a broader context beyond nervous system development, it has lately come to be appreciated that internalized RTKs retain the ability to signal and, furthermore, that the signal generated by internalized receptors can be distinct from the signaling of receptors on the cell surface.[36–38] Signaling by internalized receptors provides another level of spatiotemporal control of RTK signaling since specific sets of signaling molecules can be localized to different subcellular compartments. Moreover, signaling by internalized RTKs provides another potential interface for the discovery of pathway-biased RTK ligands, since such ligands may produce altered RTK phosphorylation, increased or decreased signal amplitude, and changes in signal duration—any of which might affect internalization and ultimate disposition of the receptor.

The identification and characterization of pathway-biased ligands for GPCRs has opened new vistas for the understanding of the 7TM receptor family, as well as for the development of novel therapeutics.[31,39–43] In addition, our understanding of the mechanism of pathway bias for GPCRs is supported by structural knowledge of key conformational transitions involved in linking ligand binding status to G protein activation and signal transduction. The ability of biased ligands to stabilize distinct receptor conformations of the serotonin 2B, vasopressin 2, and β2-adrenergic receptors has been demonstrated using crystallographic, fluorescence, and F^{19}-NMR techniques, respectively.[44–46] One important conclusion that can be drawn from the structural studies of GPCR activation by biased ligands is that small conformational movements of particular transmembrane domains appear to be sufficient for pathway activation. This is not unexpected based on earlier studies with enzymes and bacterial receptors.[47] It is quite possible, therefore, that small differences in catalytic site geometry, induced or stabilized by pathway-biased ligands, could have dramatic effects on the rate or distribution of tyrosine auto-phosphorylation or phosphorylation of docking and adapter proteins, and thereby engender signaling outcomes distinct from the natural activating ligand.

However, there is a paucity of information on the structural features connecting ligand binding and RTK dimerization to kinase domain activation. High resolution structural determinations of intact RTKs in lipid bilayers, which would facilitate understanding of the conformational changes that accompany ligand activation, are lacking. Nonetheless, it is known that RTK regions distal to the ligand binding pockets are important for kinase activation and initiation of signaling. For example, recent evidence implicating the juxtamembrane region (JM) in RTK activation comes from experiments in which the epidermal growth factor receptor (EGFR) JM was replaced with an unstructured linker. Receptor constructs containing the unstructured linker displayed normal EGF binding and ligand-induced dimerization; however, tyrosine phosphorylation in response to ligand binding was completely abolished, suggesting that the EGFR JM functions as a late-stage activation switch.[48]

In addition, dimerization studies of rat ErbB2 receptor (an EGFR family member) peptides containing the TM and JM regions indicate that dimerization, whether induced by attaching an exogenous dimerization domain or by introducing known oncogenic mutations, releases the juxtamembrane domain from a tight interaction with membrane lipids, thereby providing a potential mechanism for JM-mediated communication of ligand-(and mutation-) induced ErbB2 dimerization to receptor activation.[49] The unique mechanism of activation of the EGF family of RTKs makes interpretation of the results somewhat complicated. It remains to be determined whether the EGFR JM truly functions as a ligand-dependent kinase activation switch or whether other RTKs utilize the JM (or other cytoplasmic sequences) as an activation switch as well.

To date, there is limited precedent for ligands that produce pathway-biased signaling of RTKs. This is perhaps due to the overwhelming historical emphasis on the identification of ATP-competitive TKD inhibitors, primarily for use as chemotherapeutics.[50–52] Nevertheless, while the number of examples is limited, the pathway-biased RTK ligands that have been described provide an important proof-of-concept for this class of signaling proteins. The remainder of this chapter will take an in-depth look at several examples of pathway-biased activators and inhibitors of RTK signaling. The discovery and characterization of additional biased ligands will provide useful tools to dissect RTK signaling pathways, and potentially result in a new generation of RTK-directed medicines with improved specificity and side-effect profiles.[53,54]

BIASED LIGANDS OF RECEPTOR TYROSINE KINASES

Insulin Receptor

The insulin receptor, along with three related receptors, forms a small family of closely related receptor tyrosine kinases. The insulin receptor (IR) appears to be evolutionarily ancient with orthologs identified in primitive organisms such as cnidarians (sea anemones and jellyfish, for example) and in insects. In higher vertebrates, the IR is intimately involved in metabolic homeostasis and the control of blood glucose levels. The IR, along with the closely related insulin-like growth factor receptor 1 (IGF1R), also plays an important role in growth and development. A third receptor, the insulin-related receptor (IRR) does not bind and is not activated by insulin-like peptides; rather it has been proposed to be a renal alkali sensor involved in bicarbonate excretion.[55] The closely related IGF-2R lacks an intracellular tyrosine kinase domain and appears to be involved in the clearance of extracellular IGF by receptor-mediated endocytosis and lysosomal degradation.[56]

Unlike other RTKs, the IR and the IGF1R are present in the membrane as preformed disulfide-linked dimeric proteins consisting of two extracellular α-subunits that contain the ligand binding domain, and two intracellular β-subunits which contain the tyrosine kinase domain and binding domains (Figure 5.5). The α and β chains are synthesized as a single polypeptide that is proteolytically processed to produce the two subunits. Both the IR and IGF1R bind insulin, IGF-I and IGF-II, members of the insulin-like peptide family. Additional insulin-like peptides, the relaxins, serve as ligands for a family of seven transmembrane domain G protein-coupled receptors.[57] Even though a crystal structure of the insulin receptor

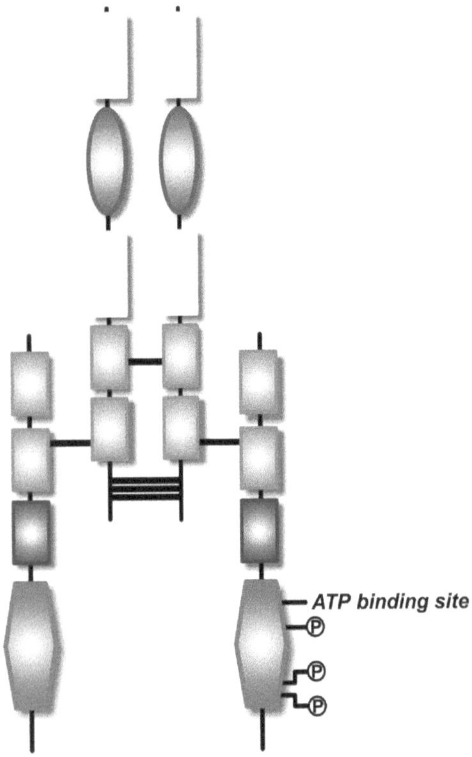

FIGURE 5.5 The modular architecture of the insulin receptor (IR). This figure shows a schematic of the dimeric ($\alpha2\beta2$) structure of the IR depicting the relative locations of the different protein modules. The heavy black horizontal lines represent the disulfide bonds joining the different subunits. The four extracellular leucine-rich repeats and two cysteine-rich domains are represented by boxes and ovals, respectively. The eight extracellular fibronectin type III repeats forming the disulfide linkages are also shown as boxes. The transmembrane and juxtamembrane domains are represented as the two boxes immediately below the fibronectin type III repeats; and the tyrosine kinase domains are shown as intracellular diamonds. The three regulatory loop phosphorylation sites discussed in the text, and the ATP binding site, are indicated.

extracellular domain bound to insulin has thus far proven elusive, it is presumed that ligand-dependent activation of the IR (and IGR1R) is initiated by a single molecule of insulin or IGF binding to the receptor with the peptide making contacts with sites on each half receptor.[58–60] Through a sequence of events that is not completely understood, peptide binding results in receptor activation and receptor auto-phosphorylation on cytoplasmic tyrosine residues. Although receptor conformational changes are undoubtedly involved in activation, the precise nature of these changes remains unknown.

As described above for the FGFR1, the insulin receptor is cis-inhibited by a series of specific protein contacts between the catalytic domain and the activation loop. The conformational change produced by insulin binding results in an initial trans-auto-phosphorylation of three tyrosine residues in the activation loop of the IR TKD and produces a 200-fold increase in receptor catalytic efficiency.[61,62] The auto-phosphorylated insulin receptor recruits the insulin receptor-specific substrate proteins, IRS1 and IRS2, and Shc,[63,64] which are in turn responsible for recruiting additional adapter proteins and orchestrating the pleiotropic cellular response to insulin, including glucose and lipid uptake, changes in gene expression and protein synthesis, and mitogenesis.[65,66]

Although they engage the same substrate proteins and overlapping sets of adapters, signaling through the IR and IGF1R has different physiological outcomes, with IGF1R signaling being essentially pro-mitogenic and IR signaling producing both cellular proliferation[67]

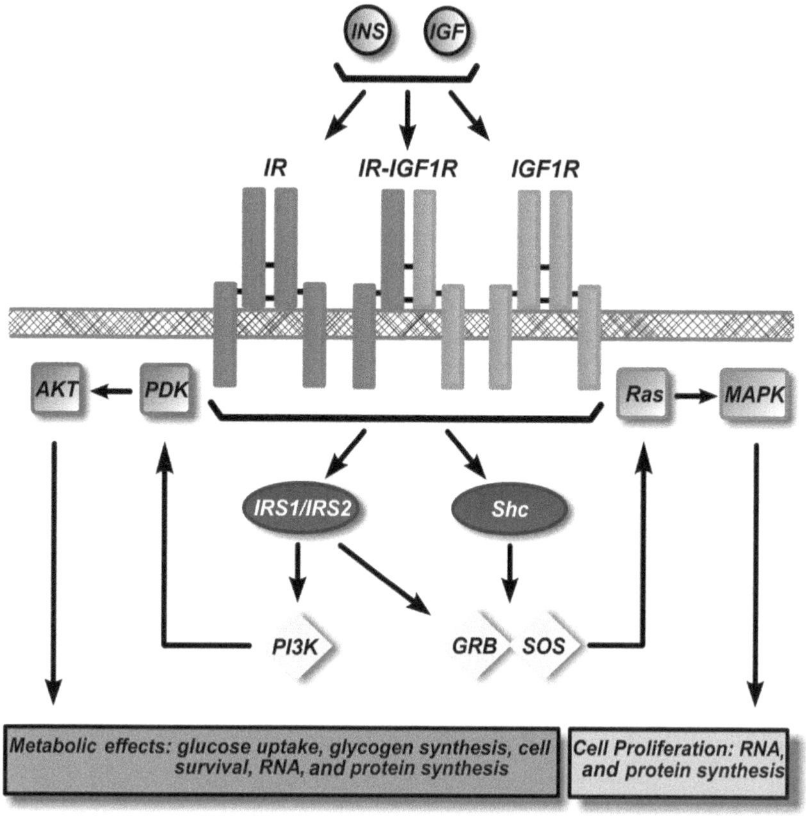

FIGURE 5.6 Simplified representation of IR signaling pathways. Binding of insulin (Ins) or IGF peptides to insulin and IGF receptor isoforms results in auto-phosphorylation and recruitment/phosphorylation of IRS1/2 and Shc receptor substrates. The substrates in turn recruit and activate additional adapter proteins (e.g. P13K, GRB, SOS) which transduce signals to metabolic (PDK and Akt) and mitogenic pathways (Ras and MAPK). GRB, growth factor receptor-bound protein; SOS, son of sevenless; MAPK, mitogen-activated protein kinase; PDK, phosphoinositide-dependent kinase-1; PI3K, phosphatidylinositide 3-kinase.

and a set of so-called metabolic effects, such as lipid and glucose uptake and glycogen synthesis (Figure 5.6). The picture is somewhat complicated by the presence of IR-IGF1R hybrid receptors,[68] as well as by other hybrid IR species, such as the recently described c-Met–IR hybrids in hepatocytes.[69] The determinants of the precise signaling outcome (metabolic vs mitogenic) is presumed to be due a combination of the expression level of various signaling components in different cell types as well as by the duration and kinetics of ligand–receptor interaction, receptor internalization, and the like.[70] For example, internalization of the IR appears to be required for phosphorylation of Shc and ERK1/2 (and therefore stimulation of cell proliferation) but not PI3K.[71] Mechanistically, it is also possible that specific conformations of the IR and IGFR produced by the different peptide ligands are responsible at least in part for signaling specificity (see Chapter 1).

Dysregulation of the insulin system results in a group of metabolic diseases known collectively as diabetes. Diabetes is characterized by hyperglycemia resulting from defects in insulin secretion, insulin action, or both. Chronic hyperglycemia resulting from diabetes is associated with long-term organ damage and failure, including the eyes, kidneys, nerves, and heart. It has been estimated that approximately 350 million individuals worldwide suffer from diabetes and that there are more than 3.4 million annual patient deaths due to consequences of uncontrolled hyperglycemia. Type I diabetes represents approximately 10% of diabetes cases and is caused by an autoimmune destruction of insulin-producing pancreatic beta cells. The majority of cases (type II diabetes) are characterized by a resistance to insulin action at one or more target tissues. Impairment of insulin secretion and defects in insulin action frequently coexist in the same patient.[72]

Existing therapy for type I diabetes consists solely of hormone administration, and insulin is sometimes prescribed in advanced cases of type II diabetes. Long-term administration of insulin is not without complications, however, including increased incidence of neoplasms.[73] There is some evidence that some of insulin's mitogenic effects are due to crosstalk of the IR and/or IGF1R with other RTK signaling networks.[74] The development of orally bioavailable small molecule insulin mimetics that can selectively and preferentially activate insulin's metabolic signaling pathways would be a welcome addition to the treatment options for these diseases. In support of this idea, it has been reported that patients with insulin-resistant type II diabetes, as well as obese non-diabetic patients, show diminished IRS1 phosphorylation and PI3K activation, but not MAPK/ERK activation, in response to an insulin challenge.[75]

Biased Signaling at the Insulin Receptor

The fungal extract demethylasterriquinone B-l, (DMAQ-Bl, L-783,281), was identified by screening insulin receptors overexpressed in CHO cells for compounds that promoted the phosphorylation of a synthetic peptide substrate.[76] DMAQ-B1 (Figure 5.7) appears to be a weak ($\approx 5\,\mu M$) direct activator of the IR tyrosine kinase domain. The compound directly stimulates IR auto-phosphorylation and IR-mediated phosphorylation of the IRS1 substrate protein.[77] DMAQ-B1 does not stimulate the catalytic activity of the IGF1R, EGFR, or platelet-derived growth factor receptor (PDGFR), indicating that it is relatively selective for the insulin receptor.[77] Furthermore, DMAQ-B1 produced a decrease in blood glucose levels following oral administration to *db/db* mice, and produced an increase in glucose tolerance and a decrease in plasma insulin levels following oral administration to *ob/ob* mice.[77]

Interestingly, in rat fibroblasts overexpressing the human IR, DMAQ-B1 is a more potent activator of Akt phosphorylation than MAPK/ERK1 phosphorylation,[78] indicating pathway bias for this compound. In support of this idea, insulin, but not DMAQ-B1, induces proliferation of vascular smooth muscle cells (VSMCs) obtained from diabetic patients. Both insulin and DMAQ-B1 were effective at stimulating glucose uptake by VSMCs.[79] DMAQ-B1 is a di-substituted quinone, which may limit its development as a chronic treatment for diabetes. Reportedly, compounds with replacements of DMAQ-B1's quinone moiety have been prepared but little information is available on their signaling properties.[80] A series of apparently unbiased 1,4 napthoquinone derivatives have been described that stimulate phosphorylation of both Akt and ERK in transfected cells,[81]

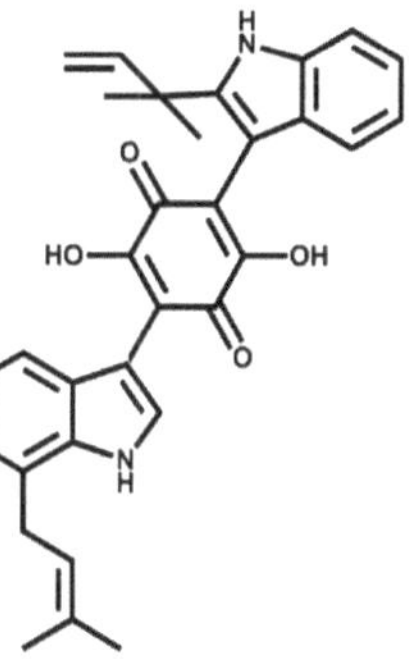

FIGURE 5.7 Chemical structure of DMAQ-B1, a biased small molecule activator of the insulin receptor.

suggesting that determinants of DMAQ-B1 signaling bias might be located in its indoyl substituents.

Biased insulin peptidomimetics have also been reported. As mentioned above, a single insulin molecule makes contacts with both half-receptors in an IR dimer. Optimized peptides containing these binding determinants have been synthesized and studied as homo- and heterodimers.[82] One of these peptides, S597 (Figure 5.8B) has been studied in some detail.[83,84] Stimulation of L6 rat myoblasts stably expressing the human insulin receptor with insulin or S597 resulted in auto-phosphorylation of the IR itself. Unlike, insulin, however, S597 did not stimulate auto-phosphorylation of IGF1R, indicating that it is specific for the insulin receptor. Interestingly, Western blot analysis of IR auto-phosphorylation in treated cells revealed that S597 produced roughly 50% of the phosphotyrosine levels induced by insulin (that is, S597 appears to be a partial agonist of IR auto-phosphorylation).

Immunoprecipitation experiments using both anti-IR and anti-phosphotyrosine antibodies suggested that the same number of insulin receptors were phosphorylated following stimulation with S597, but to a lesser degree. The degree of phosphorylation of IRS1/2 substrate proteins following S597 stimulation was similar to the level observed for IR auto-phosphorylation. Interestingly, S597 was a full agonist for Akt activation, with levels of phosphorylated Akt reaching 94% of the level produced by insulin. However, S597 was virtually inactive at stimulating ERK1/2 phosphorylation (≅20% of insulin). S597 was also able to stimulate glycogen synthesis (≅70% of insulin), but only weakly (≅30% of insulin) stimulated cell proliferation.[83] Therefore, S597 appears to be a biased agonist of the insulin receptor that possesses insulin's metabolic effects without its mitogenic activity.

As described above, ligand activation of many RTKs results in receptor dimerization and/or cell surface clustering followed by internalization of the receptor–ligand complexes.[32] It has been suggested that internalization of the IR in response to insulin stimulation is necessary for full activation of Shc but not PI3 kinase or IRS1/2.[70,71,85] Interestingly, S597 is much less efficacious in promoting internalization of the IR than is insulin, suggesting that differences in activation of the MAPK/ERK pathway may be due to differences in receptor internalization.[83] Since it appears that S597 promotes a lesser degree of receptor phosphorylation than insulin,[83] it is tempting to speculate that S597's different internalization and signaling properties are due in part to specific alterations in β-subunit phosphorylation. Indeed, it has been suggested that specific serine and threonine phosphorylation patterns on the cytoplasmic tails of 7-transmembrane G protein-coupled receptors play a pivotal role in ligand-biased signaling and in establishing a "signaling barcode" directing specific fates following GPCR internalization.[37,86–89] Nothing is yet known about the internalization of insulin receptors that have been stimulated with DMAQ-B1, and it would be interesting to determine if the biased di-substituted quinone displays reduced receptor internalization properties similar to S597.

A synthetic insulin peptidomimetic related to S597 displays different signaling properties, and an interesting, but complex, pharmacology.[90] Akin to S597, peptide S961 (Figure 5.8C) has been shown to be able to promote auto-phosphorylation of the IR and phosphorylation of Akt. At low concentrations (1–10 nM), S961 stimulated the growth of L6 cells expressing the human IR (L6-IR cells). At higher concentrations, however, S961 had no effect on the proliferation of L6-IR cells, and, at concentrations above 10 nM, was in fact an antagonist of insulin-stimulated proliferation.[90] S961 had no activity at any concentration on metabolic end points, including glycogen synthesis or lipogenesis, strongly suggesting that Akt phosphorylation is necessary but not sufficient for stimulation of insulin's metabolic effects. Therefore, S961 appears to be a biased insulin peptidomimetic as well, but one biased in favor of mitogenesis and cellular proliferation.

An interesting example of an apparently biased insulin agonist is the engineered insulin, insulin X10. Insulin is synthesized in the β-cells of the pancreas as preproinsulin. Upon cleavage of the signal sequence, the resulting molecule is transported to the Golgi where it is sequestered in secretory vesicles and assembled into hexameric insulin stabilized by Zn^{2+}.[91] Parenterally administered hexameric insulin is slowly adsorbed, which can create dosing problems for patients with type I diabetes. Insulin X10 was engineered to replace His10 in the insulin B-chain (Figure 5.8A) with aspartic acid, preventing self-assembly into multimeric forms and thereby leading to a more rapid onset of action. However, preclinical toxicity testing of insulin X10 revealed a dose-dependent increase in the incidence of mammary tumors in rats. Follow-up studies on the mechanism of action of the tumorigenicity revealed several interesting features of insulin X10 pharmacology and biology (reviewed by Hansen et al.[92]). While the exact mechanism is not fully understood, insulin X10 does display an approximately three-fold increase in binding affinity to the insulin receptor with a concomitant reduction in the rate of dissociation relative to insulin itself.[93,94] The decreased dissociation rate (i.e., prolonged association of insulin X10 and IR) is maintained following receptor internalization.[94]

Treatment of cells with insulin X10 results in different patterns of downstream effector phosphorylation/activation compared to insulin, including hyper-activation of PI3K and ERK1/2 pathways[95,96]. Interestingly, it has also been reported that insulin X10 induces a different pattern of IR tyrosine phosphorylation than does insulin, with insulin X10 favoring tyrosine kinase domain proximal sites.[92] The affinity of insulin X10 at the IGF1R, while increased relative to insulin, is still much lower than that of IGF itself.[93] Further, blockade of the IGF1R with an inhibitory antibody reduces the mitogenic effect of IGF by 70% in cultured breast carcinoma cells but only partially reduces the mitogenicity of insulin X10 (30% reduction), suggesting that most, if not all, of the insulin X10's mitogenic effects are mediated through the insulin receptor.[97] Nonetheless, it is still somewhat unclear what

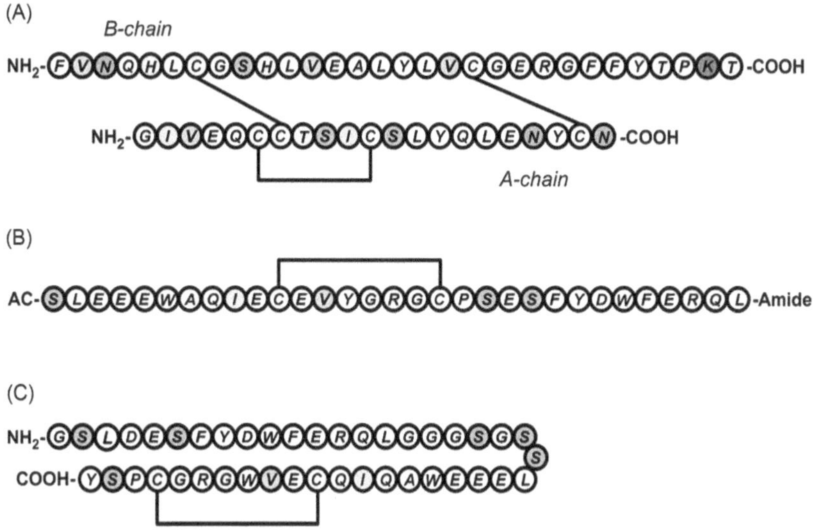

FIGURE 5.8 Primary amino acid sequences of insulin (A) and the biased insulin peptidomimetics S597 (B) and S961 (C). Positions of inter- and intramolecular disulfide bonds are shown as solid black lines.

impact the presence of different IR isoforms (A/B) or IR/IGFR hybrid receptors has on insulin X10 biology. Therefore, insulin X10 appears to be biased but in a way that emphasizes insulin's mitogenic effects rather than its metabolic ones. Again it is tempting to speculate that the observed signaling bias of insulin X10 may be related to a specific pattern of β-chain phosphorylation (i.e., an insulin receptor phosphorylation "barcode"), the precise subcellular localization and longevity of the internalized insulin X10:IR signaling complex, or both.

Neurotrophin Receptors

Neurotrophins were first identified nearly 60 years ago with the discovery of nerve growth factor (NGF) as a diffusible factor capable of promoting the growth and survival of motor and sympathetic neurons.[98–100] Over a quarter of a century would pass before the second neurotrophin, brain-derived neurotrophic factor (BDNF), was identified as a survival factor for neuronal populations that were not responsive to NGF. The two remaining mammalian neurotrophins, NT-3 and NT-4, were identified based on sequence homology to NGF and BDNF. Neurotrophins play essential roles in the development of the mammalian nervous system, maintenance of adult neuronal phenotypes, synaptic modulation and activity-dependent plasticity, memory, and injury protection and recovery.[101]

Neurotrophins are synthesized as larger pro-neurotrophin peptides and proteolytically processed to mature forms in the Golgi. They are secreted from neurons and target

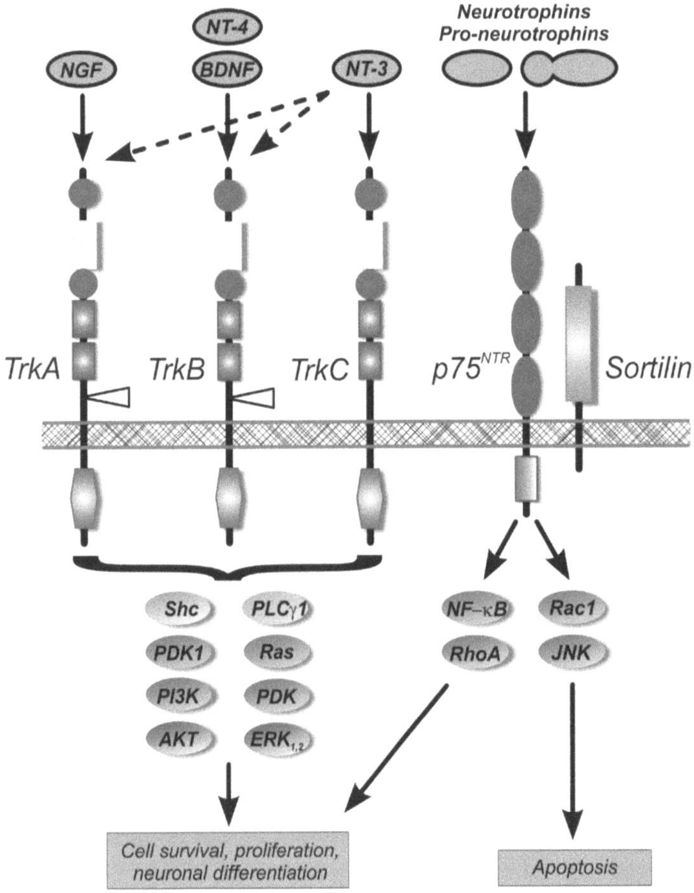

FIGURE 5.9 Schematic representation of the neurotrophin receptors and their signaling pathways. The position of a short insertion in the extracellular portion of the TrkA and TrkB receptors determining ligand specificity is indicated by the triangles. The TNF-family cytokine receptor $P75^{NTR}$ can associate with the transmembrane protein, sortilin, to form a high-affinity binding site for pro-neurotrophins. Activated (phosphorylated) Trk receptors in turn recruit the adapter proteins, PLCγ (phospholipase Cγ) and Shc, which activate a set of intracellular signaling molecules, some of which are shown in the ovals, leading to the biological response.

tissues as non-covalently linked homodimeric peptides and exert their influence by binding to and activating three receptor tyrosine kinases, the tropomyosin receptor kinases (Trks). There are three mammalian Trk receptors: TrkA, TrkB and TrkC. Structurally, the three TRK receptors share a common architecture consisting of N-terminal leucine- and cysteine-rich domains and IgG-like domains (Figure 5.9). These TRK receptors display high affinity (Kd $\approx 10^{-11}$ M) binding to their preferred neurotrophin ligands. TrkA receptors preferentially bind NGF, TrkB receptors bind BDNF and NT-4, and NT-3 preferentially activates the TrkC receptor.[102] Ligand preference of TrkA and TrkB can be influenced in part by the presence or absence of a short insertion in the extracellular juxtamembrane region. For example, TrkA or TrkB receptors lacking the juxtamembrane insertion are selectively activated by NGF or BDNF, respectively; receptor isoforms containing the insertion can additionally be activated by NT3 (TrkA)[103] or NT3 and NT4 (TrkB).[104]

In addition to the Trk receptors, there is a fourth neurotrophin receptor. The $p75^{NTR}$ receptor is a member of the tumor necrosis factor (TNF) cytokine receptor family and, like other members of the TNF receptor family, lacks a cytosolic kinase domain but instead has a so-called death domain[105] that mediates receptor signaling to intracellular transducers.[106,107] All four neurotrophins bind non-selectively to $p75^{NTR}$ with somewhat low ($\cong 10^{-8}$ M) affinity.[108] Pro-neurotrophins bind with high affinity ($\cong 10^{-10}$ M) to a cell surface complex composed of $p75^{NTR}$ and a vacuolar sorting protein called sortilin, a type 1 transmembrane protein.

Mechanisms controlling the release of non-processed neurotrophins are not well understood, but pro-neurotrophins can be released following neuronal damage or injury.[107] Binding of pro-neurotrophins to the $p75^{NTR}$–sortilin complex typically induces apoptotic cell death,[109,110] although pro-BDNF has also been found to increase hippocampal long-term depression via interactions with $p75^{NTR}$ and sortilin.[111] Co-expression of $p75^{NTR}$ and Trk receptors in a single neuron can modulate Trk receptor ligand preference, signaling properties, and functional potency.[106,107] The role of $p75^{NTR}$ in modulating Trk receptor activity is therefore complex and somewhat unpredictable. Depending upon the cellular context, activation of $p75^{NTR}$ and/or Trk receptors can produce synergistic, antagonistic, or independent responses resulting in cell survival, differentiation, or apoptosis.[112,113] Therefore, in any drug discovery strategy involving Trk receptor activation, identification of small molecule ligands that activate Trk receptors without stimulating $p75^{NTR}$ might be expected to have a therapeutic advantage over non-selective peptides or peptidomimetics.

Ligand Binding and Receptor Activation

Homodimeric neurotrophins bind to Trk receptors making contact only with a saddle formed by the C-terminal Ig-like repeat.[114–116] Trk receptor dimers appear to be stabilized by the neurotrophin dimer only, with no extensive contacts formed between the half-receptors. Peptide-induced receptor dimerization (or stabilization of preformed dimers) results in removal of pseudosubstrate inhibition of kinase activity and auto-phosphorylation of three tyrosine residues in the activation loop.[117,118] Phosphorylation of these three residues (Y676, Y680, Y681 in the human receptor, as per Genbank accession number P04629), most likely by cis-auto-phosphorylation,[119] is followed by trans-phosphorylation of additional tyrosine resides that serve as docking sites for adapter proteins, such as PLCγ and Shc, that mediate cellular pro-survival and pro-differentiation responses to neurotrophin exposure.[120–123]

Alternative splicing also generates Trk receptors harboring deletions of the kinase domain. These kinase deficient receptors can inhibit neurotrophin action by forming non-functional heterodimers with full-length receptor molecules. Alternative splicing of the TrkB gene has been shown to generate a receptor isoform lacking the cytoplasmic tyrosine kinase domain. The truncated TrkB receptor can dimerize with the full-length TrkB protein and act as an inhibitor of BDNF signaling; however, it has recently come to be appreciated that the truncated TrkB receptor can signal in its own right.[124] Finally, naturally occurring TrkC mutants harboring a short (14aa) insertion in the kinase domain have been described. These mutants display an altered response upon neurotrophin activation including decreased binding and phosphorylation of the TrkC substrates, PLCγ and Shc, and a decrease in NT-3-induced neuronal differentiation.[125]

Biased Agonism at the Tropomyosin Receptors

Due to their fundamental roles in the establishment and maintenance of nervous system organization and function, the neurotrophins are attractive candidates for therapeutic administration in a variety of human ailments. Neurotrophin therapy has been proposed for neurodegenerative diseases, such as Alzheimer's, Parkinson's, and Huntington's diseases,[126] as well as for psychiatric diseases such as depression.[127,128] In addition, it has been suggested that neurotrophin treatment has additional potential utility in the treatment of peripheral neuropathies caused by diabetes and HIV infection, and in the treatment of diabetic, vasculitic, and pressure ulcers.[129] Finally, it may be the case that neurotrophic factors are involved in a variety of non-neuronal processes including the development and maintenance of the vasculature and cardiac tissues,[130] immune cell function,[131,132] airway inflammatory disease (asthma),[133] and breast cancer.[134] To date, however, clinical trials of neurotrophin treatment have yielded disappointing results, most likely owing to a failure of peripherally administered neurotrophin to cross the blood–brain barrier, or from the short half-life these peptide display following administration.[126,129,135]

Considerable effort has been expended on identifying therapeutic molecules that do not suffer from the limitations of the neurotrophins themselves.[136–141] To serve as successful therapeutics, such synthetic activators should be characterized by pharmacokinetic and pharmacodynamic properties so as to permit sufficient target engagement following oral administration, including targets in the CNS. Synthetic activators able to engender biased signaling of Trk receptors might be expected to possess additional advantages over unbiased synthetic ligands. For example, it has long been appreciated that peripheral administration of NGF can result in increased pain sensation in humans.[142] NGF treatment of PC12 cells with NGF promotes cell survival under conditions of serum deprivation and induces a set of pro-survival and pro-differentiation genes, including the voltage-dependent sodium channel, $Na_v1.7$ (PN1) which confers electrical excitability on PC12 cells. $Na_v1.7$ gene expression is known to play a role in acute and inflammatory pain.[143] Interestingly, mutagenesis of Y791 of the TrkA receptor drastically reduces PLCγ1 binding to activated TrkA and the subsequent induction of $Na_v1.7$ gene expression.[144] The requirement of specific downstream signaling events on specific patterns of receptor auto-phosphorylation is reminiscent of the situation described above for the insulin receptor and G protein-coupled receptors.

It is not known whether biased activators of TrkA would be able to induce subtle differences in Trk receptor auto-phosphorylation. However, if such compounds can be identified, they could reasonably be expected to result in altered downstream signaling properties. Biased ligands that promoted neuronal survival without PLCγ stimulation of $Na_v1.7$ expression would conceivably have a lower incidence of on-target-induced hyperalgesia.[145,146] Also, since neurotrophin activation of Trk receptors can lead to cellular proliferation, there is the potential risk of Trk-mediated initiation or promotion of various cancers, including neuroblastoma, glioma, medulloblastoma, and prostate and breast cancer.[147] There is some reason to believe that it might be possible to separate the growth promoting/pro-survival activity of the neurotrophins from their pro-differentiation activity. For example, deletion of five conserved amino acids in the cytoplasmic portion of the juxtamembrane region of TrkA has been shown to remove NGF's ability to induce cell cycle arrest and cellular differentiation, but does not affect its ability to promote cellular survival.[148] It is therefore possible, at least in principle, to uncouple these two biological activities.

Perhaps the earliest indication of biased agonism at the tropomyosin receptors came from studies of TrkA activation by NGF and the heterologous neurotrophin, NT-3. NGF and NT-3 both bind to the TrkA receptor, but at spatially distinct sites. In primary cultures of NGF-dependent rat sympathetic neurons, activation of TrkA by the NGF or NT-3 produces different biological responses. NGF treatment of these neurons results in a dose-dependent increase in survival, neuritogenesis, cell body hypertrophy, and induction of Tα1 α-tubulin and $p75^{NTR}$ receptor mRNAs. In contrast, treatment of NGF-dependent rat sympathetic neurons with NT-3 only effectively induces neuritogenesis and induction of Tα1 mRNA; neuronal survival in the absence of NGF is only poorly mediated by NT-3.[149] The exact mechanism underlying this phenomenon is perhaps still not fully understood, but appears to involve both the time course of downstream target activation produced by the two neurotrophins, and, potentially, the ratio of TrkA to $p75^{NTR}$ receptors in the cell population. It is nevertheless clear that TrkA can be activated by different inducing signals and that these different inducers can elicit distinct sets of cellular responses.

A series of reports describing the synthesis and characterization of a series of cyclic peptidomimetics modeled on β-turn conformations of regions of NGF and NT-3 implicated in receptor binding both confirms and extends the initial observations of differential neurotrophin signaling at the TrkA receptor.[150,151] A set of cyclic peptides mimicking NGF and NT-3 β-turn regions were tested in three different assays for neurotrophic activity in cell lines expressing TrkA or TrkC receptors: receptor activation (auto-phosphorylation), survival (tropism) in serum-free medium (SFM), and neurite extension (differentiation). A few of the more interesting examples are shown in Figure 5.10A and their activities are summarized in Table 5.2. For example, the peptidomimetic 3Ac, constructed on a cyclic ether backbone, was able to stimulate TrkA and TrkC activity in all three assays. That is, 3Ac is an unbiased ligand at both receptors within the context of these experiments.

Peptidomimetics 3Ba (cyclic ether) and 1Ad (cyclic amine) appear to be unbiased TrkC ligands in that they induce TrkC auto-phosphorylation, promote survival of NIH3T3 cells expressing recombinant TrkC in SFM, and stimulate neurite outgrowth in nnr5-TrkC cells. Nnr5-TrkC cells are a variant of PC12 that have lost TrkA expression and that have been transfected with a human TrkC cDNA. However, while able to promote receptor auto-phosphorylation and induce neurite outgrowth in PC12 cells natively expressing TrkA and $p75^{NTR}$, these compounds were unable to promote survival of 3T3-TrkA cells in SFM. Peptidomimetics 3Ba and 1Ad are, in other words, biased TrkA ligands. It is interesting to note that these peptidomimetics recapitulate the results obtained with TrkA activation by NT-3, described above.[149]

The results obtained with a fourth peptidomimetic are interesting and illustrate how the complexity of the neurotrophin signaling system can make data interpretation difficult. Peptidomimetic 1Ba resembles the TrkA-biased synthetic peptides in that it induces differentiation of PC12 cells without an effect on cell survival. However, in preliminary auto-phosphorylation experiments, it also appeared that 1Ba was unable to activate receptor catalytic activity. More detailed time-course experiments revealed that while the other peptidomimetics could induce rapid and sustained (20 min) phosphorylation of TrkA, 1Ba induced rapid and transient ($<$10 min) phosphorylation. Transfection of the 3T3-TrkA cells

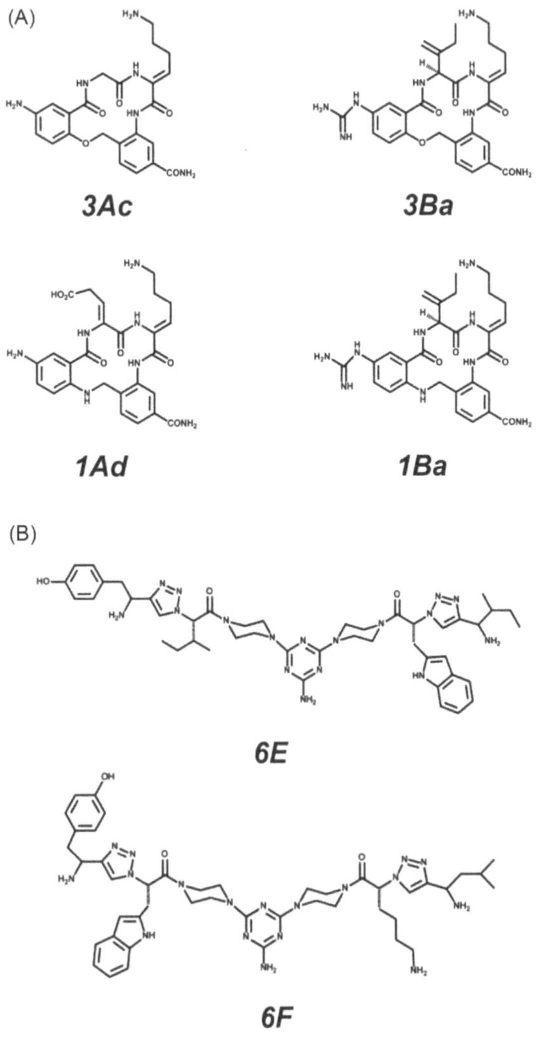

FIGURE 5.10 (A) Structures of monovalent Trk peptidomimetics. 3Ac is a non-selective and unbiased activator of TrkA and TrkC in all three assays: survival in serum-free medium (SFM); receptor auto-phosphorylation; neurite extension PC12 (TrkA+ / $p75^{NTR}$+) and nnr5 (TrkC+ / $p75^{NTR}$+) cells. Peptidomimetics 3Ba and 1Ad appear to be unbiased TrkC ligands; that is, they induce auto-phosphorylation, promote cell survival in SFM and induce neurite outgrowth in nnr5 cells. However, neither of these two peptidomimetics is able to support the survival of 3T3-TrkA cells in SFM, even though they induce auto-phosphorylation and differentiation. Mimetic 1Ba is unusual in that it promotes TrkA-mediated neurite extension without promoting survival. In auto-phosphorylation assays, 1Ba induces rapid and transient (10 min) tyrosine phosphorylation of TrkA. (B) Structures of asymmetric bivalent TrkC peptidomimetics. Molecule 6E is representative of five examples able to potentiate the trophic activity of NT-3 but had no effect on cellular differentiation either alone or in combination with NT-3. Molecule 6F is representative of a set of three examples lacking trophic activity but able to induce, as well as potentiate, NT-3-mediated cellular differentiation.

with a $p75^{NTR}$ cDNA was able to restore the sustained phosphorylation kinetics on the 1Ba-treated 3T3-TrkA cells. The native expression of $p75^{NTR}$ in PC12 cells was apparently sufficient to drive the 1Ba-induced differentiation. However, peptidomimetic 1Ba was ineffective at promoting survival of PC12 cells in SFM.

The mechanism by which $p75^{NTR}$ expression rescues kinetically long-lived TrkA auto-phosphorylation is unknown, but it does point out the underlying complexity of the neurotrophin signaling system. It is also clear that sustained activation of TrkA is in itself insufficient to confer cell survival upon serum deprivation. Finally, it is interesting to note that the peptidomimetics described above are monovalent and unlikely to dimerize. It therefore seems that Trk receptor activation can be mediated without the requirement for ligand stabilization of Trk dimers. It is possible that a fraction of Trk receptors exist in the

TABLE 5.2 Cyclic Peptidomimetic TRK Receptor Ligands (Described in the Text) were Tested in Three Assays for Neurotrophic Function: Receptor Auto-Phosphorylation; Cell Survival in Serum-Free Medium (Tropism); Neuronal Differentiation (Neurite Extension)

	Auto-phosphorylation		Tropism		Differentiation	
Compound	**3T3 TrkA**	**3T3-TrkC**	**3T3-TrkA**	**3T3-TrkC**	**PC12 (TrkA)**	**nnr5 (TrkC)**
3Ac	Sustained	Sustained	+	+	+	+
1Ad	Sustained	Sustained	–	+	+	+
3Ba	Sustained	Sustained	–	+	+	+
1Ba	Transient	Sustained	–	+	+	+

Data Summarized from Ref. 151

membrane as preformed dimers. Alternatively, it is possible that binding of a monovalent peptidomimetic to a single Trk monomer induces a conformational change that promotes dimerization with another, un-liganded, monomer. In a somewhat broader sense, these data are in agreement with recent findings on activation of the EGFR[152] and TrkB[153] receptors by monovalent peptides or small molecules.

Additional evidence for neurotrophin peptidomimetics with pathway-biased signaling activity comes from studies of a set of proteolytically-resistant, bivalent peptidomimetic compounds bearing naturally occurring amino acid side chains designed to mimic the β-turn conformations of NT-3 (Figure 5.10B).[154] Eight peptidomimetics were identified that bound selectively to TrkC. Five of these, represented by Compound 6E in Figure 5.10B, were shown to potentiate the trophic activity of a submaximal concentration of NT-3 in nnr5-TrkC cells (see above) in a survival assay following serum deprivation. However, Compound 6E, as well as the four other trophic compounds, was completely inactive in a neuronal differentiation assay (neurite extension). Three other peptidomimetics, represented by Compound 6F in Figure 5.10B, were able to induce cellular differentiation (neurite outgrowth) as well as synergistically potentiate differentiation induced by submaximal concentrations of NT-3. Compound 6F, and the other two differentiation-inducing peptidomimetics, were also able to activate TrkC and Akt, but not MAPK, as shown by tyrosine phosphorylation.[154] Akt activation is known to be important for neurotrophin-induced survival and differentiation, while MAPK activation is primarily associated with cell survival.[123] Unfortunately, the ability of Compound 6E and related peptidomimetics to induce phosphorylation of TrkC and downstream signaling intermediaries was not assessed.

Taken together these results are strongly suggestive of biased Trk signaling. Further, they are indicative that Trks, and by extension perhaps all RTKs, can adopt conformations able to promote one set of biological outcomes (neurite outgrowth) but not another (survival). However, given the complexity of the Trk signaling system and an almost bewildering number of Trk-interacting proteins,[123,155] it cannot be positively ruled out that at least some of the biased peptidomimetics described above mediate their effects by specific interactions with one or more of these Trk complexes. Further confirmation of the exact mechanism of action of these molecules should be aided by structural studies of receptor–ligand complexes.

Even though considerable progress has been made in recent years to improve the drug-like properties of peptidomimetic drugs (see, for example, Ref. 156), delivery of peptides and peptidomimetics to the CNS still poses substantial challenges.[157] Unlike native neurotrophins, or even synthetic peptidomimetics, small molecules are more likely to possess the proper pharmacokinetic and pharmacodynamic properties to allow sufficient CNS exposure following oral administration and may therefore prove efficacious in central as well as peripheral nervous system diseases. To date, there have been no definitive reports of non-peptidic, small molecule ligands of the neurotrophin receptors that display pathway-biased signaling. However, numerous examples of small molecule activators and modulators of the Trk receptors and $p75^{NTR}$ have been described.[137,139,140,158]

Recently, a series of non-peptidic small molecule TrkA agonists capable of mimicking NGF's survival and differentiation-inducing activity were reported by Scarpi et al.[159] These agonists could be distinguished from NGF on the basis of differential phosphorylation of specific TrkA tyrosine residues, raising the tantalizing possibility of phosphorylation-state sensitive biased signaling. In most cases, however, the small molecule Trk ligands thus far reported have been incompletely characterized and biased signaling produced by these molecules may therefore have gone unnoticed. This points to the need for a broad-based approach for biased ligand identification and characterization, using multiple readouts covering a range of pathway end points, preferably in primary or native expression systems. This is especially true for those allosteric small molecules displaying efficacy modulating properties, since pathway bias is a known feature of such compounds.[160]

BIASED SIGNALING BY RTK ANTAGONISTS

Growth factors and their receptor tyrosine kinases play fundamentally important roles in the growth and development of normal cells and tissues.[6,8–10] Growth factor receptor signaling is often dysregulated in human cancers due to inappropriate expression of the growth factor itself[14,161,162] or as a result of mutation or overexpression of the receptors.[162,163] It is not surprising therefore that identification and characterization of small molecule and biological inhibitors of growth factor receptor signaling for potential therapeutic utility in oncology is of keen interest to academic and pharmaceutical company researchers alike.[164,165] Adverse effects of RTK inhibitors are common[50] and it is possible that the identification of well-tolerated biased inhibitors of growth factor receptors may provide increased tolerability.

A Biased Allosteric Inhibitor of FGF Receptors

Perhaps the most convincing example of a biased RTK antagonist was identified in a traditional high throughput screen for inhibitors of the type 2 fibroblast growth factor receptor (FGFR). Fibroblast growth factors (FGFs) comprise a family of 18 secreted glycopeptides which exert their biological signaling effects via a family of four highly conserved FGFRs.[166,167] A fifth FGF receptor, FGFRL1, is conserved in all metazoans, lacks a

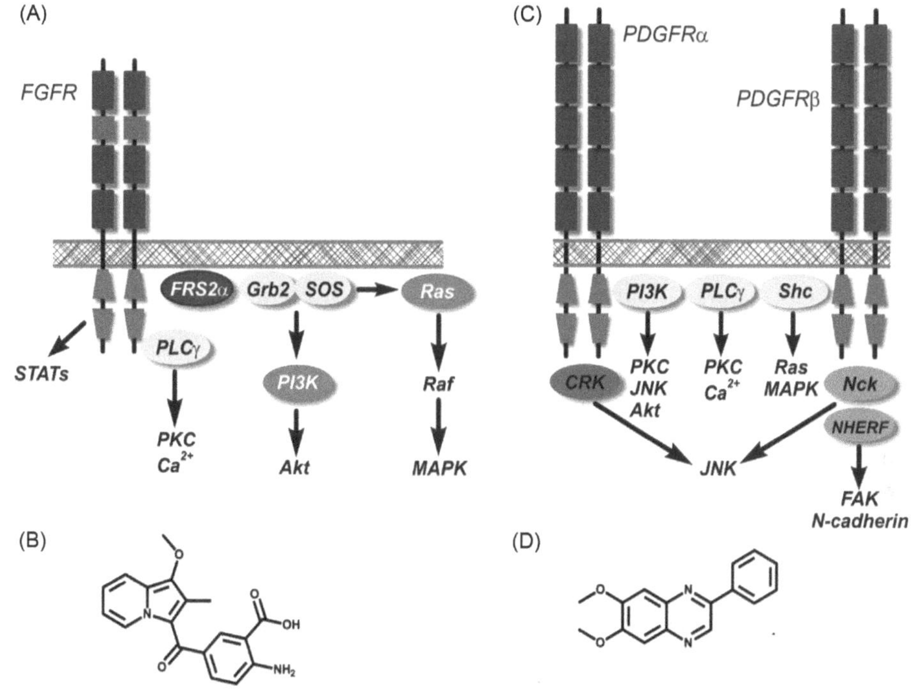

FIGURE 5.11 Biased antagonism of FGFR and PDGFR receptors. (A) Simplified representation of the FGF receptor structure and its signaling network. The three Ig-like repeats, D1–D3, are depicted as rectangles and the acid boxes as squares. Receptor activation results in the binding of substrate and adaptor proteins and in the activation of four key downstream pathways: Ras/Raf/MAPK pathway, PI3K/Akt pathway, STAT (signal transducer and activator of transcription) pathway, and PLCγ pathway. FRS2α, FGFR substrate 2α. (B) Chemical structure of the biased FGFR inhibitor, SSR128129E. (C) Simplified representation of the PDGFR structure and signaling networks. The five Ig-like repeats are depicted as rectangles; PDGFRα and PDGFRβ bind and activate a common set of adapters including PI3K, PLCγ, and Shc which in turn activate downstream signaling proteins such as Jnk, Akt, and MAPK (ERK1/2). However, each PDGF receptor also binds unique adapter proteins such as Crk (also known as p38), Nck, and the scaffold protein NHERF (Na^+/H^+ exchanger regulatory factor) resulting in individual signaling properties. (D) Chemical structure of the pan-PDGFR inhibitor, AG1296.

cytoplasmic tyrosine kinase domain, and appears to exert a negative effect on cell proliferation, but a positive effect on cell differentiation.[168] Like other RTKs, the FGFRs consist of a ligand binding extracellular domain and an intracellular tyrosine kinase domain. The extracellular portion of the FGF receptor consists of three immunoglobulin repeats, termed D1–D3, separated by a serine-rich motif known as an acid box (Figure 5.11A).

A feature unique to the FGF signaling system is the involvement of heparan sulfate proteoglycans (HSGP) in the binding, stabilization, and dimerization of the FGFR–FGF complex.[169] FGF ligands and HSGP bind to the D2 and D3 Ig repeats resulting in formation of a stable activated FGFR dimer. Receptor activation in turn results in trans-autophosphorylation of cytoplasmic tyrosines in the receptor activation loop, upon which additional tyrosines in the carboxy tail and the juxtamembrane loop are phosphorylated. The phosphorylated tyrosines in the juxtamembrane region and C-tail form binding sites for the FGFR substrate FRS2α and the signaling effector protein PLCγ1, respectively.[170,171] Binding and phosphorylation of these proteins to the activated FGFR leads to activation of MAPK, PI3K-Akt, and PKC-STAT signaling pathways resulting in increased cell survival, proliferation, migration, and differentiation. FGFs play a crucial role in embryonic morphogenesis and tissue patterning, wound healing, and the homeostatic regulation of bile acid, cholesterol, glucose, and vitamin D.[170,171] Dysregulation of the FGFR signaling system is implicated in human cancer via stimulation of tumor cell growth, angiogenesis, and migration, thereby driving interest in the identification of FGRF inhibitors for use as chemotherapeutics.[172]

The pan-FGFR antagonist SSR128129E (Figure 5.11B) was identified by high throughput screening for inhibitors of [^{125}I]FGF2 binding to the D1-D2-D3 ectodomain fragment of the FGFR1 receptor.[173] SSR128129E was almost 100 times more potent in cell-based assays (FGF2-stimulated endothelial cell proliferation, for example) than in the original binding assay (IC_{50}s of ≅30 nM and ≅2 μM, respectively. In radioligand binding assays employing full-length FGFR1, the apparent potency of SSR128129E was even further reduced such that SSR128129E was unable to inhibit FGF2 binding. This apparent disconnect between the radioligand binding and cell-based functional activities of SSR128129E is in fact characteristic of negative allosteric modulators that function primarily as efficacy, but not affinity, modulators.[174,175] SSR128129E displays other features characteristic of negative allosteric modulators including saturability of effect ("ceiling effect"). Significantly in the context of this chapter, SSR128129E exhibited a strong pathway bias.[160,175] Analysis of downstream signaling events in recombinant and native cells showed that SSR128129E was able to inhibit FGF-stimulated phosphorylation of the FGFR substrate protein, FRS2α, and its downstream target, MAPK; in contrast, phosphorylation of the other major FGFR substrate, PLCγ1, was not inhibited by SSR128129E.[173]

SSR128129E exerts its effects by binding to an extracellular site on fragments of FGFR1 containing the D2 and D3 Ig repeats, but not to fragments containing the FGF-binding domain, D2, suggesting that SSR128129E binds to a distinct site consisting of the D3 Ig repeat,[173] a hypothesis confirmed by NMR and crystallography studies.[176] Finally, it is interesting to note that in cell-based assays, SSR128129E was able to block signaling of human FGFR2-4, and of non-human FGFR paralogs, including zebra fish and *Drosophila*, indicating that the allosteric site targeted by SSR128129E has been well-conserved throughout evolution. Such evolutionary conservation strongly suggests that SSR128129E's allosteric binding site plays a fundamentally important role in regulation of FGFR activity since allosteric sites in other proteins are typically believed to not be subject to the same evolutionary constraints as orthosteric binding sites.[177] The combination of SSR128129E to function as a multi-FGFR inhibitor, coupled with a biased, allosteric mode of action, might result in a unique pharmacological profile with therapeutic utility in the treatment of FGFR-mediated disorders in proliferation, migration, or angiogenesis.[173]

Platelet-Derived Growth Factor Receptor Inhibition

A potentially biased inhibitor of platelet-derived growth factor receptor (PDGFR) activation has been reported.[178] The platelet-derived growth factors (PDGFs) are a family of four structurally-related peptides, PDGF-A–PDGF-D, that play major roles in late development, especially in blood vessel formation and organogenesis.[179–181] The four PDGF peptides form homodimers, and potentially one heterodimer, that activate two receptors, PDGFRα and PDGFRβ. The PDGF receptors belong to a small family of five related RTKs that includes Kit, Fms, and Flt3.[182] The PDGFR receptors share a conserved architecture consisting of an extracellular ligand binding portion composed of five Ig repeats (D1–D5), and an intracellular domain containing a split tyrosine kinase (Figure 5.11C).[182] The mechanisms for the regulation of PDGFR catalytic activity and its activation by PDGF ligands are still incompletely understood. It has been suggested than PDGF receptors are auto-inhibited in cis by their juxtamembrane regions and that dimerization and activation results in juxtamembrane tyrosine auto-phosphorylation, as has been observed in the related RTK, Kit.[182,183] Phosphorylation of additional tyrosine residues in the kinase linker region serves to create binding sites for specifically activated substrates and adapters, including PLCγ1, Shc, and PI3K.[184] Interestingly, the PDGFRα and the PDGFRβ are differentially regulated at the level of tyrosine phosphorylation and substrate/adapter recruitment.[185] While absolutely essential for embryonic development, up-regulation of PDGF signaling in adults is associated with various disease states, including atherosclerosis,[186] cancer,[187,188] and fibrosis.[186]

The PDGFR inhibitor, AG1296 (Figure 5.11D), was identified as an inhibitor of PDGF-BB-stimulated auto-phosphorylation of PDGF receptors endogenously expressed in Swiss 3T3 cell membranes.[189] AG1296 could inhibit PDGF-stimulated receptor auto-phosphorylation, DNA synthesis, and cell proliferation with similar potencies (0.5–3 μM). An investigation into the mechanism of action of AG1296 indicated that AG1296 neither inhibits PDGF binding nor receptor dimerization. Rather, AG1296 binds to the PDGFR kinase domain and displays a mostly mixed mode of competition with both ATP and a synthetic peptide containing one of the kinase linker region phosphorylation sites.[178] Phosphopeptide mapping of tryptic digests of partially purified, full-length PDGFRβ following PDGF stimulation suggested that AG1296 inhibited phosphorylation of tyrosine 851 (Y851) more effectively than it inhibited phosphorylation of Y751. No effect on the binding or activation of the PDGFR adapter protein, PI3K, could be detected, and effects on phosphorylation of other substrates and adapters recruited to the activated PDGFR were not reported. Nonetheless, given the differential inhibition by AG1296 on receptor auto-phosphorylation of different phosphorylation sites, it remains possible that AG1296 is in fact biased. As we have detailed in the above sections, phosphorylation of specific tyrosine residues is known to play a role in the control of downstream signaling. A more detailed examination of the effects of AG1296 on the pleiotropic signaling events mediated by PDGFs will be necessary to determine whether AG1296 is a biased antagonist of PDGFRβ signaling.

The paucity of RTK inhibitors displaying pathway-biased pharmacology is perhaps a reflection of the fact that most screening for kinase inhibitors has relied on the identification of competitive, active site inhibitors, often in biochemical assays employing isolated kinase domains.[190] We fully expect that as kinase screening strategies evolve to enable the detection of inhibitors acting at allosteric sites,[191] the identification of RTK inhibitors displaying the ability to selectively perturb some, but not all, of the signaling events activated upon ligand binding and dimerization will become more common.

CONCLUSION

Receptor tyrosine kinases are key orchestrators of virtually all aspects of cellular biology, from embryonic development to the differentiation and maintenance of the nervous system. Because of their essential roles in regulating cell growth and survival, dysregulation of RTK signaling systems is often correlated with human disease states, including proliferative disorders such as cancer, fibrosis, and neurodegenerative disease. The identification, optimization, and development of novel RTK activators and inhibitors for the treatment of human disease represent a significant portion of pharmaceutical and biotechnology drug discovery efforts.

The examples of biased RTK ligands described above indicate that pathway-bias does occur for this class of signaling proteins. These biased ligands are able to produce signaling states that are quite different from the endogenous peptide ligands, or, as in the case of the allosteric FGFR inhibitor SSR128129E, from the complete blockade of signaling that is produced by ligand or catalytic site competitive antagonists. It is further possible that the induction of these unique signaling states may offer insights into the complexities of RTK signaling, and ultimately result in a new generation of therapeutic drugs targeting RTKs. In several of the examples described above, the appearance of biased signaling was correlated with alterations in phosphorylation state, either of the RTK itself, or of one or more of its substrates or adapters. The regulatory importance of conformational changes in the TKD for determining catalytic activity is well known. And while not well understood, it seems reasonable to assume that ligand binding and receptor dimerization control kinase activation by way of inducing specific receptor conformations which are sensed by the TKD.

It is tempting to speculate further that biased activating ligands induce different receptor conformations than unbiased peptide activators such that differential phosphorylation of specific tyrosines occurs. In the case of differential phosphorylation of key regulatory RTK tyrosine residues, this could alter kinase preference for substrate and adapter phosphorylation sites, thereby increasing or decreasing the activity of certain signaling molecules and pathways. Alternatively, decreased phosphorylation of RTK tyrosines may remove docking sites, meaning that certain adapters are no longer recruited and activated. Subtle differences in the phosphorylation status of specific tyrosine residues in an RTK or in recruited docking, and adapter proteins produced by a pathway-biased ligand, may therefore have significant impacts on the nature, intensity, and duration of the evoked signal. Finally, the importance of specific phosphorylation events in orchestrating a given RTK signaling network is illustrated by the identification of more than 500 SH2 domain–phosphopeptide interactions (out of a possible 7600 interactions screened) in the insulin and FGFR signaling networks.[192] The importance of specific phosphorylation events in carboxy-tails of G protein-coupled receptors has been suggested to create a "signaling bar code" that determines not only the direction of signaling (i.e., ligand-directed bias) but subsequent internalization and desensitization events as well.

Biased ligands can have unpredictable effects on signal transduction systems, ranging from modest changes in ligand potency to dramatic shifts of ligand efficacy, depending on signaling responses considered. Even basic classifications of ligands as agonists, partial agonists, or inhibitors become assay and context dependent. For this reason, if the goal is to identify RTK activators and inhibitors with pathway bias, it becomes all the more important to move away from simple screening platforms that employ isolated TKDs and adopt more holistic approaches, including cell-based screening of full-length RTKs in a native context. The choice of cell background in this regard can be crucial. It would be desirable to use RTKs expressed at endogenous levels in the presence of a more or less native complement of docking proteins, adapters, and signal transducers.

Heterologous expression systems, while delivering high levels of expressed protein and perhaps more robust assays of RTK activation, can also alter the biological response and provide misleading results. For example, stimulation of the IR endogenously expressed in PC12 cells does not result in MAPK activation. Activation of an overexpressed IR activates MAPK and triggers cellular differentiation. Similarly, activation of the endogenous EGFR (ErbB1) receptor in PC12 cells results in transient MAPK activation, while activation of overexpressed EGFR produces sustained MAPK activation and the appearance of an EGF-dependent differentiated cell phenotype, including neurite outgrowth.[193] Therefore, even if primary screening cannot be carried out using native cell lines expressing endogenous levels of RTKs, it becomes imperative to profile active molecules in as wide a battery of assays as possible, both biochemical and cell-based signaling, in order to assess any apparent biased signaling. High-content screening methodologies coupled with high speed image analysis can be used to probe activation state and spatial organization of endogenously expressed unmodified signaling proteins in primary cells and cell lines.[194–196]

Alternatively, bioluminescence resonance energy transfer (BRET) assays can be employed to assess and quantify RTK conformational changes and RTK-adapter interactions in real-time in living cells.[197,198] BRET "toolboxes" consisting of labeled receptors and receptor interacting proteins have been constructed for GPCRs and their various signaling partners and have proven useful in the delineation of GPCR compound pharmacology.[199] Of course, care must be taken when comparing potency and efficacy readouts across a large number of assays to avoid the introduction of system or assay bias and to ensure that the data produced can be rigorously analyzed to detect true ligand-induced bias (see below).

The identification of non-competitive (allosteric) inhibitors of drug targets is recognized as increasingly important for a variety of target classes, including kinases,[200] enzymes such as proteases[201] and caspases,[202] nuclear hormone receptors,[203] and ion channels.[204] Since allosteric inhibitors do not bind to the same site as orthosteric (competitive) ligands, they do not necessarily induce and stabilize the same conformational states as their orthosteric counterparts and frequently result in pathway-biased signaling relative to the natural ligand or a truly competitive antagonist.[175] Such biased RTK antagonists would produce pathway bias by presumably allowing the receptor to adopt a "partially permissive" signaling conformation that allows the propagation of certain RTK-mediated signals but not others. For this reason, the identification of allosteric RTK inhibitors could lead to a rich source of pathway-biased ligands.

As our appreciation and understanding of signaling bias of RTKs grows, it will become more important to develop methods and conventions for quantifying biased agonism. In developing such conventions, it is important to bear in mind that the observed response of an agonist at a given signaling pathway reflects several factors: 1) "system bias," which reflects the differing coupling efficiencies of the receptor to a given signaling pathway; 2) "observational bias," which results from differing assay sensitivity and conditions; and 3) "true ligand bias", or the propensity of a given agonist to stabilize or induce particular receptor conformations associated with activation of different signaling pathways.[40] Both system bias and observational bias are independent of the agonist. Ultimately, ligand bias is the feature that can be chemically optimized, and thus is the feature that is relevant for drug discovery efforts. Therefore, biased agonism must be quantified in such a way as to eliminate as much as possible the contributions of system bias and observational bias, measuring only true ligand bias.

Additionally, it is generally recognized that the bias of a given agonist is only meaningful when compared with another agonist of the same receptor. Therefore, quantification of ligand bias must include a comparison to a reference ligand. Several potential approaches for quantifying biased agonism have been proposed. The "relative transduction ratio" is among the most widely accepted methods for quantifying ligand bias (for an in-depth review, see Ref. 40). This method is based on the operational model of agonism first described by Black and Leff[205] and takes into account both the potency (EC_{50}) and intrinsic activity (E_{max}) of each ligand for activating a given signaling pathway. Importantly, this quantification method eliminates contributions of system bias and observational bias, and quantifies ligand bias relative to a reference ligand (see Chapter 3). Because agonist-biased signaling has been much more extensively studied at GPCR targets than RTKs, most of the work on quantification of biased agonism has been done in the context of GPCRs. However, the approaches described should apply equally well to biased agonism at RTKs.

Acknowledgments

The authors are grateful to Adrian Saldanha for help in preparing the chemical structures.

References

1. Taylor SS, Kornev AP. Protein kinases: evolution of dynamic regulatory proteins. *Trends Biochem Sci* 2011;**36**(2):65–77.
2. Hunter T, Sefton BM. Transforming gene product of Rous sarcoma virus phosphorylates tyrosine. *Proc Natl Acad Sci USA* 1980;**77**(3):1311–5.
3. Eckhart W, Hutchinson MA, Hunter T. An activity phosphorylating tyrosine in polyoma T antigen immunoprecipitates. *Cell* 1979;**18**(4):925–33.
4. Manning G, Whyte DB, Martinez R, Hunter T, Sudarsanam S. The protein kinase complement of the human genome. *Science* 2002;**298**(5600):1912–34.
5. Lemmon MA, Schlessinger J. Cell signaling by receptor tyrosine kinases. *Cell* 2010;**141**(7):1117–34.
6. van der Geer P, Hunter T, Lindberg RA. Receptor protein-tyrosine kinases and their signal transduction pathways. *Annu Rev Cell Biol* 1994;**10**:251–337.
7. Manning G, Plowman GD, Hunter T, Sudarsanam S. Evolution of protein kinase signaling from yeast to man. *Trends Biochem Sci* 2002;**27**(10):514–20.
8. Lim WA, Pawson T. Phosphotyrosine signaling: evolving a new cellular communication system. *Cell* 2010;**142**(5):661–7.
9. Ullrich A, Schlessinger J. Signal transduction by receptors with tyrosine kinase activity. *Cell* 1990;**61**(2):203–12.
10. Blume-Jensen P, Hunter T. Oncogenic kinase signalling. *Nature* 2001;**411**(6835):355–65.
11. Benight NM, Waltz SE. Ron receptor tyrosine kinase signaling as a therapeutic target. *Expert Opin Ther Targets* 2012;**16**(9):921–31.

12. Ingram JL, Bonner JC. EGF and PDGF receptor tyrosine kinases as therapeutic targets for chronic lung diseases. *Curr Mol Med* 2006;**6**(4):409–21.
13. Lee YH, White MF. Insulin receptor substrate proteins and diabetes. *Arch Pharm Res* 2004;**27**(4):361–70.
14. Aaronson SA. Growth factors and cancer. *Science* 1991;**254**(5035):1146–53.
15. Huse M, Kuriyan J. The conformational plasticity of protein kinases. *Cell* 2002;**109**(3):275–82.
16. Kornev AP, Taylor SS. Defining the conserved internal architecture of a protein kinase. *Biochim Biophys Acta* 2010;**1804**(3):440–4.
17. Eswaran J, Knapp S. Insights into protein kinase regulation and inhibition by large scale structural comparison. *Biochim Biophys Acta* 2010;**1804**(3):429–32.
18. Kornev AP, Haste NM, Taylor SS, Eyck LF. Surface comparison of active and inactive protein kinases identifies a conserved activation mechanism. *Proc Natl Acad Sci USA* 2006;**103**(47):17783–8.
19. Nolen B, Taylor S, Ghosh G. Regulation of protein kinases: controlling activity through activation segment conformation. *Mol Cell* 2004;**15**(5):661–75.
20. Bae JH, Boggon TJ, Tomé F, Mandiyan V, Lax I, Schlessinger J. Asymmetric receptor contact is required for tyrosine autophosphorylation of fibroblast growth factor receptor in living cells. *Proc Natl Acad Sci USA* 2010;**107**(7):2866–71.
21. Chen H, Ma J, Li W, Eliseenkova AV, Xu C, Neubert TA, et al. A molecular brake in the kinase hinge region regulates the activity of receptor tyrosine kinases. *Mol Cell* 2007;**27**(5):717–30.
22. Niu X-L, Peters KG, Kontos CD. Deletion of the carboxyl terminus of Tie2 enhances kinase activity, signaling, and function: evidence for an autoinhibitory mechanism. *J Biol Chem* 2002;**277**(35):31768–73.
23. Hubbard SR. Juxtamembrane autoinhibition in receptor tyrosine kinases. *Nat Rev Mol Cell Biol* 2004;**5**(6):464–71.
24. Zhang X, Gureasko J, Shen K, Cole PA, Kuriyan J. An allosteric mechanism for activation of the kinase domain of epidermal growth factor receptor. *Cell* 2006;**125**(6):1137–49.
25. Furdui CM, Lew ED, Schlessinger J, Anderson KS. Autophosphorylation of FGFR1 kinase is mediated by a sequential and precisely ordered reaction. *Mol Cell* 2006;**21**(5):711–7.
26. Pawson T. Specificity in signal transduction: from phosphotyrosine-SH2 domain interactions to complex cellular systems. *Cell* 2004;**116**(2):191–203.
27. Wagner MJ, Stacey MM, Liu BA, Pawson T. Molecular mechanisms of SH2- and PTB-domain-containing proteins in receptor tyrosine kinase signaling. *Cold Spring Harb Perspect Biol* 2013;**5**(12):a008987.
28. Brummer T, Schmitz-Peiffer C, Daly RJ. Docking proteins. *FEBS J* 2010;**277**(21):4356–69.
29. Scott JD, Pawson T. Cell signaling in space and time: where proteins come together and when they're apart. *Science* 2009;**326**(5957):1220–4.
30. Oda K, Matsuoka Y, Funahashi A, Kitano H. A comprehensive pathway map of epidermal growth factor receptor signaling. *Mol Syst Biol* 2005;**1** 2005.0010.
31. Gesty-Palmer D, Luttrell LM. Refining efficacy: exploiting functional selectivity for drug discovery. *Adv Pharmacol* 2011;**62**:79–107.
32. Goh LK, Sorkin A. Endocytosis of receptor tyrosine kinases. *Cold Spring Harb Perspect Biol* 2013;**5**(5):a017459.
33. Ferguson SSG. Evolving concepts in G protein-coupled receptor endocytosis: the role in receptor desensitization and signaling. *Pharmacol Rev* 2001;**53**(1):1–24.
34. Ceresa BP, Schmid SL. Regulation of signal transduction by endocytosis. *Curr Opin Cell Biol* 2000;**12**(2):204–10.
35. Campenot RB, MacInnis BL. Retrograde transport of neurotrophins: fact and function. *J Neurobiol* 2004;**58**(2):217–29.
36. Sadowski L, Pilecka I, Miaczynska M. Signaling from endosomes: location makes a difference. *Exp Cell Res* 2009;**315**(9):1601–9.
37. Sorkin A, von Zastrow M. Endocytosis and signalling: intertwining molecular networks. *Nat Rev Mol Cell Biol* 2009;**10**(9):609–22.
38. von Zastrow M, Sorkin A. Signaling on the endocytic pathway. *Curr Opin Cell Biol* 2007;**19**(4):436–45.
39. DeWire SM, Violin JD. Biased ligands for better cardiovascular drugs: dissecting G-protein-coupled receptor pharmacology. *Circ Res* 2011;**109**(2):205–16.
40. Kenakin T, Christopoulos A. Signalling bias in new drug discovery: detection, quantification and therapeutic impact. *Nat Rev Drug Discov* 2013;**12**(3):205–16.

41. Kenakin T. New concepts in pharmacological efficacy at 7TM receptors: IUPHAR Review 2. *Br J Pharmacol* 2013;**168**(3):554–75.
42. Kenakin T. The potential for selective pharmacological therapies through biased receptor signaling. *BMC Pharmacol Toxicol* 2012;**13**:3.
43. Whalen EJ, Rajagopal S, Lefkowitz RJ. Therapeutic potential of β-arrestin- and G protein-biased agonists. *Trends Mol Med* 2011;**17**(3):126–39.
44. Wacker D, Wang C, Katritch V, Han GW, Huang XP, Vardy E, et al. Structural features for functional selectivity at serotonin receptors. *Science* 2013;**340**(6132):615–9.
45. Rahmeh R, Damian M, Cottet M, Orcel H, Mendre C, Durroux T, et al. Structural insights into biased G protein-coupled receptor signaling revealed by fluorescence spectroscopy. *Proc Natl Acad Sci USA* 2012;**109**(17):6733–8.
46. Liu JJ, Horst R, Katritch V, Stevens RC, Wuthrich K. Biased signaling pathways in beta2-adrenergic receptor characterized by 19F-NMR. *Science* 2012;**335**(6072):1106–10.
47. Koshland Jr. DE. Conformational changes: how small is big enough? *Nat Med* 1998;**4**(10):1112–4.
48. He L, Hristova K. Consequences of replacing EGFR juxtamembrane domain with an unstructured sequence. *Sci Rep* 2012;**2**:854.
49. Matsushita C, Tamagaki H, Miyazawa Y, Aimoto S, Smith SO, Sato T. Transmembrane helix orientation influences membrane binding of the intracellular juxtamembrane domain in Neu receptor peptides. *Proc Natl Acad Sci USA* 2013;**110**(5):1646–51.
50. Grant SK. Therapeutic protein kinase inhibitors. *Cell Mol Life Sci* 2009;**66**(7):1163–77.
51. Traxler P. Tyrosine kinases as targets in cancer therapy – successes and failures. *Expert Opin Ther Targets* 2003;**7**(2):215–34.
52. Noble MEM, Endicott JA, Johnson LN. Protein kinase inhibitors: insights into drug design from structure. *Science* 2004;**303**(5665):1800–5.
53. Cheng H, Force T. Why do kinase inhibitors cause cardiotoxicity and what can be done about it? *Prog Cardiovasc Dis* 2010;**53**(2):114–20.
54. Makita N, Iiri T. Tyrosine kinase inhibitor-induced thyroid disorders: a review and hypothesis. *Thyroid* 2013;**23**(2):151–9.
55. Deyev IE, Sohet F, Vassilenko KP, Serova OV, Popova NV, Zozulya SA, et al. Insulin receptor-related receptor as an extracellular alkali sensor. *Cell Metab* 2011;**13**(6):679–89.
56. Braulke T. Type-2 IGF receptor: a multi-ligand binding protein. *Horm Metab Res* 1999;**31**(2–3):242–6.
57. Bathgate RAD, Halls ML, van der Westhuizen ET, Callander GE, Kocan M, Summers RJ. Relaxin family peptides and their receptors. *Physiol Rev* 2013;**93**(1):405–80.
58. Schaffer L. A model for insulin binding to the insulin receptor. *Eur J Biochem* 1994;**221**(3):1127–32.
59. De Meyts P. The insulin receptor: a prototype for dimeric, allosteric membrane receptors? *Trends Biochem Sci* 2008;**33**(8):376–84.
60. De Meyts P, Whittaker J. Structural biology of insulin and IGF1 receptors: implications for drug design. *Nat Rev Drug Discov* 2002;**1**(10):769–83.
61. Favelyukis S, Till JH, Hubbard SR, Miller WT. Structure and autoregulation of the insulin-like growth factor 1 receptor kinase. *Nat Struct Biol* 2001;**8**(12):1058–63.
62. Cobb MH, Sang BC, Gonzalez R, Goldsmith E, Ellis L. Autophosphorylation activates the soluble cytoplasmic domain of the insulin receptor in an intermolecular reaction. *J Biol Chem* 1989;**264**(31):18701–6.
63. White MF. The insulin signalling system and the IRS proteins. *Diabetologia* 1997;**40(Suppl 2)**:S2–17.
64. Sasaoka T, Kobayashi M. The functional significance of Shc in insulin signaling as a substrate of the insulin receptor. *Endocr J* 2000;**47**(4):373–81.
65. Cohen P. The twentieth century struggle to decipher insulin signalling. *Nat Rev Mol Cell BIol* 2006;**7**(11):867–73.
66. Siddle K. Signalling by insulin and IGF receptors: supporting acts and new players. *J Mol Endocrinol* 2011;**47**(1):R1–10.
67. Ish-Shalom D, Christoffersen CT, Vorwerk P, Sacerdoti-Sierra N, Shymko RM, Naor D, et al. Mitogenic properties of insulin and insulin analogues mediated by the insulin receptor. *Diabetologia* 1997;**40(Suppl 2)**:S25–31.
68. Belfiore A, Frasca F, Pandini G, Sciacca L, Vigneri R. Insulin receptor isoforms and insulin receptor/insulin-like growth factor receptor hybrids in physiology and disease. *Endocr Rev* 2009;**30**(6):586–623.

69. Fafalios A, Ma J, Tan X, Stoops J, Luo J, Defrances MC, et al. A hepatocyte growth factor receptor (Met)-insulin receptor hybrid governs hepatic glucose metabolism. *Nat Med* 2011;**17**(12):1577–84.
70. Siddle K. Molecular basis of signaling specificity of insulin and IGF receptors: neglected corners and recent advances. *Front Endocrinol* 2012;**3**:34.
71. Leof EB. Growth factor receptor signalling: location, location, location. *Trends Cell Biol* 2000;**10**(8):343–8.
72. American Diabetes Association. Diagnosis and classification of diabetes mellitus. *Diabetes Care* 2011;**34**(**Suppl** 1):S62–9.
73. Hemkens LG, Grouven U, Bender R, Günster C, Gutschmidt S, Selke GW, et al. Risk of malignancies in patients with diabetes treated with human insulin or insulin analogues: a cohort study. *Diabetologia* 2009;**52**(9):1732–44.
74. Borisov N, Aksamitiene E, Kiyatkin A, Legewie S, Berkhout J, Maiwald T, et al. Systems-level interactions between insulin-EGF networks amplify mitogenic signaling. *Mol Syst Biol* 2009;**5**:256.
75. Cusi K, Maezono K, Osman A, Pendergrass M, Patti ME, Pratipanawatr T, et al. Insulin resistance differentially affects the PI 3-kinase- and MAP kinase-mediated signaling in human muscle. *J Clin Invest* 2000;**105** (3):311–20.
76. Salituro GM, Pelaez F, Zhang BB. Discovery of a small molecule insulin receptor activator. *Recent Prog Horm Res* 2001;**56**:107–26.
77. Zhang B, Salituro G, Szalkowski D, Li Z, Zhang Y, Royo I, et al. Discovery of a small molecule insulin mimetic with antidiabetic activity in mice. *Science* 1999;**284**(5416):974–7.
78. Webster NJG, Park K, Pirrung MC. Signaling effects of demethylasterriquinone B1, a selective insulin receptor modulator. *ChemBioChem* 2003;**4**(5):379–85.
79. Weber MA, Lidor A, Arora S, Salituro GM, Zhang BB, Sidawy AN. A novel insulin mimetic without a proliferative effect on vascular smooth muscle cells. *J Vasc Surg* 2000;**32**(6):1118–26.
80. Nankar RP, Doble M. Non-peptidyl insulin mimetics as a potential antidiabetic agent. *Drug Discov Today* 2013;**18**(15–16):748–55.
81. He K, Chan C-B, Liu X, Jia Y, Luo HR, France SA, et al. Identification of a molecular activator for insulin receptor with potent anti-diabetic effects. *J Biol Chem* 2011;**286**(43):37379–88.
82. Schäffer L, Brissette RE, Spetzler JC, Pillutla RC, Østergaard S, Lennick M, et al. Assembly of high-affinity insulin receptor agonists and antagonists from peptide building blocks. *Proc Natl Acad Sci USA* 2003;**100** (8):4435–9.
83. Jensen M, Hansen B, De Meyts P, Schaffer L, Urso B. Activation of the insulin receptor by insulin and a synthetic peptide leads to divergent metabolic and mitogenic signaling and responses. *J Biol Chem* 2007;**282**(48):35179–86.
84. Jensen M, Palsgaard J, Borup R, de Meyts P, Schaffer L. Activation of the insulin receptor (IR) by insulin and a synthetic peptide has different effects on gene expression in IR-transfected L6 myoblasts. *Biochem J* 2008;**412** (3):435–45.
85. Foti M, Ahmed Moukil M, Dudognon P, Carpentier J-L. Insulin and IGF-1 receptor trafficking and signalling. *Novartis Found Symp* 2004;**262**:125–47.
86. Butcher AJ, Prihandoko R, Kong KC, McWilliams P, Edwards JM, Bottrill A, et al. Differential G-protein-coupled receptor phosphorylation provides evidence for a signaling bar code. *J Biol Chem* 2011;**286**(13): 11506–18.
87. Gregory KJ, Hall NE, Tobin AB, Sexton PM, Christopoulos A. Identification of orthosteric and allosteric site mutations in M2 muscarinic acetylcholine receptors that contribute to ligand-selective signaling bias. *J Biol Chem* 2010;**285**(10):7459–74.
88. Tobin AB, Butcher AJ, Kong KC. Location, location, location...site-specific GPCR phosphorylation offers a mechanism for cell-type-specific signalling. *Trends Pharmacol Sci* 2008;**29**(8):413–20.
89. Nobles KN, Xiao K, Ahn S, Shukla AK, Lam CM, Rajagopal S, et al. Distinct phosphorylation sites on the β2-adrenergic receptor establish a barcode that encodes differential functions of β-arrestin. *Sci Signal* 2011;**4**(185): ra51.
90. Knudsen L, Hansen BF, Jensen P, Pedersen TÅ, Vestergaard K, Schäffer L, et al. Agonism and antagonism at the insulin receptor. *PLoS ONE* 2012;**7**(12):e51972.
91. Dunn M. Zinc–ligand interactions modulate assembly and stability of the insulin hexamer – a review. *Biometals* 2005;**18**(4):295–303.
92. Hansen BF, Kurtzhals P, Jensen AB, Dejgaard A, Russell-Jones D. Insulin X10 revisited: a super-mitogenic insulin analogue. *Diabetologia* 2011;**54**(9):2226–31.

93. Drejer K, Kruse V, Dahl Larsen U, Hougaard P, Bjørn S, Gammeltoft S. Receptor binding and tyrosine kinase activation by insulin analogues with extreme affinities studied in human hepatoma HepG2 cells. *Diabetes* 1991;**40**(11):1488–95.
94. Hamel FG, Siford GL, Fawcett J, Chance RE, Frank BH, Duckworth WC. Differences in the cellular processing of AspB10 human insulin compared with human insulin and LysB28ProB29 human insulin. *Metabolism* 1999;**48**(5):611–7.
95. Sciacca L, Cassarino MF, Genua M, Pandini G, Moli R, Squatrito S, et al. Insulin analogues differently activate insulin receptor isoforms and post-receptor signalling. *Diabetologia* 2010;**53**(8):1743–53.
96. Oleksiewicz MB, Bonnesen C, Hegelund AC, Lundby A, Holm G-MN, Jensen MB, et al. Comparison of intracellular signalling by insulin and the hypermitogenic AspB10 analogue in MCF-7 breast adenocarcinoma cells. *J Appl Toxicol* 2011;**31**(4):329–41.
97. Milazzo G, Sciacca L, Papa V, Goldfine ID, Vigneri R. ASPB10 insulin induction of increased mitogenic responses and phenotypic changes in human breast epithelial cells: evidence for enhanced interactions with the insulin-like growth factor-I receptor. *Mol Carcinog* 1997;**18**(1):19–25.
98. Levi-Montalcini R. The saga of the nerve growth factor. *Neuroreport* 1998;**9**(16):R71–83.
99. Cowan WM. Viktor Hamburger and Rita Levi-Montalcini: the path to the discovery of nerve growth factor. *Annu Rev Neurosci* 2001;**24**(1):551–600.
100. Shooter EM. Early days of the nerve growth factor proteins. *Annu Rev Neurosci* 2001;**24**(1):601–29.
101. Minichiello L. TrkB signalling pathways in LTP and learning. *Nat Rev Neurosci* 2009;**10**(12):850–60.
102. Barbacid M. Neurotrophic factors and their receptors. *Curr Opin Cell Biol* 1995;**7**(2):148–55.
103. Clary DO, Reichardt LF. An alternatively spliced form of the nerve growth factor receptor TrkA confers an enhanced response to neurotrophin 3. *Proc Natl Acad Sci USA* 1994;**91**(23):11133–7.
104. Strohmaier C, Carter BD, Urfer R, Barde YA, Dechant G. A splice variant of the neurotrophin receptor trkB with increased specificity for brain-derived neurotrophic factor. *EMBO J* 1996;**15**(13):3332–7.
105. Liepinsh E, Ilag LL, Otting G, Ibanez CF. NMR structure of the death domain of the p75 neurotrophin receptor. *EMBO J* 1997;**16**(16):4999–5005.
106. Dechant G, Barde YA. The neurotrophin receptor p75(NTR): novel functions and implications for diseases of the nervous system. *Nat Neurosci* 2002;**5**(11):1131–6.
107. Ibáñez CF, Simi A. p75 neurotrophin receptor signaling in nervous system injury and degeneration: paradox and opportunity. *Trends Neurosci* 2012;**35**(7):431–40.
108. Bothwell M. Functional interactions of neurotrophins and neurotrophin receptors. *Annu Rev Neurosci* 1995;**18**:223–53.
109. Teng HK, Teng KK, Lee R, Wright S, Tevar S, Almeida RD, et al. ProBDNF induces neuronal apoptosis via activation of a receptor complex of NTR and sortilin. *J Neurosci* 2005;**25**(22):5455–63.
110. Nykjaer A, Lee R, Teng KK, Jansen P, Madsen P, Nielsen MS, et al. Sortilin is essential for proNGF-induced neuronal cell death. *Nature* 2004;**427**(6977):843–8.
111. Woo NH, Teng HK, Siao CJ, Chiaruttini C, Pang PT, Milner TA, et al. Activation of NTR by proBDNF facilitates hippocampal long-term depression. *Nat Neurosci* 2005;**8**(8):1069–77.
112. Bandtlow C, Dechant G. From cell death to neuronal regeneration, effects of the p75 neurotrophin receptor depend on interactions with partner subunits. *Sci STKE* 2004;**2004**(235):pe24.
113. Lu B, Pang PT, Woo NH. The yin and yang of neurotrophin action. *Nat Rev Neurosci* 2005;**6**(8):603–14.
114. Ultsch MH, Wiesmann C, Simmons LC, Henrich J, Yang M, Reilly D, et al. Crystal structures of the neurotrophin-binding domain of TrkA, TrkB and TrkC. *J Mol Biol* 1999;**290**(1):149–59.
115. Wehrman T, He X, Raab B, Dukipatti A, Blau H, Garcia KC. Structural and mechanistic insights into nerve growth factor interactions with the TrkA and p75 receptors. *Neuron* 2007;**53**(1):25–38.
116. Wiesmann C, Ultsch MH, Bass SH, de Vos AM. Crystal structure of nerve growth factor in complex with the ligand-binding domain of the TrkA receptor. *Nature* 1999;**401**(6749):184–8.
117. Middlemas DS, Meisenhelder J, Hunter T. Identification of TrkB autophosphorylation sites and evidence that phospholipase C-gamma 1 is a substrate of the TrkB receptor. *J Biol Chem* 1994;**269**(7):5458–66.
118. Bertrand T, Kothe M, Liu J, Dupuy A, Rak A, Berne PF, et al. The crystal structures of TrkA and TrkB suggest key regions for achieving selective inhibition. *J Mol Biol* 2012;**423**(3):439–53.
119. Iwasaki Y, Nishiyama H, Suzuki K, Koizumi S. Sequential cis/trans autophosphorylation in TrkB tyrosine kinase. *Biochemistry* 1997;**36**(9):2694–700.

120. Stephens RM, Loeb DM, Copeland TD, Pawson T, Greene LA, Kaplan DR. Trk receptors use redundant signal transduction pathways involving SHC and PLC-β1 to mediate NGF responses. *Neuron* 1994;**12**(3):691–705.
121. Alsina FC, Ledda F, Paratcha G. New insights into the control of neurotrophic growth factor receptor signaling: Implications for nervous system development and repair. *J Neurochem* 2012;**123**(5):652–61.
122. Chao MV. Neurotrophins and their receptors: a convergence point for many signalling pathways. *Nat Rev Neurosci* 2003;**4**(4):299–309.
123. Reichardt LF. Neurotrophin-regulated signalling pathways. *Philos Trans R Soc Lond B Biol Sci* 2006;**361**(1473):1545–64.
124. Fenner BM. Truncated TrkB: beyond a dominant negative receptor. *Cytokine Growth Factor Rev* 2012;**23**(1–2):15–24.
125. Guiton M, Gunn-Moore FJ, Glass DJ, Geis DR, Yancopoulos GD, Tavaré JM. Naturally occurring tyrosine kinase inserts block high affinity binding of phospholipase C and Shc to TrkC and neurotrophin-3 signaling. *J Biol Chem* 1995;**270**(35):20384–90.
126. Allen SJ, Watson JJ, Shoemark DK, Barua NU, Patel NK. GDNF, NGF and BDNF as therapeutic options for neurodegeneration. *Pharmacol Ther* 2013;**138**(2):155–75.
127. Rantamäki T, Castrén E. Targeting TrkB neurotrophin receptor to treat depression. *Expert Opin Ther Targets* 2008;**12**(6):705–15.
128. Kozisek ME, Middlemas D, Bylund DB. Brain-derived neurotrophic factor and its receptor tropomyosin-related kinase B in the mechanism of action of antidepressant therapies. *Pharmacol Ther* 2008;**117**(1):30–51.
129. Aloe L, Rocco ML, Bianchi P, Manni L. Nerve growth factor: from the early discoveries to the potential clinical use. *J Transl Med* 2012;**10**:239.
130. Caporali A, Emanueli C. Cardiovascular actions of neurotrophins. *Physiol Rev* 2009;**89**(1):279–308.
131. Vega JA, Garcia-Suarez O, Hannestad J, Perez-Perez M, Germana A. Neurotrophins and the immune system. *J Anat* 2003;**203**(1):1–19.
132. Scuri M, Samsell L, Piedimonte G. The role of neurotrophins in inflammation and allergy. *Inflamm Allergy Drug Targets* 2010;**9**(3):173–80.
133. Freund-Michel V, Frossard N. The nerve growth factor and its receptors in airway inflammatory diseases. *Pharmacol Ther* 2008;**117**(1):52–76.
134. Hondermarck H. Neurotrophins and their receptors in breast cancer. *Cytokine Growth Factor Rev* 2012;**23**(6):357–65.
135. Mocchetti I, Brown M. Targeting neurotrophin receptors in the central nervous system. *CNS Neurol Disord Drug Targets* 2008;**7**(1):71–82.
136. Beglova N, Maliartchouk S, Ekiel I, Zaccaro MC, Saragovi HU, Gehring K. Design and solution structure of functional peptide mimetics of nerve growth factor. *J Med Chem* 2000;**43**(19):3530–40.
137. Longo FM, Massa SM. Small-molecule modulation of neurotrophin receptors: a strategy for the treatment of neurological disease. *Nat Rev Drug Discov* 2013;**12**(7):507–25.
138. Skaper SD. Peptide mimetics of neurotrophins and their receptors. *Curr Pharm Des* 2011;**17**(25):2704–18.
139. Obianyo O, Ye K. Novel small molecule activators of the Trk family of receptor tyrosine kinases. *Biochim Biophys Acta* 2013;**1834**(10):2213–8.
140. Webster NJ, Pirrung MC. Small molecule activators of the Trk receptors for neuroprotection. *BMC Neurosci* 2008;**9(Suppl 2)**:S1.
141. Saragovi HU, Zaccaro MC. Small molecule peptidomimetic ligands of neurotrophin receptors, identifying binding sites, activation sites and regulatory sites. *Curr Pharm Des* 2002;**8**(24):2201–16.
142. Dyck PJ, Peroutka S, Rask C, Burton E, Baker MK, Lehman KA, et al. Intradermal recombinant human nerve growth factor induces pressure allodynia and lowered heat-pain threshold in humans. *Neurology* 1997;**48**(2):501–5.
143. Nassar MA, Stirling LC, Forlani G, Baker MD, Matthews EA, Dickenson AH, et al. Nociceptor-specific gene deletion reveals a major role for Nav1.7 (PN1) in acute and inflammatory pain. *Proc Natl Acad Sci USA* 2004;**101**(34):12706–11.
144. Choi D-Y, Toledo-Aral JJ, Segal R, Halegoua S. Sustained signaling by phospholipase C-γ mediates nerve growth factor-triggered gene expression. *Mol Cell Biol* 2001;**21**(8):2695–705.
145. Shu X-Q, Mendell LM. Neurotrophins and hyperalgesia. *Proc Natl Acad Sci USA* 1999;**96**(14):7693–6.

146. Siniscalco D, Giordano C, Rossi F, Maione S, de Novellis V. Role of neurotrophins in neuropathic pain. *Curr Neuropharmacol* 2011;**9**(4):523–9.
147. Krüttgen A, Schneider I, Weis J. The Dark Side of the NGF Family: Neurotrophins in Neoplasias. *Brain Pathol* 2006;**16**(4):304–10.
148. Meakin SO, MacDonald IS. A novel juxtamembrane deletion in rat TrkA blocks differentiative but not mitogenic cell signaling in response to nerve growth factor. *J Neurochem* 1998;**71**(5):1875–88.
149. Belliveau DJ, Krivko I, Kohn J, Lachance C, Pozniak C, Rusakov D, et al. NGF and neurotrophin-3 both activate TrkA on sympathetic neurons but differentially regulate survival and neuritogenesis. *J Cell Biol* 1997;**136**(2):375–88.
150. Peleshok J, Saragovi HU. Functional mimetics of neurotrophins and their receptors. *Biochem Soc Trans* 2006;**34**(Pt 4):612–7.
151. Zaccaro MC, Lee HB, Pattarawarapan M, Xia Z, Caron A, L'Heureux P-J, et al. Selective small molecule peptidomimetic ligands of TrkC and TrkA receptors afford discrete or complete neurotrophic activities. *Chem Biol* 2005;**12**(9):1015–28.
152. Liu P, Cleveland TE, Bouyain S, Byrne PO, Longo PA, Leahy DJ. A single ligand is sufficient to activate EGFR dimers. *Proc Natl Acad Sci USA* 2012;**109**(27):10861–6.
153. Massa SM, Yang T, Xie Y, Shi J, Bilgen M, Joyce JN, et al. Small molecule BDNF mimetics activate TrkB signaling and prevent neuronal degeneration in rodents. *J Clin Invest* 2010;**120**(5):1774–85.
154. Chen D, Brahimi F, Angell Y, Li Y-C, Moscowicz J, Saragovi HU, et al. Bivalent peptidomimetic ligands of TrkC are biased agonists and selectively induce neuritogenesis or potentiate neurotrophin-3 trophic signals. *ACS Chem Biol* 2009;**4**(9):769–81.
155. Schecterson LC, Bothwell M. Neurotrophin receptors: Old friends with new partners. *Dev Neurobiol* 2010;**70**(5):332–8.
156. Bellmann-Sickert K, Beck-Sickinger AG. Peptide drugs to target G protein-coupled receptors. *Trends Pharmacol Sci* 2010;**31**(9):434–41.
157. Patel MM, Goyal BR, Bhadada SV, Bhatt JS, Amin AF. Getting into the brain: approaches to enhance brain drug delivery. *CNS Drugs* 2009;**23**(1):35–58.
158. Price RD, Milne SA, Sharkey J, Matsuoka N. Advances in small molecules promoting neurotrophic function. *Pharmacol Ther* 2007;**115**(2):292–306.
159. Scarpi D, Cirelli D, Matrone C, Castronovo G, Rosini P, Occhiato EG, et al. Low molecular weight, non-peptidic agonists of TrkA receptor with NGF-mimetic activity. *Cell Death Dis* 2012;**3**:e339.
160. Urban JD, Clarke WP, von Zastrow M, Nichols DE, Kobilka B, Weinstein H, et al. Functional selectivity and classical concepts of quantitative pharmacology. *J Pharmacol Exp Ther* 2007;**320**(1):1–13.
161. Witsch E, Sela M, Yarden Y. Roles for growth factors in cancer progression. *Physiology (Bethesda)* 2010;**25**(2):85–101.
162. Goustin AS, Leof EB, Shipley GD, Moses HL. Growth factors and cancer. *Cancer Res* 1986;**46**(3):1015–29.
163. Zhang X, Nie D, Chakrabarty S. Growth factors in tumor microenvironment. *Front Biosci (Landmark Ed)* 2010;**15**:151–65.
164. Takeuchi K, Ito F. Receptor tyrosine kinases and targeted cancer therapeutics. *Biol Pharm Bull* 2011;**34**(12):1774–80.
165. Arora A, Scholar EM. Role of tyrosine kinase inhibitors in cancer therapy. *J Pharmacol Exp Ther* 2005;**315**(3):971–9.
166. Itoh N, Ornitz DM. Fibroblast growth factors: from molecular evolution to roles in development, metabolism and disease. *J Biochem* 2011;**149**(2):121–30.
167. Beenken A, Mohammadi M. The FGF family: biology, pathophysiology and therapy. *Nat Rev Drug Discov* 2009;**8**(3):235–53.
168. Trueb B. Biology of FGFRL1, the fifth fibroblast growth factor receptor. *Cell Mol Life Sci* 2011;**68**(6):951–64.
169. Schlessinger J, Plotnikov AN, Ibrahimi OA, Eliseenkova AV, Yeh BK, Yayon A, et al. Crystal structure of a ternary FGF–FGFR–heparin complex reveals a dual role for heparin in FGFR binding and dimerization. *Mol Cell* 2000;**6**(3):743–50.
170. Eswarakumar VP, Lax I, Schlessinger J. Cellular signaling by fibroblast growth factor receptors. *Cytokine Growth Factor Rev* 2005;**16**(2):139–49.

171. Turner N, Grose R. Fibroblast growth factor signalling: from development to cancer. *Nat Rev Cancer* 2010;**10**(2):116–29.
172. Daniele G, Corral J, Molife LR, de Bono JS. FGF receptor inhibitors: role in cancer therapy. *Curr Oncol Rep* 2012;**14**(2):111–9.
173. Bono F, De Smet F, Herbert C, De Bock K, Georgiadou M, Fons P, et al. Inhibition of tumor angiogenesis and growth by a small-molecule multi-FGF receptor blocker with allosteric properties. *Cancer Cell* 2013;**23**(4):477–88.
174. Ehlert FJ. Estimation of the affinities of allosteric ligands using radioligand binding and pharmacological null methods. *Mol Pharmacol* 1988;**33**(2):187–94.
175. Leach K, Sexton PM, Christopoulos A. Allosteric GPCR modulators: taking advantage of permissive receptor pharmacology. *Trends Pharmacol Sci* 2007;**28**(8):382–9.
176. Herbert C, Schieborr U, Saxena K, Juraszek J, De Smet F, Alcouffe C, et al. Molecular mechanism of SSR128129E, an extracellularly acting, small-molecule, allosteric inhibitor of FGF receptor signaling. *Cancer Cell* 2013;**23**(4):489–501.
177. May LT, Leach K, Sexton PM, Christopoulos A. Allosteric modulation of G protein-coupled receptors. *Ann Rev Pharmacol Toxicol* 2007;**47**:1–51.
178. Kovalenko M, Ronnstrand L, Heldin CH, Loubtchenkov M, Gazit A, Levitzki A, et al. Phosphorylation site-specific inhibition of platelet-derived growth factor beta-receptor autophosphorylation by the receptor blocking tyrphostin AG1296. *Biochemistry* 1997;**36**(21):6260–9.
179. Betsholtz C. Biology of platelet-derived growth factors in development. *Birth Defects Res C Embryo Today* 2003;**69**(4):272–85.
180. Fredriksson L, Li H, Eriksson U. The PDGF family: four gene products form five dimeric isoforms. *Cytokine Growth Factor Rev* 2004;**15**(4):197–204.
181. Andrae J, Gallini R, Betsholtz C. Role of platelet-derived growth factors in physiology and medicine. *Genes Dev* 2008;**22**(10):1276–312.
182. Chen P-H, Chen X, He X. Platelet-derived growth factors and their receptors: Structural and functional perspectives. *Biochim Biophys Acta* 2013;**1834**(10):2176–86.
183. Mol CD, Lim KB, Sridhar V, Zou H, Chien EY, Sang BC, et al. Structure of a c-kit product complex reveals the basis for kinase transactivation. *J Biol Chem* 2003;**278**(34):31461–4.
184. Tallquist M, Kazlauskas A. PDGF signaling in cells and mice. *Cytokine Growth Factor Rev* 2004;**15**(4):205–13.
185. Rosenkranz S, Kazlauskas A. Evidence for distinct signaling properties and biological responses induced by the PDGF receptor α and β subtypes. *Growth Factors* 1999;**16**(3):201–16.
186. Raines EW. PDGF and cardiovascular disease. *Cytokine Growth Factor Rev* 2004;**15**(4):237–54.
187. Board R, Jayson GC. Platelet-derived growth factor receptor (PDGFR): A target for anticancer therapeutics. *Drug Resist Updat* 2005;**8**(1–2):75–83.
188. Dai Y. Platelet-derived growth factor receptor tyrosine kinase inhibitors: a review of the recent patent literature. *Expert Opin Ther Pat* 2010;**20**(7):885–97.
189. Kovalenko M, Gazit A, Böhmer A, Rorsman C, Rönnstrand L, Heldin C-H, et al. Selective platelet-derived growth factor receptor kinase blockers reverse sis-transformation. *Cancer Res* 1994;**54**(23):6106–14.
190. von Ahsen O, Bomer U. High-throughput screening for kinase inhibitors. *ChemBioChem* 2005;**6**(3):481–90.
191. Lebakken CS, Reichling LJ, Ellefson JM, Riddle SM. Detection of Allosteric Kinase Inhibitors by Displacement of Active Site Probes. *J Biomol Screen* 2012;**17**(6):813–21.
192. Liu BA, Engelmann BW, Jablonowski K, Higginbotham K, Stergachis AB, Nash PD. SRC homology 2 domain binding sites in insulin, IGF-1 and FGF receptor mediated signaling networks reveal an extensive potential interactome. *Cell Commun Signal* 2012;**10**(1):27.
193. Marshall CJ. Specificity of receptor tyrosine kinase signaling: transient versus sustained extracellular signal-regulated kinase activation. *Cell* 1995;**80**(2):179–85.
194. Ghosh RN, Grove L, Lapets O. A quantitative cell-based high-content screening assay for the epidermal growth factor receptor-specific activation of mitogen-activated protein kinase. *Assay Drug Devel Technol* 2004;**2**(5):473–81.
195. Mayr LM, Bojanic D. Novel trends in high-throughput screening. *Curr Opin Pharmacol* 2009;**9**(5):580–8.
196. Wang J, Xie X. Development of a quantitative, cell-based, high-content screening assay for epidermal growth factor receptor modulators. *Acta Pharmacol Sin* 2007;**28**(10):1698–704.

197. Siddiqui S, Cong WN, Daimon CM, Martin B, Maudsley S. BRET biosensor analysis of receptor tyrosine kinase functionality. *Front Endocrinol* 2013;**4**:46.
198. Tan PK, Wang J, Littler P-LH, Wong KK, Sweetnam TA, Keefe W, et al. Monitoring interactions between receptor tyrosine kinases and their downstream effector proteins in living cells using bioluminescence resonance energy transfer. *Mol Pharmacol* 2007;**72**(6):1440–6.
199. Schann S, Bouvier M, Neuville P. Technology combination to address GPCR allosteric modulator drug-discovery pitfalls. *Drug Discov Today Technol* 2013;**10**(2):e261–7.
200. Fang Z, Grutter C, Rauh D. Strategies for the selective regulation of kinases with allosteric modulators: exploiting exclusive structural features. *ACS Chem Biol* 2013;**8**(1):58–70.
201. Merdanovic M, Monig T, Ehrmann M, Kaiser M. Diversity of allosteric regulation in proteases. *ACS Chem Biol* 2013;**8**(1):19–26.
202. Häcker H-G, Sisay MT, Gütschow M. Allosteric modulation of caspases. *Pharmacol Ther* 2011;**132**(2):180–95.
203. Huang P, Chandra V, Rastinejad F. Structural overview of the nuclear receptor superfamily: Insights into physiology and therapeutics. *Annu Rev Physiol* 2010;**72**:247–72.
204. Dilly S, Lamy C, Marrion NV, Liegeois JF, Seutin V. Ion-channel modulators: more diversity than previously thought. *ChemBioChem* 2011;**12**(12):1808–12.
205. Black JW, Leff P. Operational models of pharmacological agonism. *Proc R Soc Lond B Biol Sci* 1983;**220**(1219):141–62.

CHAPTER

6

Conformational Mechanisms of Signaling Bias of Ion Channels

James Herrington[1] *and Brian J. Arey*[2]

[1]Department of Exploratory Biology and Genomics, Research and Development, Bristol-Myers Squibb Co., Wallingford, CT, USA, [2]Department of Cardiovascular Drug Discovery Biology, Research and Development, Bristol-Myers Squibb Co., Hopewell, NJ, USA

OUTLINE

B. Arey (Ed): *Biased Signaling in Physiology, Pharmacology and Therapeutics*
DOI: http://dx.doi.org/10.1016/B978-0-12-411460-9.00006-9

INTRODUCTION

Cells are extremely complex entities that are in constant communication with their environment in order to adapt and respond to the frequent changes in their surroundings. Evolution has selected several mechanisms for cells to perceive their environment, the most notable of which are receptors, but other mechanisms also exist, including the presence of cell surface ion channels. Ion channels are evolutionarily ancient and may represent one of the earliest forms of intracellular signaling, present in both prokaryotes and eukaryotes. Ion channels are membrane proteins that open and close in response to external stimuli, allowing passage of ions across the cell membrane, and are key regulators of the voltage across the cell membrane. Since their primary function is to regulate the passage of hydrophilic ions across the amphipathic cell membrane, ion channels are transmembrane proteins comprised of subunits with discreet functional domains that are characterized by their hydrophilic/lipophilic properties.[1]

Ion channels are a diverse and physiologically important class of membrane proteins that are significantly represented in the genomes of both plants and animals. Analysis of the sequence of the human genome has revealed the existence of approximately 230 putative genes encoding ion channels.[2] These proteins have a rich pharmacological history since many of the natural remedies used throughout prehistory and ancient times have been found to be directed toward this class of proteins. Indeed, the first receptor identified by Jean-Pierre Changeux was the nicotinic acetylcholine receptor. Through detailed studies of naturally occurring and synthetic compounds, we now recognize the existence of molecules that span the gamut of pharmacological activities: those that open channels, others that close them, some that block the ion conduction pore, and yet others that more subtly regulate their behavior (modulators).

There is also considerable diversity in the types of molecules that interact with channels, ranging from various metal ions to large proteins. The origin of these molecules is similarly diverse and includes tropical plants and the venoms of various predatory animals. Physiologically, channels control diverse processes such as neurotransmitter release, muscle contraction, and hormone secretion, among others. For this reason, ion channels are the targets of many therapeutic drugs. Estimates suggest that 5–10% of existing drugs have their therapeutic effect by targeting ion channels.[3,4] Interest in ion channels as drug targets is growing as new physiological roles and intracellular signaling pathways for channels are discovered and technological advances have enabled the routine measurement of channel activity with sufficient throughput.[5,6] The critical physiological importance of channels is also demonstrated by the diseases that arise from their mutation. These monogenic inherited diseases caused by the mutation of a single gene encoding an ion channel are referred to as channelopathies. To date, more than 60 channelopathies have been described.[7] Some of these diseases have revealed novel therapeutic approaches to treatment.

Ion channels can be thought of as molecular machines, catalyzing the movement of up to 10^8 ions per second across membranes in a highly regulated and specific manner. In addition, ion channels transition from open to closed very rapidly (in microseconds). Over the last 60 years, our knowledge of the processes of gating and ion movement through the channel has grown considerably. One inherent requirement for the ability to open and close so rapidly is the existence of multiple conformational states of the channel protein. Thus, ion channels are considered allosteric receptors and, in fact, have been instrumental in understanding the mechanism of allosteric ligand binding to proteins.[8] Further, given the wealth of pharmacological data on these proteins, we now have a much better understanding of how binding of one molecule to a channel can influence the binding of another (synergy), and how function is affected. In this way, channels are classical examples of how biased signaling occurs, often with important physiological consequences and potentially therapeutic implications. In this chapter, we will focus on the nature of allosterism as it relates to ion channels and provide examples of how channels can signal in a biased fashion. We will also highlight future opportunities for exploiting conformationally-dependent ligand–channel interactions for therapeutic benefit. As a starting point, a review of ion channel types, their basic properties, and their conformational dynamics will provide a framework for the discussion.

ION CHANNEL OVERVIEW

All ion channels have a few common features: a narrow ion conduction pathway, a water-filled pore cavity, and a gate (Figure 6.1). The ion conduction pathway, or selectivity filter, regulates which ions can pass through the channel. The water-filled pore is the passageway for ions through the membrane. The gate controls passage of ions through the pore and opens and closes through conformational changes in the channel protein in response to stimuli. In general, channels fall into two broad classes: those gated by ligands (the ligand-gated family, LGICs) and those gated by membrane voltage (the voltage-gated family).

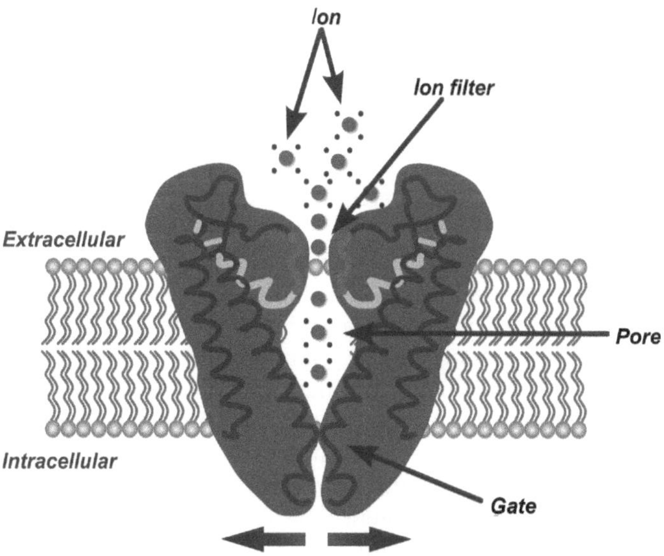

FIGURE 6.1 Schematic diagram of the basic structural elements of an ion channel. A schematic diagram of the major features of a prototypical ion channel including the ion selectivity filter, the water-filled pore, and the channel gate. Ions are depicted as circles moving through the filter in a single-file manner. Water molecules are represented by black dots. The illustration is based upon the structure of the bacterial potassium channel, KcsA.[193]

Ligand-gated Ion Channel Superfamily

Ligand-gated ion channels bind neurotransmitters and open in response to ligand binding. These channels control synaptic transmission between two neurons or between a neuron and a muscle. One subfamily encompasses the Cys-loop channels, so named because of a large extracellular domain containing Cys loops.[9,10] Members of this family include the channels that bind acetylcholine (the nicotinic acetylcholine receptor), GABA (the $GABA_A$ receptor), the 5HT3 receptor, and glycine receptors. Five subunits assemble to form the functional pentameric channel. Most are composed of two alpha subunits and three other subunits (beta, gamma, or delta) but some consist of five alpha subunits. Each subunit contains a large amino-terminal extracellular domain with a characteristic disulfide bond formed by a pair of cysteine residues (the Cys loop) and a transmembrane region formed by four helical segments (M1–M4). The channel pore is formed primarily by one of the helical segments (M4) from two of the subunits. The orthosteric ligand binding site is formed by the extracellular domain at the interface between subunits.

Another class of ligand-gated channel encompasses the tetrameric receptors for glutamate.[11] Glutamate is the major neurotransmitter in the mammalian brain and is largely responsible for excitatory synaptic transmission. The ionotropic glutamate receptor has three major structural domains: a large amino-terminal extracellular domain (ATD), an extracellular ligand binding domain (LBD), and a transmembrane region consisting of two helical segments per subunit. Binding of glutamate causes the "clamshell"- like LBD to close, resulting in opening of the pore. Several subtypes of ionotropic glutamate receptors exist, and can be broadly segregated based on their pharmacological sensitivity to glutamate-like molecules such as NMDA, kainate, and AMPA. As is typical for ion channel pharmacology, the distinct profile of these synthetic agonists in neuronal preparations foreshadowed the molecular identification of the three subfamilies of ionotropic glutamate receptor.

Yet another class of ligand-gated channel is characterized by a trimeric subunit arrangement. Each subunit has a large extracellular loop and two transmembrane segments. Members of this family include the acid-sensing ion channels (ASICs) that are gated by protons.[12,13] P2X channels, which are gated by extracellular ATP, are another example.[14,15]

Voltage-gated Ion Channel Superfamily

The other large superfamily includes those channels gated by membrane voltage.[16] This family includes the voltage-gated sodium (Na^+), calcium (Ca^{2+}) and potassium (K^+) channel subfamilies. The basic architectural unit is six transmembrane segments (S1–S6). The six transmembrane units can be further subdivided into a voltage-sensing domain (S1–S4) and a pore domain (S5–S6). Voltage-sensing arises from a series of positively charged residues (arginines and lysines) with the S4 segment. For the voltage-gated potassium (K_V) channels, each subunit has six transmembrane segments and the functional unit is a tetramer. The voltage-gated Na^+ and Ca^{2+} channels are large polypeptides where each effective "subunit" is linked together in a single protein. Also included in this family are the transient-receptor potential (TRP) channels.[17] TRP channels are weakly voltage-gated due to the reduced number of positive charges in the S4 segment. The TRP channels, however, are gated by a variety of molecules and often by temperature.

Other Ion Channel Families

In addition to the ligand-gated and voltage-gated superfamilies, there are other channels that don't fit easily into either superfamily. For example, many channels are not specifically gated by a known external stimulus but rather undergo opening (and closing) transitions without any apparent stimulus. An example includes the inwardly rectifying potassium channels. These channels essentially lack the voltage-sensor domain (S1–S4 segments) of voltage-gated channels and possess the pore domain composed of two transmembrane segments. Like K_V channels, each subunit has a single pore domain and the basic functional unit is a tetramer. The inwardly rectifying channel also possesses a large intracellular domain structure that gives rise to unique features. Another example is the twin-pore K^+ channels, with two pore domains linked together into a single polypeptide and the functional unit is a dimer.[18] Although these channels can exist as constitutively open, they are capable of being gated by diverse external stimuli.

Another channel family is the amiloride-sensing sodium channel of epithelial cells (ENaC). This channel is involved in sodium transport across epithelia. Three subunits are known to exist and the functional unit is thought be a heteromeric trimer and thus may belong to the ASIC/P2X family of channels. Each subunit has an intracellular N-terminus region followed by a transmembrane domain, a large extracellular loop, a second transmembrane segment, and a C-terminal intracellular tail. ENaC channels are related to the degenerins of *Caenorhabditis elegans.*

Other Classification Schemes

Channels can also be characterized by other properties in addition to their gating mechanism. One such property is their ion selectivity; that is, which ions do they allow to pass through their open pore and which do they exclude. For example, some ion channels are highly selective for cations over anions (or vice versa). In addition, cationic selective channels come in different degrees of selectivity for various cations; for example, there are non-selective cation channels which don't readily discriminate between, for example, sodium and potassium. An example of this type is the nicotinic acetylcholine receptor channel.

At the other extreme, some cation channels are exquisitely selective for one cation over another. For example, voltage-gated calcium channels are 1000-fold selective for Ca^{2+}over Na^+. This degree of selectivity is required from a physiological standpoint since in the extracellular space sodium is found at nearly 100-fold higher concentrations compared with Ca^{2+}. How does the channel do this given that a sodium ion has an ionic radius (0.9 Å) that is actually smaller than Ca^{2+} (0.99 Å)? The answer lies in specific binding sites in the pore of the channel that exclude the smaller Na^+ ion (reviewed in Hille[1]). Another distinguishing feature of ion channels is the size of the currents generated by single channels. The current flowing through an open single channel is very small, in the range of picoamperes (10^{-12} amperes). However, the size of the unitary charge of a monovalent ion is incredibly small (10^{-19} coulombs). Thus, a single channel requires a turnover of 6×10^6 ions per second to pass 1 pA of current! How does a channel achieve such high transit rates? The definitive answer came to light with the emergence of channel structures solved by X-ray crystallography.

ION CHANNEL STRUCTURES

Our understanding of the conformational changes involved in ion channel opening and closing has grown considerably in the last decade, as crystal structures of various bacterial and mammalian channels have emerged. Although we don't yet have the structure of a specific ion channel captured in a native membrane environment in both the open and closed conformations, the structures solved to date have revealed a wealth of information about ion selectivity and the conformational changes that underlie gating. Let's first consider a prototypical channel, a bacterial potassium channel.

Potassium Channel Structure

The first ion channel structure to be solved was the bacterial potassium channel, KcsA, by Rod MacKinnon's laboratory in 1998.[19] Improvement in the resolution of the structure of KcsA yielded a more complete understanding of the chemistry of ion selectivity and the energetics of ion conduction.[20] Notably, the channel is a tetramer with four-fold symmetry (Figure 6.2A). The structure has an overall shape of an inverted teepee with a narrow selectivity filter that allows potassium ions to transit in single file nature, a large water-filled inner vestibule, and a gate formed by the crossing of alpha helices at the intracellular end of the protein (Figure 6.2B). The selectivity filter is lined by carbonyl oxygen atoms which coordinate potassium ions about 7.5 Å apart and exclude smaller sodium ions. High transit rates are achieved by electrostatic repulsion of the potassium ions in the filter. The positions of the water-filled cavity and helix dipoles help to overcome the energetic penalty of having

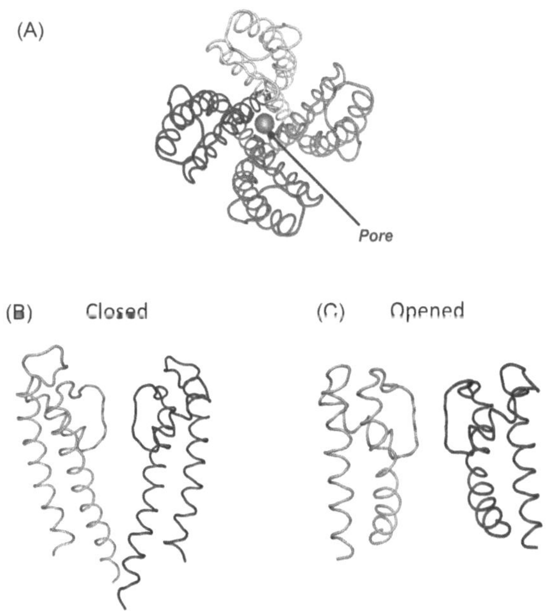

FIGURE 6.2 Structures of bacterial potassium channels. Molecular models of the bacterial ion channels, KcsA and MthK, based upon the crystal structures and depicting the movement of the intracellular molecular gate upon opening. (A) shows the structure of KcsA as seen from the extracellular side. A potassium ion is shown in the pore. Each of the four subunits is depicted. Note the four-fold symmetry. (B) shows the structure of KcsA as seen from the side. Note that the intracellular gate is closed. For clarity, only two of the four subunits are shown. (C) shows the structure of MthK. The gate is open. Only two of the four subunits are shown and the RCK domains have been omitted. *The figure is adapted from MacKinnon.*[193] *Images were created with the software program Cn3D. MMDB IDs are 12521 and 19970 for KcsA and MthK, respectively.*

an ion in the center of the bilayer. This channel was clearly captured in a closed state, where potassium is prevented from exit by the intracellular gate.

The next potassium channel structure that emerged, again from MacKinnon's lab, was of a bacterial calcium-activated potassium channel, MthK.[21] In this structure, the helices are splayed about halfway across the membrane, at a glycine residue, such that the intracellular gate is open (Figure 6.2C). The opening is large, about 10 Å, which explains why some large cations can block potassium channels from the inside when the channel is open.[22] The striking differences between these two ion channel structures suggest that channels undergo large conformational changes upon gating.

The three-dimensional crystal structures of KcsA and MthK marked a new era of ion channel science, where functional studies could be interpreted on structural grounds. In 2003, MacKinnon shared the Nobel Prize in Chemistry for these contributions to the structural understanding of ion channels. Despite these seminal contributions, we were missing structural information on channels gated by voltage. Recall that voltage-gated channels have four additional segments (S1–S4) compared with the simple two transmembrane segments in KcsA. Indeed, the question of the structural movements involved in voltage-dependent gating remained a mystery. Not surprisingly, MacKinnon's laboratory provided the first picture of a voltage-gated channel.[23,24] As before, the channel, termed KvAP, was of bacterial origin.

There are several key features of the KvAP structure. The overall organization is highly conserved. KvAP has four-fold symmetry and its pore region, selectivity filter, and intracellular gate are strikingly similar to that of KcsA. The S1–S4 segments form a structure, referred to as a voltage-sensor paddle, that surrounds the pore. The paddle is connected to the pore by flexible hinges.[23] Movement of the paddle, with its positive charges, in response to a change in the membrane voltage is thought to pull on the gate, opening the channel. Studies with Fab fragments targeting this region[24] and biotin-avidin accessibility experiments[25] show that indeed this domain is highly mobile in the membrane and capable of moving 15–20 Å across the membrane. The exact nature of this movement in the native membrane environment is a topic of intense study and debate. The structure of a mammalian voltage-gated potassium channel, Kv1.2, in complex with an accessory beta subunit followed and demonstrated a similar basic architecture as KvAP.[26] Recently, the structures of bacterial voltage-gated sodium channels NavAb and NavRh have been solved and also demonstrate this basic arrangement.[27,28]

The structure of KvAP suggested that the paddle domain exists as an autonomous unit, linked to the pore by a flexible hinge. Indeed, the paddles can be transferred from one channel to another, yielding functional channels.[29,30] Interestingly, the paddle domain exists in a phosphatase, yielding a voltage-dependent enzyme, and highlights the true module nature of this functional unit.[31]

The voltage-sensing paddle domain is of interest from a signaling perspective for several reasons. Within a subfamily of voltage-gated channels, the paddle domain is less conserved than the pore domain, making it a potential unique binding site for drugs. Indeed, small molecule inhibitors and peptides from venomous animals target this domain (see below). Some of these inhibitors have exquisite selectivity across channels. As you might expect, binding of these agents does not block ion conduction through the pore. Rather, they modify the energetics of gating. Thus, they confer a biased signaling effect, modifying the activity of the channel at some membrane voltages but not other voltages. We will discuss the so-called gating modifiers in subsequent sections.

Many voltage-gated ion channels undergo inactivation. However, little is known about the structural changes that underlie the inactivation process. For potassium channels, there appears to be two types of inactivation. In one, termed N-type, an intracellular loop of the channel physically occludes the open channel pore, much in the same way as large cations can block the pore by accessing the cavity from the intracellular side. In another type, termed C-type, the selectivity of the channel changes, suggesting the selectivity filter is a dynamic structure that changes conformation during gating.[32,33] Structural studies on KcsA confirm this mechanism.[34]

Voltage-gated sodium channels also undergo fast inactivation by a mechanism similar to N-type inactivation of potassium channels. In this case, the key residues are isoleucine, phenylalanine and methionine (IFM), residing on the intracellular loop between domains III and IV. Voltage-gated calcium channels also inactivate. In some cases, this is a strictly voltage-dependent process. The underlying structural changes of voltage-dependent inactivation of calcium channels are not understood. In other cases, the process is calcium dependent and involves a calmodulin interaction with specific domains of the channel. The diversity of inactivation mechanisms highlights a unique feature of voltage-gated ion channels: Specific stimuli can have distinct and often prolonged effects on channel availability, thereby influencing the excitability of individual cells and cellular networks.

Ligand-gated Channel Structure

The structures of several ligand-gated ion channels have been solved. These include the P2X(4) channel,[35] the acetylcholine receptor (binding site only),[36,37] an ASIC channel,[38] the glutamate-gated cation channel, GluA2,[39] and the invertebrate glutamate-gated chloride channel, GluCl.[40] There are several striking differences between these structures, including the nature of the orthosteric binding site. In trimeric P2X and pentameric Cys-loop receptors, the agonist binding site is located between subunits. In the case of GluA2, agonist binding occurs within individual subunits. In this chapter, we will focus on GluR2 and GluCl where there is considerable information of the conformational dynamics involved in channel gating.

GluA2 is a member of the AMPA class of glutamate-gated ion channels. GluA2 is highly expressed in the brain and most native AMPA receptors are heteromeric complexes consisting of dimers of GluA2 and either GluA1, GluA3, or GluA4. The structure of a homomeric rat GluA2 channel was solved by the laboratory of Eric Gouaux.[39] The AMPA receptor GluA2 has three basic domains: a large N-terminal domain (ATD), the LBD, and the transmembrane domain (TMD) forming the channel (Figure 6.3). Although the functional channel is a tetramer, the symmetrical arrangement of subunits is strikingly unique. The extracellular ATDs and LBDs are organized as a pair of dimers with two-fold symmetry. However, the TMDs forming the ion channel have four-fold symmetry. This "symmetry mismatch" yields two conformationally distinct dimer pairs, with each one coupling differently to the channel region. This arrangement is likely conserved for other ionotropic glutamate receptors that are heteromers, such as the NMDA receptor which is composed of NR1 and NR2 subunits.

The ATD of GluA2 has a clamshell-like shape in which the amino-terminal portion of the sequence defines most of one lobe and the carboxyl-terminal region composes most of the second lobe.[41] The ATD functions in receptor trafficking and assembly. No known

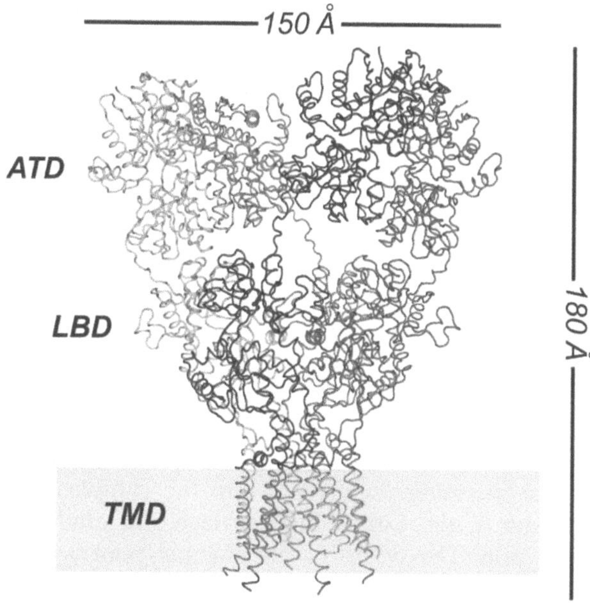

FIGURE 6.3 Structure of a mammalian ligand-gated ion channel. The structure of the homomeric AMPA receptor, GluA2, is depicted based upon the crystal structure as seen from the side. Each of the four subunits is shown. The position of the amino-terminal domains (ATD), the ligand binding domains (LBD), and the transmembrane domains (TMD) are noted. *The figure is adapted from Sobolevsky et al.*[39] *Images were created with Cn3D software. MMDB ID is 78644.*

ligands bind to this region in AMPA receptors. However, in NMDA receptors, the region is the site for binding of several modulators of channel function, including zinc, protons, polyamines, and small phenylethanolamine molecules such as ifenprodil (see below). The LBD of GluA2 also has a clamshell-like structure.[42] Binding of glutamate causes closure of the clamshell. Thus, closure of the clamshell is communicated through linkers to the transmembrane channel, opening the gate. The structure of the LBD with various agonists and antagonists bound has shed considerable light on the mechanism of agonism. Quite simply, the extent of activation of the channel depends on the degree of closure of the clamshell. Partial agonists lead to less domain closure than full agonists like glutamate. Antagonists bind and stabilize the open conformation of the clamshell, preventing opening of the channel.[42] This mechanism appears to be generalized for ionotropic glutamate receptors as similar findings have been described for NMDA receptors.[43,44]

The transmembrane domain of each subunit of GluA2 has three transmembrane segments (M1, M3, and M4) and a central pore-like helix (M2). In notable contrast to potassium channels, the GluA2 channel has a broad cytoplasmic face and a narrow extracellular top. The channel gate is formed by the crossing of the M3 helices near the extracellular side. The narrow ion conduction pathway is formed by highly conserved residues in the M3 helix. Channel opening is linked to agonist binding by the connection of the LBD to the M3 helix. Binding of glutamate causes closure of the clamshell and a rotation, increasing the separation between the LBD and the TMD. This movement pulls on the M3 helices, causing an opening of the gate. Thus, the energy of agonist binding results in a major conformational change in the LBD, which is then applied to the mechanical work of opening the channel.[39]

Ligand-gated ion channels often undergo a process of agonist-dependent desensitization. With continuous exposure to agonist, the channel adopts a closed conformation which has high affinity for the agonist. For some ligand-gated channels, this process is extremely rapid and nearly complete. For example, the AMPA receptor opens for only a few milliseconds before desensitizing. Structural studies with GluR2 have shed light on this process.[39,45,46] In the current model, activation of the receptor by clamshell closure produces a strain on the linkers between the LBD and TMD. To release this strain, the LBD dimer dissociates and rotates, allowing the linkers to adopt a relaxed (closed state-like) conformation. In this way, the gate closes despite the fact that the LBD has agonist bound and the clamshell is in the closed conformation. For recovery from desensitization to occur, agonist must dissociate before the receptor resumes its resting state with the LBDs existing as dimers.

The invertebrate glutamate (Glu)-gated Cl^- channel (GluCl) is pentameric like the ACh receptor and is a member of the greater LGIC family but is permeable to anions such as Cl^-, unlike the vertebrate Glu receptors that are cation selective. From a functional perspective, the invertebrate Glu neurotransmitter behaves as an inhibitory input, opposing neuronal excitability and repolarizing or hyperpolarizing the membrane potential.[40] Crystal structures of this channel, in complex with Cl^-, have revealed many insights into the structure of the pore of these channels. As one might expect, the pore extends from the extracellular surface through the membrane-spanning region and into the cytoplasm. It takes on a general funnel shape, being wider at the extracellular surface and gradually narrowing as it reaches the cytoplasmic region. The width of the channel pore widens rapidly again as it reaches its cytoplasmic base.

The extracellular N-terminus of each subunit of the GluCl channel is comprised of a series of anti-parallel β-sheets with a small α-helix at the extreme N-terminus

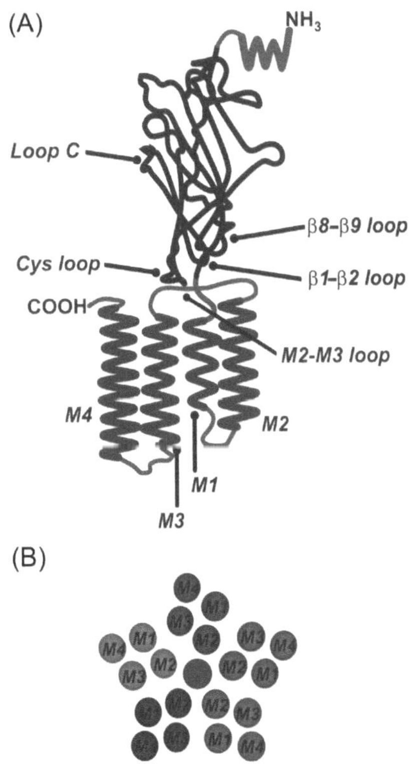

FIGURE 6.4 Schematic diagram of the pentameric *C. elegans* glutamate-gated Cl^- channel. Illustration of the structure of the glutamate-gated Cl^- channel as representative of the pentameric LGIC family based upon the published crystal structure of Hibbs and Gouaux. (A) depicts notable domains and features of the channel subunit structure, including the important Cys loop and loop C of the extracellular domain. The membrane-spanning α-helices are labeled M1–M4. Not shown for clarity are the β-sheets of the extracellular domain. (B) illustrates the juxtaposition of the transmembrane helices of the five subunits that comprise the holochannel as seen from above. The pore formed by the arrangement of the M2 domains of each subunit is represented in the center by a circle. Note the pentameric shape of the holochannel.

(Figure 6.4A). The important Loop C is found on the external surface of the extracellular N-terminal domain in each subunit. The membrane-spanning region is comprised of four linked α-helices (labeled M1–M4). Interestingly, the amino acid residues lining the pore of the channel do not possess formal charges but are electrostatically neutral. The positive charge potential that arises at the base of the pore has been shown to arise from the orientation of peptide dipoles that arise from the tilt of the M2 α-helix for each subunit.

The Cl^- binding site of the GluCl channel has been resolved by soaking the GluCl channel crystals with iodide.[40] This technique has revealed a putative site of coordination of Cl^- within the base of the channel but still within the transmembrane domain. This site of coordination is provided by positive amino acid residues from each M2 helix of the subunits. Three amino acids (Pro–Ala–Ile) from the M2 helices create the positive electrostatic interaction required to coordinate Cl^- ions.[40] The prolines from adjacent subunit M2 helices form a pocket of electropositive charges that bind Cl^- ion[40] (Figure 6.4B). These residues are responsible for determining ion selectivity of the GluCl channel.[47] Similar sequences have also been found to determine selectivity for related ion channels (e.g., glycine receptors).[48] Interestingly, eukaryotic cation channels of the LGIC family have been found to replace alanine with glutamate, thereby reversing the electrostatic charge at this critical position of LGIC channels and filling the electropositive pocket with its carboxylic acid side chain.[47] Thus, anion selectivity in GluCl channels is determined by a constrictive pore (smaller diameter compared with cation channels that are devoid of Pro at this site) and an electropositive potential contributed by the amino-terminus of the M2 α-helix of each subunit,[40] the result of which is to concentrate Cl^- ions at the cytoplasmic surface of the GluCl receptor pore.

CONFORMATIONAL DYNAMICS

We have seen that the major role of ion channels is to regulate the movement of ions across the cell membrane and that the gating of ion channels is dependent on the tertiary structure of the channel. In terms of biological function, ion channels behave as all proteins do, through conformational-dependent responses to exogenous stimuli. For ion channels, these stimuli can take on numerous forms, such as membrane voltage, mechanical deformation (e.g., stretch), exogenous ligands, or pH. Despite their differences in stimuli, all ion channels respond by regulating the flow of ions across a cell membrane. The resultant ion flux can cause a change in the cell's membrane potential or alter intracellular ion concentrations. Ultimately, these changes can trigger the release of stored vesicles, alter enzyme activities, and regulate gene expression. Therefore, one can consider the flow of ions, and the resultant change in membrane potential and/or ion concentration, as the primary signaling mechanism of ion channels; that is, ion channels regulate cellular processes through passage of ions.

For ligand-gated channels, it seems obvious that agonist binding triggers channel opening. Despite this simplicity, the precise changes in channel conformation leading from agonist binding to channel opening are certainly more complex. Early studies of the neuromuscular junction in the 1950s led to the formulation of a two-state theory to explain the action of acetylcholine on the endplate ACh receptor. This model, proposed by del Castillo and Katz,[49] suggested that binding of agonist changed the channel from a closed state to an open state. This linear, two-state model predicts that in the absence of agonist, all channels are closed. Further, the agonist cannot dissociate from the active (opened) receptor; that is, the affinity of the active state for agonist was infinite.

The role of protein conformation in the activation process of ion channels was first postulated by Changeux in his landmark work on the nicotinic acetylcholine receptor (see Chapter 1). This hypothesis was based upon his earlier work with Monod and Wyman[50] on enzymes and is one theory to explain how conformation of ion channels determines function. This model of conformational dynamics, called the concerted model (also referred to as the MWC model),[51,52] suggests that proteins (e.g., enzymes or receptors) are dynamic entities and that the functional subunits that comprise these proteins can be found in either a relaxed or tensed state. The key imperative in this model is that it is assumed that all parts of the protein are found within the same conformation at any one point in time; that is, the protein is either relaxed or tense but the subunits do not differ in their conformational state. Thus, a modulator affects the function of the protein acting at a topographically distinct site on the protein by inducing or maintaining either the relaxed or tensed state in one subunit of the protein that is conferred to the other parts of the protein.

Alternatively, the sequential model of allostery put forth by Koshland et al.[53] suggests that in the absence of ligand, receptors are freely flowing between available, energetically allowable conformations (KNF model). A given ligand induces a given, active or inactive, conformation of the protein. This is also referred to as the induced fit model. Interestingly, in early iterations of their MWC model, Changeux and colleagues utilized certain aspects of the induced-fit model as a basis for explaining their own.[54] It is important to point out that the MWC model also conflicted with the classical understanding of enzyme inhibition through steric hindrance at a common binding site.

The MWC model of allosteric regulation of proteins proposed two very important concepts concerning complex protein conformation. The first is that complex proteins have a given quaternary structure with a given symmetry and finite arrangement of subunits (we now know that monomeric proteins also exist in differing conformations in solution). The second concept is that subunit oligomers undergo reversible transitions between conformations in the absence of ligand that ultimately impact the quaternary structure of the protein.[54] The implication is that ligand-stabilized conformations are within the range of natural conformations available to the protein and not induced by the ligand. Further understanding of protein conformational dynamics has revealed that a hybrid model that combines the basic principles of the MWC and KNF models is probably a more accurate representation of how allosteric ligands affect protein conformation. Interestingly, Changeux and colleagues understood relatively early that this model of protein conformational dynamics had application in understanding receptor–ligand interactions.[54,55] As we have seen in previous chapters, ligands, whether allosteric or orthosteric, act to stabilize conformations of the protein, thereby regulating the function of the protein.

Ion channels have been a focus of drug discovery as many medicinal plants and natural remedies affect ion channel function. Studies of the conformational dynamics, using crystal structures and three-dimensional modeling, have revealed key aspects of interactions between ligands and binding sites on ion channels from many different families. One of the most studied families in this regard is the pentameric ligand-activated ion channels.[56]

Initial attempts to understand the site of ligand binding in this family of receptors utilized affinity and photoaffinity labeling techniques to probe the amino acid residues of the ion channel important to, or in close proximity to, the ligand. First attempts in this regard understandably used crude affinity probes attached to ligands. These probes relied on electrophilic groups (e.g., maleimide or diazide) attached to known ligands such as tetraethylammonium or phenyltrimethylammonium iodide. The electrophilic groups react with amino acids with nucleophilic side chains such as cysteine, tyrosine, aspartate, and glutamate. Unfortunately, these probes were too weak to allow efficient labeling of amino acids on the ion channel.[56]

A more powerful technique utilizing photosensitive moieties attached to known ligands proved to be a suitable approach to labeling amino acids within the channel for study. Using this strategy, probing the acetylcholine receptor (AChR, a Na^+, and K^+ permeant ion channel) from the electric organ of the ray *Torpedo electroplax* with p-(dimethylamino) benzenediazonium fluoroborate (DDF) revealed key residues within the α1 subunit. Additional studies using photoaffinity probes designed from other AChR antagonists subsequently suggested that the acetylcholine (ACh) binding site spanned multiple subunits. Further evaluation of other LGICs, including the $GABA_A$ receptor,[57] confirmed these observations and led to a model of ligand binding in this ion channel family. The model hypothesized that the ligand binding region is composed of a pocket lined by loops formed by the amino acid sequences in these subunits that are complimentary. A positive face of the binding pocket was proposed that contained loops A, B, and C, and a negative face comprised of loops D, E, and F.[56]

The first crystal structure to reveal the nature of the three-dimensional organization of the ACh binding site was solved by Smit et al.[58] using the secreted ACh binding protein of *Lymnea stagnalis*. This structure revealed that the ACh binding protein is an oligomeric protein comprised of five identical subunits aligned along an axis with five-fold symmetry. Each subunit is comprised of an N-terminal α-helix, two 3_{10} helices, and a core of 10 β-strands in a β-sandwich configuration. The loops (so-called Cys loops) for each subunit are formed by an intrasubunit cysteine–cysteine disulfide linkage. ACh binding sites occur at sites at the interface between each subunit as had been hypothesized by mutational and photoaffinity studies. Subsequent solution of crystal structures of the extracellular domain of the nicotinic AChR confirmed these observations in terms of the site of orthosteric ACh binding.[59]

The conformational movements that occur in response to orthosteric ligand binding at the LGICs have been deduced from the differences in structures of the ACh binding protein in complex with agonists and antagonists. The structure of the ACh binding protein bound to various toxins (e.g., snail venom conotoxin,[60] snake venom neurotoxin, and cobratoxin[61]) that act as antagonists to ACh binding revealed that in the presence of these molecules, loop C is displaced outward. In the presence of nicotinic agonists, loop C moves inward to wrap around or "cap" the agonist, thus securing it in the binding pocket.[37] It has been hypothesized that this movement of loop C in the presence of agonist is key to the allosteric activation of channel opening since fixation of this loop in the capped configuration leads to extended channel opening.[62]

Through experiments using mutants, Mukhtasimova et al.[62] have found that "trapping" the ACh receptor in various intermediate states shows that the ACh receptor is a conformationally fluid structure that transitions between open and closed states. Surprisingly, they have demonstrated that the ability of different agonists to elicit the transition from closed to open states of the receptor is independent of agonist efficacy, confirming earlier work.[63,64] However, they observed that in the presence of ligand, the ACh receptor rapidly transitions to one of two primed conformational states: one that leads to relatively brief channel openings and one that leads to temporally longer channel openings. The ability to induce the long-lived open conformation is proportional to agonist efficacy.[62] They have proposed a model of activation based upon their observations and that of others (Figure 6.5). From an evolutionary perspective this type of agonist-mediated allosteric activation of LGIC activation may be conserved from early on, as comparisons of prokaryotic and eukaryotic crystal structures have revealed a similar interfacial binding pocket of glutamate-activated chloride channels from *C. elegans* with similar counterparts in bacteria (GLIC),[40,65] and chimeric receptors constructed from prokaryotic and eukaryotic channels demonstrate similar activities.[66]

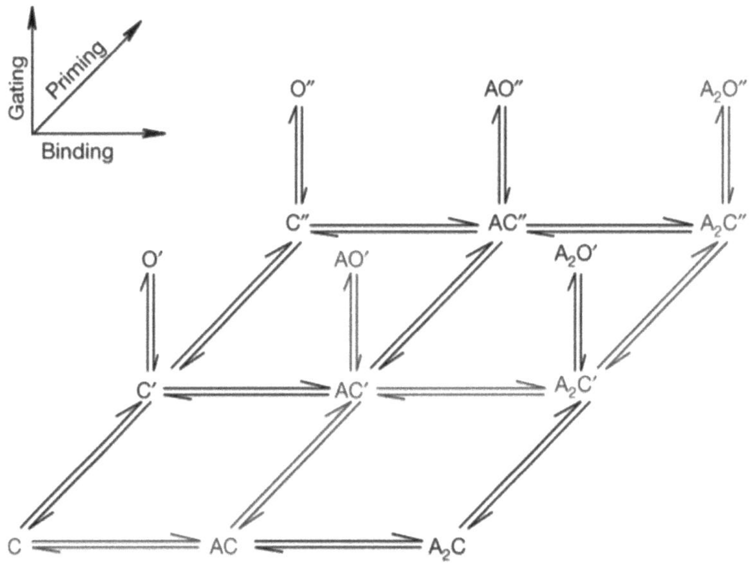

FIGURE 6.5 Priming model of the opening of the ACh receptor as proposed by Mukhtasimova et al.[62] Based upon channel kinetics of mutant mammalian ACh receptors in the presence of agonist, a conformational-dependent model of open probability in the presence of two agonists was developed based upon the model of del Castillo and Katz. Letter designations refer to the conformational state of the channel as closed (C), bound to agonist (A), or open (O). States depicted by prime (′) refer to different primed conformational states. Successful paths to primed and open states confirmed by observed rate constants are shown. In the absence of agonist, closed conformations have negligible impact on the open probability. Binding models that follow the paths bind/prime/prime/bind or bind/bind/prime/prime did not result in predicted rate constants consistent with experimental values. Only the model represented by bind/prime/bind/prime resulted in appropriate rate constants. This suggests that the ACh receptor requires a well-defined and sequential binding of agonist. Binding of each agonist is followed by a primed state that ultimately stabilizes the correct open conformation. *Figure reproduced with permission of the Nature Publishing Group, London, UK.*

ALLOSTERIC MODULATION OF ION CHANNELS

As both voltage-gated and ligand-gated channels are conformationally dynamic proteins, it follows that transitions between those states can be modulated allosterically by a ligand binding event. As a consequence, cellular events downstream of ion conduction through the pore can be biased by ligand–channel interactions. In this section, we will discuss several examples of allosteric modulation of channels. First, a few words are warranted about the mechanism of allosterism and the terminology as it relates to ion channels.

Allosterism is used throughout the literature very loosely and generally to describe "any mechanism in which a protein can exist in two (or more) distinct conformations, which differ in their affinity for a ligand…and an allosteric regulator is anything that binds better to one conformation than the other (i.e, almost everything)".[67] The term "allosteric" ligand is used to describe molecules that modulate the receptor through sites that are distinct from the traditional agonist "orthosteric" binding site. That is, "orthosteric" and "allosteric" are used to differentiate between canonical agonist binding domains and sites located elsewhere. However, both types of ligands can produce global changes in protein conformation that influence receptor function.[68] Nevertheless, the terms are useful and are cemented in the literature. Regarding binding of allosteric ligands, basic allosteric theory predicts that binding of agonist at the orthosteric site influences the binding of an allosteric modulator, and vice-versa.[69]

Allosteric ligands are generally classified in terms of the functional effects they exert on the receptor; that is, positive allosteric modulators (PAMs) increase function whereas negative allosteric modulators (NAMs) inhibit function. In the case of PAMs, these ligands are generally thought to be unable to produce receptor activation on their own, but facilitate the response of the receptor to the orthosteric ligand by modifying the energy barriers associated with transitions between functional conformations (see Chapter 1). However, this distinction is often blurred as some PAMs produce activation of receptors independent of the traditional agonist. Examples of agonist-independent activation of receptors by PAMs exist for GABA(A) receptors[70] and alpha-7 nicotinic ACh receptors.[71] This is possible since the Boltzmann distribution law predicts that unliganded "spontaneous" channel openings can occur, albeit very rarely.[72] A PAM will alter the resting distribution of functional channel states, stabilizing the spontaneous opening of the receptor in the absence of the "orthosteric" agonist. Thus, activation of a receptor by a PAM independent of the traditional agonist is entirely consistent with allosteric theory.

Allosteric Modulation of Ligand-gated Channels

Allosteric modulators have been described for many of the members of the ligand-gated superfamily and, in some cases, are important therapeutics. Here we shall review examples of allosteric modulators of GABA(A) receptors, nicotinic ACh receptors, and ionotropic glutamate receptors. From a mechanistic point of view, these are the best understood and serve to illustrate the complexity possible for allosteric modulation in this family of ion channels.

GABA(A) receptors bind to gamma-aminobutyric acid (GABA), the principal inhibitory neurotransmitter in the mammalian brain. Binding of GABA induces opening of the channel and the flow of chloride ions. These receptors are important drug targets for sleep disorders, anxiety, and epilepsy.[73] As a pentameric channel, the complexity of subunit combinations is significant. Subtype-specific modulators hold promise for the development of novel therapeutics for anxiety and sleep induction with improved side-effect profiles.[74,75] Several positive allosteric modulators (PAMs) of GABA(A) receptors have been identified. For example, benzodiazepines such as diazepam bind to a specific subtype of GABA(A) receptor.[76] Discovered by Roche chemist Leo Sternbach in the 1950s, these were the first class of allosteric modulators targeting a membrane receptor to be used clinically and remain the most widely prescribed class of drugs to promote sleep.

The binding site for benzodiazepines is in the amino-terminal extracellular domain, at the interface of the α and γ subunits.[77] At the single channel level, benzodiazepines cause an increase in the frequency of the channel opening in "bursts" without changing the duration of bursts.[78] The overall effect is to increase the affinity of the receptor for GABA. The striking finding here is that the change in affinity is modest, about two- to four-fold, at saturating concentrations of diazepam.[70,79] Thus, even a relatively small change in agonist affinity can produce profound functional effects on the receptor and be therapeutically relevant. Barbiturates, like phenobarbital, also potentiate GABA(A) currents but by a different mechanism. Acting at a transmembrane site on the β-subunit,[80,81] pentobarbital increases burst duration without altering burst frequency.[78] These actions underlie the anti-seizure activity of this class of drugs. Thus, two clinically important allosteric modulators of the GABA(A) receptor have distinct mechanisms of action, via different sites, yielding very different behavioral results.

Allsoteric modulation of the nicotinic ACh receptor has been studied extensively. Several classes of positive allosteric modulators of the nAChR have been identified (reviewed in Changeux[8]). In the central nervous system, nAChRs are of two general types: heteromeric (containing predominately alpha-4 subunit) and homomeric (containing alpha-7 subunit). Partial agonists of alpha-4-containing receptors are used clinically for smoking cessation. Preclinical data suggest that activation of alpha-7 receptors may be useful for the treatment of schizophrenia and the cognitive deficits in Alzheimer's disease. As such, there is considerable interest in identifying positive allosteric modulators that selectively target the alpha-7 receptor.

Numerous chemotypes have been identified as modulators of nAChR (reviewed in Williams[68]). Alpha-7 PAMs have been classified as type I or type II based on the properties of modulation.[82] Type I PAMs increase alpha-7 current without affecting receptor kinetics while type II PAMs increase alpha-7 current and profoundly alter channel kinetics. Specifically, type II PAMs slow receptor desensitization and channel closing after removal of agonist ("deactivation"). Unfortunately, the definitive binding site(s) of alpha-7 PAMs has remained elusive. Some type I PAMs appear to bind to the N-terminal extracellular domain at subunit interfaces while others appear to require transmembrane regions.[83–85] Interestingly, the structure of the type I PAM galantamine bound to the ACh binding protein from *Aplysia* reveals binding at the subunit interface.[86] Type II PAMs such as PNU-120596 require transmembrane domains for potentiation.[87] Despite the interest in this overall approach, the therapeutic utility of either of these classes of alpha-7 PAMs has yet to be demonstrated.

Allosteric modulators of ionotropic glutamate receptors are numerous and provide diverse examples of both positive and negative mechanisms of modulation. Importantly, many of these modulators come with binding site knowledge based on X-ray crystallographic studies of extracellular domains. Consider, for example, the homomeric AMPA receptor GluA2. Several small molecules have been shown to bind at the LBD dimer interface. For example, the PAMs aniracetam and CX614 bind adjacent to a hinge region of the "clamshell," stabilizing the closed (glutamate bound) conformation of the clamshell. In this way, these compounds slow channel deactivation by increasing the affinity of the receptor for glutamate.[88] In contrast, compounds such as cyclothiazide act as PAMs by stabilizing the LBD dimer, thereby inhibiting receptor desensitization.[45,46] The LBDs of the NMDA receptor (GluN1 and GluN2) are also sensitive to allosteric modulation, but in the negative direction. The molecule TCN-201 acts as a negative allosteric modulator (NAM) of glycine by binding to a site at the dimer interface.[89] PAMs of the NMDA receptor have been identified as well, such as pregnenolone sulfate, but appear to bind at a site distinct from the LBD.[90]

In addition to the LBD, the amino-terminal domain (ATD) of NMDA receptors is a site of allosteric action. Zinc, protons, polyamines, and small phenylethanolamine molecules such as ifenprodil act as NAMs of NMDA receptor function, decreasing the probability of channel opening. Zinc is a potent NAM at NMDA receptors containing GluN2A subunit and acts by reducing channel opening probability.[91] Zinc binds within the clamshell cleft of GluN2B.[92] Ifenprodil binds at the ATD subunit interface of GluN1 and GluN2B causing clamshell closure.[93] These allosteric actions are communicated over surprisingly long distances. Specifically, binding of molecules to the ATD is communicated to the channel region, over 100 Å away, ultimately leading to a reduced channel opening.

Allosteric Modulation of Voltage-gated Channels

Voltage-gated ion channels are not typically thought of as being targets of allosteric modulation. However, many molecules are known to bind to sites distinct from the ion conduction pathway and alter voltage-dependence of channel gating. In this section, we will highlight a few examples of this type of modulation.

Peptides isolated from the venoms of predatory organisms are a rich source of ion channel modulators. Some of these peptides are simple channel blockers and act by binding to the pore region of the channel, preventing ion conduction. Charybdotoxin, isolated from the venom of the scorpion *Leiurus quinquestriatus hebraeus*, is a classic example of this mechanism where mutagenesis and structural studies have yielded a detailed picture of peptide docking and channel block.[94,95] Other peptides act more discretely, by modifying the voltage-dependence of channel gating. Examples of these include the alpha and beta scorpion toxins targeting sodium channels.[96] Alpha scorpion toxins act by slowing inactivation whereas beta scorpion toxins alter channel activation. The beta scorpion toxins are prototypical "gating modifier" peptides that alter the voltage dependence of channel opening. Both alpha and beta scorpion toxins are thought to act by a voltage sensor trapping mechanism where one of the S4 segments is held in the active conformation by the peptide.[96] Examples of this type of modulation also exist for Cav[97] and Kv channels.[98] The Kv gating modifier peptides, isolated from tarantulas, have been studied the most extensively and are known to bind to the voltage-sensing domain (VSD) and make movement of the VSD energetically unfavorable. In this way, they act as channel inhibitors, requiring a larger depolarization to achieve the same degree of channel opening without inhibitor. However, subtle changes in the channel-peptide interaction can switch this inhibition to activation.[99] Notably, these peptides interact with the voltage sensor within the lipid membrane and behave as cargo that can remain bound while the channel opens and closes.[100,101]

Modulation of gating of voltage-gated channels is not limited to peptides. Several small molecules are known to similarly act to modify channel gating. For sodium channels, these include the alkaloids from tree frogs and the insecticidal pyrethrins from plant leaves.[1] These compounds are functional channel activators but bind at a site distinct from the scorpion toxins. However, small molecule inhibitors of sodium channels targeting one of the VSDs (VSD4) have been recently identified that show subtype specificity and hold promise as potential therapeutics.[102] Small molecule modulators of Kv channels are also known. Retigabine activates Kv7 channels by binding to the activation gate.[103,104] Interestingly, another Kv7 channel opener, NH29, acts at the voltage-sensing domain. NH29 is thought to trap the voltage sensor in the active conformation similar to scorpion toxins.[105] Hence, these examples suggest that allosteric modulation of voltage-gated ion channels by small molecule agents may be more common than once thought.

Clearly, allosteric modulation of ion channels is diverse in terms of sites of action and mechanism of modulation. The rapid growth in this area in the last 10 years suggests that new molecules and sites of action will undoubtedly emerge in the near future. Ideally, some of these molecules and sites can be taken advantage of for the treatment of disorders where targeting a specific ion channel in a nuanced fashion has therapeutic advantages from either an efficacy or safety perspective.

BIASED SIGNALING BY ION CHANNELS

As we have seen in earlier chapters, conformational dynamics induced by the physical interaction between ligand and receptor is a universal theme across classes of receptors. Similarly, it seems that ligand-specific activation of signaling signatures (biased signaling) appears to be a common theme across many receptor classes. It is hard to imagine how ion selectivity could potentially be altered in a ligand-dependent manner in ion channels. However, emerging literature points to just that. Indeed, it is now recognized that channel specificity for permeant ions can change in response to alterations within the quaternary structure of the channel or in response to changes in the tertiary structure of given subunits. This has been noted for voltage-gated,[106] mechanosensitive,[107] and ligand-gated channels.[108,109] In addition, effects of associated proteins can regulate the dynamics of channel function to a given ion. Lastly, some ion channels have been demonstrated to regulate other receptors or activate traditional signaling enzymes (e.g., ERK, PI3K, Rho, etc.) in addition to regulating ion flux and membrane potential. Given the importance of conformational changes in receptor signaling that have been detailed thus far as well as in ion channel function, it seems likely that ion channels may also possess the potential for signaling bias in response to agonists, whether orthosteric or allosteric.

Biased Signaling by Alteration of Ion Selectivity

For many years, ion selectivity of channels was thought to be a permanent feature determined by the structural organization of the contributing subunits. This was the case, despite the elucidation of the roles of conformational dynamics and allosterism in determining/regulating ion channel function. In recent years, however, a number of ion channels across different families have been found to possess differential selectivity to ions. As noted above, dynamic changes in ion selectivity occur during C-type inactivation of Kv channels (reviewed in Yellen[110]). During this type of inactivation, the selectivity filter changes structure, increasing the permeability of the channel to sodium.[32,33] These changes in selectivity are relatively subtle and may not serve a signaling function. However, the most dramatic example of dynamic changes in ion selectivity comes from the ATP-gated P2X family of LGICs. Current evidence suggests these changes in selectivity can serve important biased signaling functions that impact physiological processes.

The P2X receptor was first identified as an ATP-dependent permeability to nucleotides and other large molecules in mast cells,[111] which was named the P2Z receptor. Following its subsequent cloning in the mid 1990s, it was renamed $P2X_7$. In the initial description of its activity in response to ATP, a time-dependent increase in membrane permeability to nucleotides and other phosphorylated species was noted. Indeed, it was observed that longer exposure of the receptor to ATP led to a time-dependent increase in the permeability to larger molecules. In cloning $P2X_7$, Surprenant et al.[112,113] demonstrated that activity of this channel was biphasic; that is, transient exposure of agonist (ATP) elicited cation permeability (e.g., Na^+ or K^+), but sustained exposure of agonist induced permeability to increasingly larger molecules. These data demonstrated that the $P2X_7$ receptor has the capability to alter its permeability in a time-dependent and ligand-dependent fashion. In addition, since even large molecules (>500 Da) were able to pass through the channel, this implied that the pore structure was remarkably elastic. This process of time-dependent changes in permeability is referred to as pore dilation.

P2X receptors are trimeric proteins comprised of seven potential subunits within the P2X receptor family, $P2X_{1-7}$. The amino and carboxy termini of each subunit are intracellular. The C-terminus (between 31 and 242 amino acids) is much longer than the short amino terminus and provides for most of the divergence in terms of sequence identity across the family of proteins.[109,112] These receptors contain two lipophilic, transmembrane domains that are connected by a rather large extracellular loop (60% of the total structure). The extracellular loop contains five disulfide bridges provided by ten cysteines that are conserved among the seven P2X subunits.[109,114] Three subunits are combined to form a functional P2X receptor channel. P2X receptors can be either homotrimers or heterotrimers. In the quaternary structure of the P2X receptors, the second transmembrane domain from each subunit aligns to create the pore of the channel.[114,115] Key amino acid residues within these domains form the gate of the channel.[114–116]

It has been shown that the first transmembrane domain plays little or no role in regulating ion passage.[114,117,118] Studies using chimeric[119] or mutant[120] channels have revealed that three to four key residues within the second transmembrane domain of each subunit form the constricted gate region of these channels and that these residues are critical for the dynamic changes in ion permeability that are characteristic of this family of ion channels.[119] The sequence of these key residues varies somewhat between the P2X channel subunits and across species but the general topology of these residues seems to be conserved.[119] Interestingly, not all P2X receptors have shown the ability to dynamically change their permeability.[108,113] Indeed, only $P2X_2$, $P2X_4$, and $P2X_7$ have been reported to have this ability. Of equal importance is the observation that pore dilation does not occur in all cell types.[114,121] Taken together, these data suggest that pore dilation of some P2X receptors is physiologically important.

Pore dilation by P2X receptors is important in immune function. Current data suggest that $P2X_7$ is involved in allowing passage of agonists for NLRP3 (Nod-like receptor 3) into the inflammatory cells activated at the site of injury. Subsequent activation of NLRP3 located on inflammasomes within immune cells leads to the secretion of interleukin-1β (IL1β), an important inflammatory mediator which recruits other immune cells to the site.[122] Furthermore, activation of the dilated pore of some P2X receptors is associated with activation of caspase 3 but not caspase 1.[121] In addition, $P2X_4$ and $P2X_7$ have been found to be important to kidney function at multiple sites across the nephron.[123] Most notably, $P2X_4$ has been reported to regulate Na^+ reabsorption via modulation of renal epithelium in the thick ascending loop.[123] Additionally, $P2X_4$ receptors have been suggested to regulate sodium reabsorption through modulation of ENaC channels in renal epithelial cells.[123] Interestingly, some effects of $P2X_7$ and $P2X_4$ in the kidney have also been linked to stimulation of inflammasome activity.[123,124] $P2X_7$ receptors have also been implicated in mediating the effects of nerve damage in chronic pain[121] through their effects on local IL1β release. For these reasons, P2X receptors have been studied as potential targets for the development of selective antagonists for treatment of inflammatory diseases and as analgesics for treatment of chronic pain.[121]

Small molecule agonists and antagonists of P2X receptors have been known for some time. However, many of these display non-selective properties within the P2X receptor family and at other receptor types. Many of these first generation compounds were based upon the structure of ATP and were purinergic analogs. Recent research has discovered novel chemotypes that are capable of selectively blocking $P2X_7$ receptors.[121,125] In addition, some compounds have been found to be allosteric modulators of $P2X_7$ function. One notable example in this regard is ivermectin.

Ivermectin is a macrocylic lactone that activates a glutamate-gated chloride channel of nematode and arthropod parasites. It is used widely as an anthelmintic in veterinary medicine and to treat river blindness in humans. However, at higher concentrations it acts as a PAM at P2X receptors. Originally thought to be a $P2X_4$ selective compound,[126,127] ivermectin has been found to allosterically potentiate ATP-induced currents in HEK293 cells expressing the human $P2X_7$ receptor and in human monocyte-derived macrophages that endogenously express $P2X_7$ receptors.[127] These effects were blocked by the $P2X_7$ selective antagonists, A438079 or AZ10606120. However, ivermectin did not affect the ability to induce pore dilation, or the associated ATP-induced current decay, suggesting that ivermectin acts as a positive allosteric modulator at the $P2X_7$ receptor[127] that is biased for the acute ATP-induced cation currents but does not affect pore dilation. Thus, ivermectin would seem to be a biased modulator of $P2X_7$ receptor signaling. The effects of ivermectin on the $P2X_7$ receptor are reminiscent of those reported for calmidazolium (a known calcium channel blocker), which was found to block ATP-induced cation currents in P2X receptors but had no effect on pore dilation.[128] These data suggest that the cation currents elicited by ATP at P2X receptors can be wholly separated from the pore dilation process seen in some of these channels. Taken together, they imply the potential for biased agonism/antagonism at P2X receptors.

Another family of ion channels that display dynamic ion selectivity is the transient receptor potential (TRP) family of channels. These channels are responsive to exogenous forces, such as heat, cold, and mechanical force (e.g., stretch), and a variety of chemical agents. Some TRP channels display time-dependent dynamic ion selectivity following activation.[107,129,130] TRPV1 is one member of this family of proteins that is expressed abundantly in sensory neurons. TRPV1 is activated by many factors including capsaicin (the active ingredient in chili peppers), protons, prostaglandins, and heat. The TRPV1 channel is a non-selective cation channel that is permeable to monovalent cations but displays preference for Ca^{2+}. Studies have found that in HEK293 cells expressing TRPV1 channels, the agonist capsaicin elicits a time-dependent increase in permeability of the large cation N-methyl-D-glucamine (NMDG) relative to Na^+, while also displaying a time-dependent decrease in permeability relative to Na^+.[107] Using an excluded volume calculation method, these investigators estimated that these time-dependent changes in permeability were due to an increase in the diameter of the channel pore from approximately 10 Å to 12 Å.[107]

Furthermore, it has been observed that the preference for Ca^{2+} in TRPV1 channels dynamically changes in response to agonists such as capsaicin.[107] The TRPV1 channel displays a near 10-fold preference for Ca^{2+} as compared with other monovalent cations such as Na^+.[131] However, over time this preference is significantly reduced. Therefore, these channels display a biphasic current in the presence of capsaicin, a transient current, and a later current associated with increased permeability to larger cations and Na^+. Perhaps more importantly, these effects were found to be agonist dependent.[107] Agonists such as capsaicin and N-arachidonoyl dopamine (NADA) evoke notable biphasic currents, albeit with differential magnitudes. However, other agonists such as piperine and camphor elicit the transient current but not the later current; still others induce the later currents without the transient current.[107] The sensitivity of the TRPV1 receptor to capsaicin is modulated by activation of protein kinase C, suggesting intracellular regulation of channel function. These data strongly support the notion of biased signaling in ion channels in that the TRPV1 receptor is capable of activating different currents (signaling) in response to physiological ligands (and that this can be modulated by the intracellular milieu) and that the nature and magnitude of the currents evoked is ligand dependent.[107] Interestingly, similar characteristics have also been noted for the TRPV3 receptor.[129]

Activation of Intracellular Signaling Enzymes

In addition to regulating membrane potential, ion channels have also been shown to activate downstream intracellular mediators, although less is known concerning this aspect of ion channel function. This is perhaps best described for Ca^{2+} channels, which have been associated with activation of intracellular kinase signaling pathways such as the ERK (extracellular regulated kinase) pathways and the Rho/Ras kinase pathways. Interestingly, serine-threonine kinases and tyrosine kinases are well known for their ability to trans-regulate ion channel function through phosphorylation following activation by other receptor classes such as GPCRs or growth factor receptors.[132] These activities act to either enhance or reduce the cell's sensitivity to stimuli. However, there are recent reports of downstream activation of kinase pathways following stimulation of some P2X and TRP channels.

Within the P2X family, the $P2X_3$ receptor has been suggested as a key mediator of inflammatory pain stimuli at the level of the dorsal root ganglion neuron.[133,134] $P2X_3$ receptor activation by ATP was found to be responsible for the enhanced inflammatory response of primary sensory afferent neurons via activation of $ERK_{1,2}$ signaling and this was correlated *in vivo* with a reduced sensitivity to joint pain following inflammatory stimulation.[133,134] Furthermore, blockade of either ERK phosphorylation[134,135] or $P2X_3$ activation (TNP-ATP)[133] reduces sensitivity to pain in animal models of inflammatory joint diseases such as arthritis.

These data suggest that the P2X receptors are physiological and pathophysiologic mediators of inflammatory pain via ERK activation and that modulation of their ability to stimulate ERK may be a therapeutically relevant target for treatment of joint pain.[133] Since activation of ERK signals is dependent upon the Ca^{2+} conductance of these channels, ligands that block or modulate the selectivity filter of P2X receptors such as calmidazolium (see above) may represent useful pharmacological tools by selectively blocking the Ca^{2+} permeability of P2X channels while allowing passage of other ions through the dilated pore. Pharmacological modulation of ion selectivity would represent a novel approach to affecting channel function.

Members of the TRP channel family have also been found to activate classical intracellular signaling enzymes. For example, TRPV4 has been identified as a key mechanotransducing receptor in vascular endothelial cells. Vascular endothelial cells respond to mechanical forces such as stretch and the frictional shear forces applied by flowing blood.[136] These forces significantly impact endothelial cell function. Indeed, atherosclerotic plaque formation occurs preferentially in regions of low shear, turbulent blood flow but not in those regions of the vascular bed exposed to high shear, laminar blood flow.[136] Endothelial cells express many members of the TRP family of ion channels, in addition to TRPV4.[137,138] TRPV4 is a key mediator of flow-induced vasodilation of blood vessels, suggesting an important role as a mechanotransducing receptor within vascular endothelial cells.[139]

Similar to TRPV1, TRPV4 is a non-selective cation channel that prefers Ca^{2+} over Na^{+}.[140,141] Cyclical stretching of endothelial cells *in vitro* by flexing the floor of the culture dish where the cells are adhered results in activation of ionic currents, ERK phosphorylation, and PI3K/AKT phosphorylation.[142] At the level of the whole cell, these stimuli lead to thickening of cytoskeletal stress fibers and the characteristic reorientation of the endothelial cell anti-parallel to the main axis of the strain.[142] This is a key feature of healthy endothelial cells and helps to maintain a strong barrier between circulating cells of the blood and the interstitial space.[136] A role for TRPV4 in this process may exist since the dynamic changes in endothelial cell cytoskeletal components can be completely blocked by the application of gadolinium chloride (a non-selective TRPV4 channel blocker) or knockdown of TRPV4 expression using specific small interfering RNAs. It has been suggested that activation of these signaling pathways plays a crucial role in activating stress fiber formation and cellular remodeling in response to mechanical force in endothelial cells.[142] Others have hypothesized that shear forces can activate ion channels to affect ERK signaling in endothelial cells via voltage-dependent Na^+ channels.[143]

Recently, a related TRP family member, the transient receptor potential melastatin-like 7 (TRPM7) channel, was found to have two distinct functions within the protein complex. This channel contains both an ion channel pore that is permeant to cations and a serine-threonine kinase domain (Figure 6.6). Such channels are referred to as chanzymes to describe their bifunctionality.[144] The kinase domain is structurally related to protein kinase A and has sequence homology to the α-kinase family of serine-threonine kinases.[145] The channel function is activated by low pH and phospholipase C-linked receptors.[146] This channel has been observed to be involved in regulating apoptosis via Fas receptor activation in T-cells. It has been found that the apoptotic activity of the TRPM7 chanzyme is dissociable from the activity of the serine-threonine kinase domain, suggesting that the kinase domain has an independent function.

Desai et al.[146] have shown that caspases acting intracellularly can cleave the kinase domain from the remaining channel structure, leading to release of the active kinase and actually enhancing the conductance of the channel and regulation of Fas receptor localization.[146] In their studies, immunoneutralization of either caspase 3 or caspase 8 blocked the cleavage of TRPM7 *in vitro*. Cleavage occurs at aspartate 1510 in the carboxy-terminus of the TRPM7 protein and it seems that the enhanced channel conductance induced by this cleavage is relatively specific since truncations of the channel at other sites within this region do not have this effect.

Additionally, Desai et al.[146] demonstrate that the increase in TRPM7 conductance is solely due to the cleavage of the C-terminus and not activation of the kinase activity. Creation of a cleavage-resistant TRPM7 by mutating D1510 to alanine produces a channel that does not participate in Fas-induced cell apoptosis. The physiological importance of this has been revealed in knockout mice, where Fas receptor knockout mice and TRPM7 knockout mice display a similar pathological phenotype. These data suggest that members of this family of unique channels have the capability to selectively signal through multiple pathways that are functionally distinct (ion flux and phosphorylation). It is not yet clear

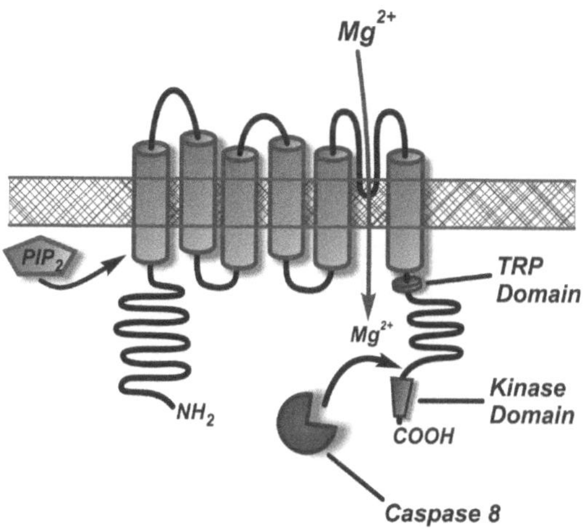

FIGURE 6.6 Schematic diagram of the TRPM chanzyme. The TRPM chanzyme is a channel that also contains an intracellular kinase domain. Structural features of the TRPM chanzyme are illustrated. Note the six transmembrane helices and the intracellular TRP domain. The TRPM channel can be modulated by PIP2. The channel is permeant to cations, primarily Mg^{2+}. The kinase domain is shown at the intracellular carboxy-terminus. Under some conditions, caspases can cleave the active kinase domain from the channel protein. This has the result of increasing channel conductance.

whether the caspase-induced cleavage of the TRPM7 channel is stimulus dependent or what activities lie downstream of the liberated kinase domain, but these data provide further insight into the plasticity of ion channels and the possibilities for biased signaling.

Biased Signaling By Calcium Influx Though Channels

Calcium ions are key second messengers in cells, impacting such diverse processes as neurotransmitter release, muscle contraction, hormone secretion, and the regulation of gene transcription. Several excellent reviews of the cellular actions of ions are available elsewhere.[147,148] Here, we shall focus on how Ca^{2+} flux through ion channels can lead to biased cellular signaling. Biased signaling by Ca^{2+} arises from the spatial and temporal organization of changes in the concentration of the ion. That is, Ca^{2+} elevations in the cytosol due to the opening of an ion channel often result in a localized, transient change that is not reflected in the rest of the cell. In this way, "microdomains" of Ca^{2+} can have specialized functions that often result in distinct cellular programs compared with more global changes in concentration within the cell.[149]

For instance, Ca^{2+} influx through presynaptic voltage-gated calcium (Cav) channels is key to triggering neurotransmitter release from neurons. At these sites, there is an association between the Cav channel and the exocytotic machinery. For example, one of the intracellular loops of Cav2.1 and Cav2.2 channels interacts with the SNARE proteins, syntaxin and SNAP25, on the synaptic membrane and with synaptotagmin on the vesicle membrane.[150] This close interaction ensures that influx of Ca^{2+} through these channels during the action potential triggers the fusion of nearby docked vesicles. Interestingly, the association of Cav2 channels and SNARE proteins may provide more than a source of localized Ca^{2+} ions. In some neurons, release can be detected independent of changes of Ca^{2+} but dependent on voltage.[151,152] The voltage-dependent release is sensitive to disruption of the Cav2.1–SNARE interaction, suggesting that voltage-dependent conformational changes in the Cav2 channel are coupled to increased vesicle fusion.[151]

Calcium entry via ion channels also shows evidence of biased signaling to the nucleus. Calcium influx through Cav1.2 channels activates the mitogen-activated protein kinase (MAPK) pathway leading to the phosphorylation of cyclic AMP response element-binding protein (CREB). CREB is a transcription factor that regulates genes involved in numerous functions, including synaptic plasticity. Surprisingly, Ca^{2+}elevation produced by entry through other channels does not activate this pathway. The specificity for Cav1.2 arises from association of calmodulin with these channels. Binding of calmodulin to a specific motif on the Cav1.2 channel in a Ca^{2+}-dependent manner activates the MAPK pathway by a mechanism not yet fully understood. Hence, localized elevations in Ca^{2+} from the opening of a specific Cav channel leads to the initiation of a specialized transcriptional program in the nucleus. Another related example is Ca^{2+} signaling in T-cells. In these cells, activation of the T-cell receptor (TCR) leads to the release of calcium from intracellular stores, whose depletion activates the store-operated plasma membrane current, I(CRAC). The periodic activation of ICRAC channels leads to oscillations in intracellular calcium.[153,154] Oscillations in Ca^{2+} are much more efficient in triggering gene transcription than sustained elevations, and specificity is encoded in the frequency of the oscillations.[155]

Channel Functions beyond Ion Conduction

In addition to their central role in generating and regulating electrical activity of cells, there is a growing literature that suggests ion channels can serve cellular functions in addition to ion conduction (reviewed in detail elsewhere[156]).

The earliest indication of channel function independent of conduction came from the study of excitation-contraction coupling in skeletal muscle. In these cells, depolarization activates the calcium channel, Cav1.1, in the plasma membrane. Opening of the channel leads to depolarization and muscle contraction even in the absence of external Ca^{2+}. Thus, calcium flux through the Cav1.1 channel is not required for contraction. These observations suggest that the Ca^{2+} associated with contraction must come from intracellular stores. Indeed, it is now well recognized that Ca^{2+} released from the sarcoplasmic reticulum in muscle tissue occurs via the ryanodine receptor (a calcium channel). In these cells, the cytoplasmic domain of the Cav1.1 channel is physically coupled to the ryanodine receptor so that the voltage-dependent conformational change in Cav1.1 is communicated to the ryanodine receptor, opening the channel.[157,158]

Certain members of the TRP channel family provide other examples where ion conduction can be separated from other functions. As discussed above, members of the TRPM subfamily have a large cytoplasmic carboxy-terminus that functions as an enzyme. For example, the carboxy-termini of TRPM6 and TRPM7 have a kinase domain capable of phosphorylating the channel and other substrates as well.[159,160]

Auxillary subunits of ion channels can have very specialized functions, often distinct from regulating ion conduction. A good example is the beta subunit of Kv channels. These are generally cytoplasmic auxillary subunits that bind to the pore-forming alpha subunit to modulate gating and channel trafficking.[161] Structural analysis of one such beta subunit, Kvβ2, suggests it may function as an oxidoreductase enzyme.[162,163] However, it is still unclear if the subunit functions as an enzyme in physiological systems.[161] The beta subunit of voltage-gated sodium (Nav) channels has a function that is quite different to that of the beta subunit of Kv channels. The beta subunit of Nav channels has a single transmembrane domain with a large immunoglobulin-like extracellular domain. The IgG-like domain acts as an adhesion molecule, mediating cell–cell adhesion through binding to other adhesion molecules as well as through homotypic binding to beta subunits on adjacent cells. The beta subunit has a cytoplasmic C-terminus that binds cytoskeletal proteins such as ankyrin G.[164] These interactions are likely important for the high density localization of Nav channels at specific sites, such as the nodes of Ranvier.

Another interesting role of ion channels that is potentially distinct from ion conduction is the regulation of cellular proliferation. Specifically, expression of several types of ion channels is correlated with cell proliferation and is up-regulated in cells derived from tumors. For example, expression of the channel Kv10.1 (also referred to as Eag) is up-regulated in certain cancers.[165,166] Interestingly, these channels appear to modulate the MAPK pathway and proliferation by a voltage-dependent mechanism independent of ion conduction through the channel.[167] Another example is the two-pore channel, KCNK9, which is amplified in colorectal cancers.[168] In this case, the oncogenic activity of KCNK9 appears to be dependent on potassium conduction through the channel.[169] Nav channels have also been implicated in cell proliferation and cancer.[170] Taken together, the data suggest a role for ion channels in regulation of proliferation. However, it remains unclear whether dysregulation of channel expression is causative in certain cancers or the result of other genomic changes. In any event, it remains possible that changes in the expression of certain channels may be markers of some cancers and that modulation of channel function may be therapeutically useful.

Biased Signaling Arising from Modulation of Ion Channel Function

Ion channels, like most signaling proteins, are the target of extensive modulation by a variety of signaling mechanisms. The literature in this field is extensive and the mechanisms are diverse and include phosphorylation, RNA editing, alternative splicing, protein–protein interactions, and many others. Some of the better understood examples will be highlighted here to demonstrate the complexity and impact of this modulation.

Ion channel function is often regulated by phosphorylation of the pore-forming subunit or accessory subunits. Numerous examples of channel regulation by phosphorylation exist for both voltage-gated channels and ligand-gated channels (reviewed in Levitan[171]). The earliest examples of channel regulation was the demonstration that cardiac calcium channel currents are enhanced by β-adrenergic agonists.[172] Catecholamines released by sympathetic nerves increase cardiac contractility by increasing Ca^{2+} influx thorough Cav1.2 channels (reviewed in Catterall[173]). Phosphorylation of residues in the C-terminus of Cav1.2 by protein kinase A and casein kinase II is largely responsible for the up-regulation of activity in response to β-adrenergic stimulation. Mice with channels lacking these residues have impaired cardiac function and exercise capacity, suggesting that phosphorylation of Cav1.2 at these sites is required for the sympathetic activation of the "fight or flight" response.[174]

Another example of voltage-gated channel regulation by phosphorylation is the potassium channel, Kv2.1. For this channel, considerable detail is known about the sites and functional effects of phosphorylation.[175] The native Kv2.1 channel is phosphorylated at multiple intracellular residues and the degree of phosphorylation is regulated by neuronal activity. Phosphorylation produces a graded functional effect on the gating of the channel and, ultimately, the firing properties of the neurons that express the channel.[176] Importantly, the regulation is bidirectional, where neuronal activity induces dephosphorylation and suppression of activity leads to hyperphosphorylation of specific residues.[177] For ligand-gated channels, phosphorylation of AMPA receptors is a well-studied example. Phosphorylation of AMPA receptors on the intracellular C-terminus is important for regulation of receptor trafficking.[178] Similar examples exist for the kainate and NMDA subtypes. In many of these cases, the precise residues have been identified, the kinases responsible have been determined, and regulation by neuronal activity has been demonstrated (reviewed in Traynelis[11]).

The AMPA receptor is an excellent example of regulation of channel function by RNA editing and alternative splicing. The AMPA receptor subunit, GluA2 mRNA, is edited to change a glutamine in the transmembrane domain to an arginine. This change results in a receptor that has low Ca^{2+} permeability.[179] Most of the receptors in the CNS are the edited form, thereby restricting the triggering of Ca^{2+}-dependent cellular signaling to activation of NMDA receptors rather than AMPA receptors. Alternative splicing of AMPA receptors also impacts function. For example, the interchangeable exons "flip" and "flop" encode a region of the LBD and confer properties of receptor desensitization, deactivation, and sensitivity to allosteric modulators.[180,181] Numerous other examples of alternative splicing of channel mRNAs exist in the literature. Notably, alternative splicing of neuronal calcium channel alpha subunits controls channel kinetics as well as modulation by G protein signaling pathways.[182]

Regulation of ion channels by G protein signaling cascades is rich in examples and variety. In addition to the regulation of Cav1.2 by phosphorylation discussed earlier, there are two other mechanisms that merit highlighting. Regulation of the inwardly rectifying potassium channel by acetylcholine is a physiologically important mechanism for control of heart rate by parasympathetic input. The current, originally referred to as $I_{K(ACh)}$, is now known to be composed of GIRK1/GIRK4 subunits. Regulation of this channel occurs by a membrane delimited pathway[183] and, after much debate, was shown to be due to direct interaction of G protein $\beta\gamma$ subunits with the channel.[184] Structural studies are consistent with a membrane-delimited mechanism and show four $\beta\gamma$ subunits bound per channel.[185] Modulation of Cav channels in sympathetic neurons also occurs by a direct $\beta\gamma$ mechanism.[186,187] Interestingly, modulation of these Cav channels by $\beta\gamma$ subunits is due to a change in the voltage-dependence of activation[188] and is itself susceptible to regulation by protein kinases.[189,190] In addition to direct regulation by $\beta\gamma$ subunits, neuronal Cav channels are modulated by another membrane-delimited pathway. This time the messenger is a lipid rather than a protein and the process is independent of voltage. Phosphatidylinositol 4,5-bisphosphonate (PIP(2)) supports the function of these Cav channels. Activation of G protein-coupled receptors coupled to phospholipase C causes depletion of membrane PIP(2), resulting in inhibition of channel function.[191] A similar mechanism explains the muscarinic modulation of the voltage-gated potassium channel, termed M-current, formed by KCNQ2/KCNQ3 subunits.[192]

CONCLUSION

Overall, the known effects of conformational-dependent activation of ion channels by ligands and stimuli across receptor families—as well as ligand-dependent modulation of ion selectivity within some ion channels and activation of signaling pathways separate from ion conductance—make a strong case for the potential of biased signaling within some ion channels. Further work is needed, however, to understand the commonality of this mechanism across the larger family of ion channels.

References

1. Hille B. *Ion channels of excitable membranes*. 3rd ed. Sunderland, MA: Sinauer Associates; 2001.
2. Jegla TJ, Zmasek CM, Batalov S, Nayak SK. Evolution of the human ion channel set. *Comb Chem High Throughput Screen* 2009;**12**(1):2–23.
3. Drews J. Drug discovery: a historical perspective. *Science* 2000;**287**(5460):1960–4.
4. Zheng CJ, Han LY, Yap CW, Ji ZL, Cao ZW, Chen YZ. Therapeutic targets: progress of their exploration and investigation of their characteristics. *Pharmacol Rev* 2006;**58**(2):259–79.

5. Dunlop J, Bowlby M, Peri R, Vasilyev D, Arias R. High-throughput electrophysiology: an emerging paradigm for ion-channel screening and physiology. *Nat Rev Drug Discov* 2008;**7**(4):358–68.
6. Bagal SK, Brown AD, Cox PJ, Omoto K, Owen RM, Pryde DC, et al. Ion channels as therapeutic targets: a drug discovery perspective. *J Med Chem* 2013;**56**(3):593–624.
7. Ashcroft FM. From molecule to malady. *Nature* 2006;**440**(7083):440–7.
8. Changeux JP, Edelstein SJ. Allosteric receptors after 30 years. *Neuron* 1998;**21**(5):959–80.
9. Thompson AJ, Lester HA, Lummis SC. The structural basis of function in Cys-loop receptors. *Q Rev Biophys* 2010;**43**(4):449–99.
10. Miller PS, Smart TG. Binding, activation and modulation of Cys-loop receptors. *Trends Pharmacol Sci* 2010;**31**(4):161–74.
11. Traynelis SF, Wollmuth LP, McBain CJ, Menniti FS, Vance KM, Ogden KK, et al. Glutamate receptor ion channels: structure, regulation, and function. *Pharmacol Rev* 2010;**62**(3):405–96.
12. Sherwood TW, Frey EN, Askwith CC. Structure and activity of the acid-sensing ion channels. *Am J Physiol Cell Physiol* 2012;**303**(7):C699–710.
13. Wemmie JA, Taugher RJ, Kreple CJ. Acid-sensing ion channels in pain and disease. *Nat Rev Neurosci* 2013;**14**(7):461–71.
14. Coddou C, Yan Z, Obsil T, Huidobro-Toro JP, Stojilkovic SS. Activation and regulation of purinergic P2X receptor channels. *Pharmacol Rev* 2011;**63**(3):641–83.
15. Baconguis I, Hattori M, Gouaux E. Unanticipated parallels in architecture and mechanism between ATP-gated P2X receptors and acid sensing ion channels. *Curr Opin Struct Biol* 2013;**23**(2):277–84.
16. Yu FH, Catterall WA. The VGL-chanome: a protein superfamily specialized for electrical signaling and ionic homeostasis. *Sci STKE* 2004;**2004**(253):re15.
17. Ramsey IS, Delling M, Clapham DE. An introduction to TRP channels. *Annu Rev Physiol* 2006;**68**:619–47.
18. Lesage F, Lazdunski M. Molecular and functional properties of two-pore-domain potassium channels. *Am J Physiol Renal Physiol* 2000;**279**(5):F793–801.
19. Doyle DA, Morais Cabral J, Pfuetzner RA, Kuo A, Gulbis JM, Cohen SL, et al. The structure of the potassium channel: molecular basis of K^+ conduction and selectivity. *Science* 1998;**280**(5360):69–77.
20. Zhou M, Morais-Cabral JH, Mann S, MacKinnon R. Potassium channel receptor site for the inactivation gate and quaternary amine inhibitors. *Nature* 2001;**411**(6838):657–61.
21. Jiang Y, Lee A, Chen J, Cadene M, Chait BT, MacKinnon R. Crystal structure and mechanism of a calcium-gated potassium channel. *Nature* 2002;**417**(6888):515–22.
22. Jiang Y, Lee A, Chen J, Cadene M, Chait BT, MacKinnon R. The open pore conformation of potassium channels. *Nature* 2002;**417**(6888):523–6.
23. Jiang Y, Lee A, Chen J, Ruta V, Cadene M, Chait BT, et al. X-ray structure of a voltage-dependent K^+ channel. *Nature* 2003;**423**(6935):33–41.
24. Jiang Y, Ruta V, Chen J, Lee A, MacKinnon R. The principle of gating charge movement in a voltage-dependent K^+ channel. *Nature* 2003;**423**(6935):42–8.
25. Ruta V, Chen J, MacKinnon R. Calibrated measurement of gating-charge arginine displacement in the KvAP voltage-dependent K^+ channel. *Cell* 2005;**123**(3):463–75.
26. Long SB, Campbell EB, Mackinnon R. Crystal structure of a mammalian voltage-dependent Shaker family K^+ channel. *Science* 2005;**309**(5736):897–903.
27. Payandeh J, Scheuer T, Zheng N, Catterall WA. The crystal structure of a voltage-gated sodium channel. *Nature* 2011;**475**(7356):353–8.
28. Zhang X, Ren W, DeCaen P, Yan C, Tao X, Tang L, et al. Crystal structure of an orthologue of the NaChBac voltage-gated sodium channel. *Nature* 2012;**486**(7401):130–4.
29. Alabi AA, Bahamonde MI, Jung HJ, Kim JI, Swartz KJ. Portability of paddle motif function and pharmacology in voltage sensors. *Nature* 2007;**450**(7168):370–5.
30. Bosmans F, Martin-Eauclaire MF, Swartz KJ. Deconstructing voltage sensor function and pharmacology in sodium channels. *Nature* 2008;**456**(7219):202–8.
31. Murata Y, Iwasaki H, Sasaki M, Inaba K, Okamura Y. Phosphoinositide phosphatase activity coupled to an intrinsic voltage sensor. *Nature* 2005;**435**(7046):1239–43.

32. Starkus JG, Kuschel L, Rayner MD, Heinemann SH. Ion conduction through C-type inactivated Shaker channels. *J Gen Physiol* 1997;**110**(5):539–50.
33. Kiss L, LoTurco J, Korn SJ. Contribution of the selectivity filter to inactivation in potassium channels. *Biophys J* 1999;**76**(1 Pt 1):253–63.
34. Cuello LG, Jogini V, Cortes DM, Perozo E. Structural mechanism of C-type inactivation in K(+) channels. *Nature* 2010;**466**(7303):203–8.
35. Kawate T, Michel JC, Birdsong WT, Gouaux E. Crystal structure of the ATP-gated P2X4 ion channel in the closed state. *Nature* 2009;**460**(7255):592–8.
36. Brejc K, van Dijk WJ, Klaassen RV, Schuurmans M, van Der Oost J, Smit AB, et al. Crystal structure of an ACh-binding protein reveals the ligand-binding domain of nicotinic receptors. *Nature* 2001;**411**(6835):269–76.
37. Hansen SB, Sulzenbacher G, Huxford T, Marchot P, Taylor P, Bourne Y. Structures of Aplysia AChBP complexes with nicotinic agonists and antagonists reveal distinctive binding interfaces and conformations. *EMBO J* 2005;**24**(20):3635–46.
38. Gonzales EB, Kawate T, Gouaux E. Pore architecture and ion sites in acid-sensing ion channels and P2X receptors. *Nature* 2009;**460**(7255):599–604.
39. Sobolevsky AI, Rosconi MP, Gouaux E. X-ray structure, symmetry and mechanism of an AMPA-subtype glutamate receptor. *Nature* 2009;**462**(7274):745–56.
40. Hibbs RE, Gouaux E. Principles of activation and permeation in an anion-selective Cys-loop receptor. *Nature* 2011;**474**(7349):54–60.
41. Jin R, Singh SK, Gu S, Furukawa H, Sobolevsky AI, Zhou J, et al. Crystal structure and association behaviour of the GluR2 amino-terminal domain. *EMBO J* 2009;**28**(12):1812–23.
42. Armstrong N, Gouaux E. Mechanisms for activation and antagonism of an AMPA-sensitive glutamate receptor: crystal structures of the GluR2 ligand binding core. *Neuron* 2000;**28**(1):165–81.
43. Inanobe A, Furukawa H, Gouaux E. Mechanism of partial agonist action at the NR1 subunit of NMDA receptors. *Neuron* 2005;**47**(1):71–84.
44. Kussius CL, Popescu GK. NMDA receptors with locked glutamate-binding clefts open with high efficacy. *J Neurosci* 2010;**30**(37):12474–9.
45. Sun Y, Olson R, Horning M, Armstrong N, Mayer M, Gouaux E. Mechanism of glutamate receptor desensitization. *Nature* 2002;**417**(6886):245–53.
46. Armstrong N, Jasti J, Beich-Frandsen M, Gouaux E. Measurement of conformational changes accompanying desensitization in an ionotropic glutamate receptor. *Cell* 2006;**127**(1):85–97.
47. Keramidas A, Moorhouse AJ, Schofield PR, Barry PH. Ligand-gated ion channels: mechanisms underlying ion selectivity. *Prog Biophys Mol Biol* 2004;**86**(2):161–204.
48. Lee DJS, Keramidas A, Moorhouse AJ, Schofield PR, Barry PH. The contribution of proline 250 (P-2′) to pore diameter and ion selectivity in the human glycine receptor channel. *Neurosci Lett* 2003;**351**(3):196–200.
49. del Castillo J, Katz B. Interaction at end-plate receptors between different choline derivatives. *Proc R Soc Lond B Biol Sci* 1957;**146**(924):369–81.
50. Monod J, Wyman J, Changeux J-P. On the nature of allosteric transitions: a plausible model. *J Mol Biol* 1965;**12**:88–118.
51. Changeux JP. On the allosteric properties of L-threonine deaminase. V. The allosteric transition. *Bull Soc Chim Biol (Paris)* 1965;**47**:267–80.
52. Changeux JP. On the allosteric properties of L-threonine deaminase. VI. General Discussion. *Bull Soc Chim Biol (Paris)* 1965;**47**:281–300.
53. Koshland Jr. DE, Némethy G, Filmer D. Comparison of experimental binding data and theoretical models in proteins containing subunits. *Biochemistry* 1966;**5**(1):365–85.
54. Changeux J-P, Edelstein SJ. Allosteric mechanisms of signal transduction. *Science* 2005;**308**(5727):1424–8.
55. Changeux J-P, Thiery J, Tung Y, Kittel C. On the cooperativity of biological membranes. *Proc Natl Acad Sci USA* 1967;**57**(2):335–41.
56. Lemoine D, Jiang R, Taly A, Chataigeneau T, Specht A, Grutter T. Ligand-gated ion channels: new insights into neurological disorders and ligand recognition. *Chem Rev* 2012;**112**(12):6285–318.
57. Smith GB, Olsen RW. Identity of a [^{3}H]muscimol photoaffinity substrate in the bovine γ-amino butyric acid$_A$ receptor α subunit. *J Biol Chem* 1994;**269**(32):20380–7.

58. Smit AB, Syed NI, Schapp D, van Minnen J, Klumperman J, Kits KS, et al. A glia-derived acetylcholine-binding protein that modulates synaptic transmission. *Nature* 2001;**411**(6835):261–8.
59. Dellisanti CD, Yao Y, Stroud JC, Wang Z-Z, Chen L. Crystal structure of the extracellular domain of nAChR α1 bound to α-bungarotoxin at 1.94 Å resolution. *Nat Neurosci* 2007;**10**(8):953–62.
60. Celie PHN, Kasheverov IE, Mordvintsev DY, Hogg RC, van Nierop P, van Elk R, et al. Crystal structure of nicotinic acetylcholine receptor homolog AChBP in complex with an α-conotoxin PnIA variant. *Nat Struct Mol Biol* 2005;**12**(7):582–8.
61. Bourne Y, Talley TT, Hansen SB, Taylor P, Marchot P. Crystal structure of a Cbtx–AChBP complex reveals essential interactions between snake α-neurotoxins and nicotinic receptors. *EMBO J* 2005;**24**(8):1512–22.
62. Mukhtasimova N, Lee WY, Wang H-L, Sine SM. Detection and trapping of intermediate states priming nicotinic receptor channel opening. *Nature* 2009;**459**(7245):451–4.
63. Sine SM, Steinbach JH. Acetylcholine receptor activation by a site-selective ligand: nature of brief open and closed states in BC3H-1 cells. *J Physiol (London)* 1986;**370**:357–79.
64. Sine SM, Steinbach JH. Activation of acetylcholine receptors on clonal mammalian BC3H-1 cells by low concentration of agonist. *J Physiol (London)* 1986;**373**:129–62.
65. Zimmermann I, Dutzler R. Ligand activation of the prokaryotic pentameric ligand-gated ion channel ELIC. *PLoS Biol* 2011;**9**(6):e1001101.
66. Duret G, Van Renterghem C, Weng Y, Prevost M, Moraga-Cid G, Huon C, et al. Functional prokaryotic–eukaryotic chimera from the pentameric ligand-gated ion channel family. *Proc Natl Acad Sci* 2011;**108**(29):12143–8.
67. Colquhoun D. Binding, gating, affinity and efficacy: the interpretation of structure–activity relationships for agonists and of the effects of mutating receptors. *Br J Pharmacol* 1998;**125**(5):924–47.
68. Williams DK, Wang J, Papke RL. Positive allosteric modulators as an approach to nicotinic acetylcholine receptor-targeted therapeutics: advantages and limitations. *Biochem Pharmacol* 2011;**82**(8):915–30.
69. Kenakin T. Allosteric theory: taking therapeutic advantage of the malleable nature of GPCRs. *Curr Neuropharmacol* 2007;**5**(3):149–56.
70. Campo-Soria C, Chang Y, Weiss DS. Mechanism of action of benzodiazepines on $GABA_A$ receptors. *Br J Pharmacol* 2006;**148**(7):984–90.
71. Gill JK, Savolainen M, Young GT, Zwart R, Sher E, Millar NS. Agonist activation of alpha7 nicotinic acetylcholine receptors via an allosteric transmembrane site. *Proc Natl Acad Sci USA* 2011;**108**(14):5867–72.
72. Colquhoun D, Lape R. Perspectives on: conformational coupling in ion channels: allosteric coupling in ligand-gated ion channels. *J Gen Physiol* 2012;**140**(6):599–612.
73. Macdonald RL, Olsen RW. GABAA receptor channels. *Annu Rev Neurosci* 1994;**17**:569–602.
74. Wafford KA, Macaulay AJ, Fradley R, O'Meara GF, Reynolds DS, Rosahl TW. Differentiating the role of gamma-aminobutyric acid type A (GABAA) receptor subtypes. *Biochem Soc Trans* 2004;**32**(Pt3):553–6.
75. Whiting PJ. GABA-A receptors: a viable target for novel anxiolytics? *Curr Opin Pharmacol* 2006;**6**(1):24–9.
76. Trincavelli ML, Da Pozzo E, Daniele S, Martini C. The GABAA-BZR complex as target for the development of anxiolytic drugs. *Curr Top Med Chem* 2012;**12**(4):254–69.
77. Sigel E. Mapping of the benzodiazepine recognition site on GABA(A) receptors. *Curr Top Med Chem* 2002;**2**(8):833–9.
78. Twyman RE, Rogers CJ, Macdonald RL. Differential regulation of gamma-aminobutyric acid receptor channels by diazepam and phenobarbital. *Ann Neurol* 1989;**25**(3):213–20.
79. Gielen MC, Lumb MJ, Smart TG. Benzodiazepines modulate GABAA receptors by regulating the preactivation step after GABA binding. *J Neurosci* 2012;**32**(17):5707–15.
80. Amin J. A single hydrophobic residue confers barbiturate sensitivity to gamma-aminobutyric acid type C receptor. *Mol Pharmacol* 1999;**55**(3):411–23.
81. Serafini R, Bracamontes J, Steinbach JH. Structural domains of the human GABAA receptor 3 subunit involved in the actions of pentobarbital. *J Physiol* 2000;**524(Pt 3)**:649–76.
82. Gronlien JH, Hakerud M, Ween H, Thorin-Hagene K, Briggs CA, Gopalakrishnan M, et al. Distinct profiles of alpha7 nAChR positive allosteric modulation revealed by structurally diverse chemotypes. *Mol Pharmacol* 2007;**72**(3):715–24.
83. Bertrand D, Bertrand S, Cassar S, Gubbins E, Li J, Gopalakrishnan M. Positive allosteric modulation of the alpha7 nicotinic acetylcholine receptor: ligand interactions with distinct binding sites and evidence for a prominent role of the M2–M3 segment. *Mol Pharmacol* 2008;**74**(5):1407–16.

84. Collins T, Millar NS. Nicotinic acetylcholine receptor transmembrane mutations convert ivermectin from a positive to a negative allosteric modulator. *Mol Pharmacol* 2010;**78**(2):198–204.
85. Gronlien JH, Ween H, Thorin-Hagene K, Cassar S, Li J, Briggs CA, et al. Importance of M2–M3 loop in governing properties of genistein at the alpha7 nicotinic acetylcholine receptor inferred from alpha7/5-HT3A chimera. *Eur J Pharmacol* 2010;**647**(1–3):37–47.
86. Hansen SB, Taylor P. Galanthamine and non-competitive inhibitor binding to ACh-binding protein: evidence for a binding site on non-alpha-subunit interfaces of heteromeric neuronal nicotinic receptors. *J Mol Biol* 2007;**369**(4):895–901.
87. Young GT, Zwart R, Walker AS, Sher E, Millar NS. Potentiation of alpha7 nicotinic acetylcholine receptors via an allosteric transmembrane site. *Proc Natl Acad Sci USA* 2008;**105**(38):14686–91.
88. Jin R, Clark S, Weeks AM, Dudman JT, Gouaux E, Partin KM. Mechanism of positive allosteric modulators acting on AMPA receptors. *J Neurosci Nurs* 2005;**25**(39):9027–36.
89. Hansen KB, Ogden KK, Traynelis SF. Subunit-selective allosteric inhibition of glycine binding to NMDA receptors. *J Neurosci* 2012;**32**(18):6197–208.
90. Jang MK, Mierke DF, Russek SJ, Farb DH. A steroid modulatory domain on NR2B controls N-methyl-D-aspartate receptor proton sensitivity. *Proc Natl Acad Sci USA* 2004;**101**(21):8198–203.
91. Paoletti P, Ascher P, Neyton J. High-affinity zinc inhibition of NMDA NR1-NR2A receptors. *J Neurosci* 1997;**17**(15):5711–25.
92. Karakas E, Simorowski N, Furukawa H. Structure of the zinc-bound amino-terminal domain of the NMDA receptor NR2B subunit. *EMBO J* 2009;**28**(24):3910–20.
93. Karakas E, Simorowski N, Furukawa H. Subunit arrangement and phenylethanolamine binding in GluN1/GluN2B NMDA receptors. *Nature* 2011;**475**(7355):249–53.
94. MacKinnon R, Heginbotham L, Abramson T. Mapping the receptor site for charybdotoxin, a pore-blocking potassium channel inhibitor. *Neuron* 1990;**5**(6):767–71.
95. Banerjee A, Lee A, Campbell E, Mackinnon R. Structure of a pore-blocking toxin in complex with a eukaryotic voltage-dependent K(+) channel. *Elife* 2013;**2**:e00594.
96. Catterall WA, Cestele S, Yarov-Yarovoy V, Yu FH, Konoki K, Scheuer T. Voltage-gated ion channels and gating modifier toxins. *Toxicon* 2007;**49**(2):124–41.
97. McDonough SI. Gating modifier toxins of voltage-gated calcium channels. *Toxicon* 2007;**49**(2):202–12.
98. Swartz KJ. Tarantula toxins interacting with voltage sensors in potassium channels. *Toxicon* 2007;**49**(2):213–30.
99. Milescu M, Lee HC, Bae CH, Kim JI, Swartz KJ. Opening the Shaker K^+ channel with hanatoxin. *J Gen Physiol* 2013;**141**(2):203–16.
100. Phillips LR, Milescu M, Li-Smerin Y, Mindell JA, Kim JI, Swartz KJ. Voltage-sensor activation with a tarantula toxin as cargo. *Nature* 2005;**436**(7052):857–60.
101. Milescu M, Vobecky J, Roh SH, Kim SH, Jung HJ, Kim JI, et al. Tarantula toxins interact with voltage sensors within lipid membranes. *J Gen Physiol* 2007;**130**(5):497–511.
102. McCormack K, Santos S, Chapman ML, Krafte DS, Marron BE, West CW, et al. Voltage sensor interaction site for selective small molecule inhibitors of voltage-gated sodium channels. *Proc Natl Acad Sci USA* 2013;**110**(29):E2724–32.
103. Wuttke TV, Seebohm G, Bail S, Maljevic S, Lerche H. The new anticonvulsant retigabine favors voltage-dependent opening of the Kv7.2 (KCNQ2) channel by binding to its activation gate. *Mol Pharmacol* 2005;**67**(4):1009–17.
104. Lange W, Geissendorfer J, Schenzer A, Grotzinger J, Seebohm G, Friedrich T, et al. Refinement of the binding site and mode of action of the anticonvulsant retigabine on KCNQ K+ channels. *Mol Pharmacol* 2009;**75**(2):272–80.
105. Peretz A, Pell L, Gofman Y, Haitin Y, Shamgar L, Patrich E, et al. Targeting the voltage sensor of Kv7.2 voltage-gated K + channels with a new gating-modifier. *Proc Natl Acad Sci USA* 2010;**107**(35):15637–42.
106. Noskov SY, Berneche S, Roux B. Control of ion selectivity in potassium channels by electrostatic and dynamic properties of carbonyl ligands. *Nature* 2004;**431**(7010):830–3.
107. Chung M-K, Guler AD, Caterina MJ. TRPV1 shows dynamic ionic selectivity during agonist stimulation. *Nat Neurosci* 2008;**11**(5):555–64.
108. Khakh BS, Bao XR, Labarca C, Lester HA. Neuronal P2X transmitter-gated cation channels change their ion selectivity in seconds. *Nat Neurosci* 1999;**2**(4):322–30.

109. Khakh BS, Lester HA. Dynamic selectivity filters in ion channels. *Neuron* 1999;**23**(4):653–8.
110. Yellen G. The moving parts of voltage-gated ion channels. *Q Rev Biophys* 1998;**31**(3):239–95.
111. Cockcroft S, Gomperts BD. ATP induces nucleotide permeability in rat mast cells. *Nature* 1979;**279**(5713):541–2.
112. Surprenant A, Rassendren F, Kawashima E, North RA, Buell G. The cytolytic P2Z receptor for extracellular ATP identified as a P2X receptor (P2X7). *Science* 1996;**272**(5262):735–8.
113. Virginio C, MacKenzie A, North RA, Surprenant A. Kinetics of cell lysis, dye uptake and permeability changes in cells expressing the rat P2X7 receptor. *J Physiol* 1999;**519**(2):335–46.
114. North RA. Molecular physiology of P2X receptors. *Physiol Rev* 2002;**82**(4):1013–67.
115. Rassendren F, Buell G, Newbolt A, North RA, Surprenant A. Identification of amino acid residues contributing to the pore of a P2X receptor. *EMBO J* 1997;**16**(12):3446–54.
116. Egan TM, Cox JA, Voigt MM. Molecular structure of P2X receptors. *Curr Opin Med Chem* 2004;**4**:821–9.
117. Li M, Chang T-H, Silberberg SD, Swartz KJ. Gating the pore of P2X receptor channels. *Nat Neurosci* 2008;**11**(8):883–7.
118. Shinozaki Y, Sumitomo K, Tsuda M, Koizumi S, Inoue K, Torimitsu K. Direct observation of ATP-induced conformational changes in single P2X(4) receptors. *PLoS Biol* 2009;**7**(5):e1000103.
119. Sun C, Heid ME, Keyel PA, Salter RD. The second transmembrane domain of P2X7 contributes to dilated pore formation. *PLoS ONE* 2013;**8**(4):e61886.
120. Kracun S, Chaptal V, Abramson J, Khakh BS. Gated access to the pore of a P2X receptor: Structural implications for closed–open transitions. *J Biol Chem* 2010;**285**(13):10110–21.
121. Donnelly-Roberts DL, Jarvis MF. Discovery of selective P2X7 receptor antagonists offers new insights into P2X7 receptor function and indicates a role in chronic pain states. *Br J Pharmacol* 2007;**151**(5):571–9.
122. Schroder K, Tschopp J. The inflammasomes. *Cell* 2010;**140**(6):821–32.
123. Birch RE, Schweibert EM, Peppiatt-Wildman CM, Wildman SS. Emerging key roles for P2X receptors in the kidney. *Front Physiol* 2013;**4**:262.
124. Ji X, Naito Y, Weng H, Endo K, Ma X, Iwai N. P2X7 deficiency attenuates hypertension and renal injury in deoxycorticosterone acetate-salt hypertension. *Am J Physiol Renal Physiol* 2012;**303**(8):F1207–15.
125. Guile SD, Alcaraz L, Birkinshaw TN, Bowers KC, Ebden MR, Furber M, et al. Antagonists of the P2X7 Receptor. From lead identification to drug development. *J Med Chem* 2009;**52**(10):3123–41.
126. Priel A, Silberberg SD. Mechanism of ivermectin facilitation of human P2X4 receptor channels. *J Gen Physiol* 2004;**123**(3):281–93.
127. Norenberg W, Sobottka H, Hempel C, Plotz T, Fischer W, Schmalzing G, et al. Positive alloseric modulation by ivermectin of human but not murine $P2X_7$ receptors. *Br J Pharmacol* 2012;**167**(1):48–66.
128. Virginio C, Church D, North RA, Surprenant A. Effects of divalent cations, protons and calmidazolium at the rat P2X7 receptor. *Neuropharmacology* 1997;**36**(9):1285–94.
129. Chung M-K, Güler AD, Caterina MJ. Biphasic currents evoked by chemical or thermal activation of the heat-gated ion channel, TRPV3. *J Biol Chem* 2005;**280**(16):15928–41.
130. Banke TG, Chaplan SR, Wickenden AD. Dynamic changes in the TRPA1 selectivity filter lead to progressive but reversible pore dilation. *Am J Physiol Cell Physiol* 2010;**298**(6):C1457–68.
131. Caterina MJ, Julius D. The vanilloid receptor: a molecular gateway to the pain pathway. *Annu Rev Neurosci* 2001;**24**:487–517.
132. Vay L, Gu C, McNaughton PA. The thermo-TRP ion channel family: properties and therapeutic implications. *Br J Pharmacol* 2012;**165**(4):787–801.
133. Seino D, Tokunaga A, Tachibana T, Yoshiya S, Dai Y, Obata K, et al. The role of ERK signaling and the P2X receptor on mechanical pain evoked by movement of inflamed knee joint. *Pain* 2006;**123**(1):193–203.
134. Dai Y, Fukuoka T, Wang H, Yamanaka H, Obata K, Tokunaga A, et al. Contribution of sensitized P2X receptors in inflamed tissue to the mechanical hypersensitivity revealed by phosphorylated ERK in DRG neurons. *Pain* 2004;**108**(3):258–66.
135. Cruz CD, Neto FL, Castro-Lopes J, McMahon SB, Cruz F. Inhibition of ERK phosphorylation decreases nociceptive behaviour in monoarthritic rats. *Pain* 2005;**116**(3):411–9.
136. Hahn C, Schwartz MA. Mechanotransduction in vascular physiology and atherogenesis. *Nat Rev Mol Cell Biol* 2009;**10**(1):53–62.

137. Kwan H-Y, Huang Y, Yao X. TRP channels in endothelial function and dysfunction. *Biochim Biophys Acta* 2007;**1772**(8):907–14.
138. Watanabe H, Murakami M, Ohba T, Takahashi Y, Ito H. TRP channel and cardiovascular disease. *Pharmacol Ther* 2008;**118**(3):337–51.
139. Köhler R, Heyken W-T, Heinau P, Schubert R, Si H, Kacik M, et al. Evidence for a functional role of endothelial transient receptor potential V4 in shear stress-induced vasodilatation. *Arterioscler Thromb Vasc Biol* 2006;**26**(7):1495–502.
140. Voets T, Prenen J, Vriens J, Watanabe H, Janssens A, Wissenbach U, et al. Molecular determinants of permeation through the cation channel TRPV4. *J Biol Chem* 2002;**277**(37):33704–10.
141. Watanabe H, Davis JB, Smart D, Jerman JC, Smith GD, Hayes P, et al. Activation of TRPV4 channels (hVRL-2/mTRP12) by phorbol derivatives. *J Biol Chem* 2002;**277**(16):13569–77.
142. Thodeti CK, Matthews B, Ravi A, Mammoto A, Ghosh K, Bracha AL, et al. TRPV4 channels mediate cyclic strain-induced endothelial cell reorientation through integrin-to-integrin signaling. *Circ Res* 2009;**104**(9):1123–30.
143. Traub O, Ishida T, Ishida M, Tupper JC, Berk BC. Shear stress-mediated extracellular signal-regulated kinase activation is regulated by sodium in endothelial cells: potential role for a voltage-dependent sodium channel. *J Biol Chem* 1999;**274**(29):20144–50.
144. Nadler MJS, Hermosura MC, Inabe K, Perraud A-L, Zhu Q, Stokes AJ, et al. LTRPC7 is an MgATP-regulated divalent cation channel required for cell viability. *Nature* 2001;**411**(6837):590–5.
145. Yamaguchi H, Matsushita M, Nairn AC, Kuriyan J. Crystal structure of the atypical protein kinase domain of a TRP channel with phosphotransferase activity. *Mol Cell* 2001;**7**(5):1047–57.
146. Desai BN, Krapivinsky G, Navarro B, Krapivinsky L, Carter Brett C, Febvay S, et al. Cleavage of TRPM7 releases the kinase domain from the ion channel and regulates its participation in Fas-induced apoptosis. *Dev Cell* 2012;**22**(6):1149–62.
147. Berridge MJ. Calcium signalling remodelling and disease. *Biochem Soc Trans* 2012;**40**(2):297–309.
148. Clapham DE. Calcium signaling. *Cell* 2007;**131**(6):1047–58.
149. Berridge MJ. Calcium microdomains: organization and function. *Cell Calcium* 2006;**40**(5–6):405–12.
150. Catterall WA. Structure and regulation of voltage-gated channels. *Annu Rev Cell Dev Biol* 2000;**16**:521–55.
151. Mochida S, Yokoyama CT, Kim DK, Itoh K, Catterall WA. Evidence for a voltage-dependent enhancement of neurotransmitter release mediated via the synaptic protein interaction site of N-type channels. *Proc Natl Acad Sci USA* 1998;**95**(24):14523–8.
152. Zhang C, Zhou Z. Ca($^{2+}$)-independent but voltage-dependent secretion in mammalian dorsal root ganglion neurons. *Nat Neurosci* 2002;**5**(5):425–30.
153. Lewis RS, Cahalan MD. Mitogen-induced oscillations of cytosolic Ca^{2+} and transmembrane Ca^{2+} current in human leukemic T cells. *Cell Regul* 1989;**1**(1):99–112.
154. Dolmetsch RE, Lewis RS. Signaling between intracellular Ca^{2+} stores and depletion-activated Ca^{2+} channels generates $[Ca^{2+}]i$ oscillations in T lymphocytes. *J Gen Physiol* 1994;**103**(3):365–88.
155. Dolmetsch RE, Xu K, Lewis RS. Calcium oscillations increase the efficiency and specificity of gene expression. *Nature* 1998;**392**(6679):933–6.
156. Kaczmarek LK. Non-conducting functions of voltage-gated ion channels. *Nat Rev Neurosci* 2006;**7**(10):761–71.
157. Rios E, Pizarro G. Voltage sensor of excitation-contraction coupling in skeletal muscle. *Physiol Rev* 1991;**71**(3):849–908.
158. Paolini C, Fessenden JD, Pessah IN, Franzini-Armstrong C. Evidence for conformational coupling between two calcium channels. *Proc Natl Acad Sci USA* 2004;**101**(34):12748–52.
159. Runnels LW, Yue L, Clapham DE. TRP-PLIK, a bifunctional protein with kinase and ion channel activities. *Science* 2001;**291**(5506):1043–7.
160. Schmitz C, Dorovkov MV, Zhao X, Davenport BJ, Ryazanov AG, Perraud AL. The channel kinases TRPM6 and TRPM7 are functionally nonredundant. *J Biol Chem* 2005;**280**(45):37763–71.
161. Pongs O, Schwarz JR. Ancillary subunits associated with voltage-dependent K^+ channels. *Physiol Rev* 2010;**90**(2):755–96.
162. Gulbis JM, Mann S, MacKinnon R. Structure of a voltage-dependent K^+ channel beta subunit. *Cell* 1999;**97**(7):943–52.

163. Gulbis JM, Zhou M, Mann S, MacKinnon R. Structure of the cytoplasmic beta subunit-T1 assembly of voltage-dependent K^+ channels. *Science* 2000;**289**(5476):123–7.
164. Patino GA, Isom LL. Electrophysiology and beyond: multiple roles of Na + channel beta subunits in development and disease. *Neurosci Lett* 2010;**486**(2):53–9.
165. Pardo LA, del Camino D, Sanchez A, Alves F, Bruggemann A, Beckh S, et al. Oncogenic potential of EAG K (+) channels. *EMBO J* 1999;**18**(20):5540–7.
166. Patt S, Preussat K, Beetz C, Kraft R, Schrey M, Kalff R, et al. Expression of ether a go-go potassium channels in human gliomas. *Neurosci Lett* 2004;**368**(3):249–53.
167. Hegle AP, Marble DD, Wilson GF. A voltage-driven switch for ion-independent signaling by ether-a-go-go K+ channels. *Proc Natl Acad Sci USA* 2006;**103**(8):2886–91.
168. Mu D, Chen L, Zhang X, See LH, Koch CM, Yen C, et al. Genomic amplification and oncogenic properties of the KCNK9 potassium channel gene. *Cancer Cell* 2003;**3**(3):297–302.
169. Pei L, Wiser O, Slavin A, Mu D, Powers S, Jan LY, et al. Oncogenic potential of TASK3 (Kcnk9) depends on K+ channel function. *Proc Natl Acad Sci USA* 2003;**100**(13):7803–7.
170. Brackenbury WJ. Voltage-gated sodium channels and metastatic disease. *Channels (Austin)* 2012;**6**(5):352–61.
171. Levitan IB. Modulation of ion channels by protein phosphorylation and dephosphorylation. *Annu Rev Physiol* 1994;**56**:193–212.
172. Reuter H. The dependence of slow inward current in Purkinje fibres on the extracellular calcium-concentration. *J Physiol* 1967;**192**(2):479–92.
173. Catterall WA. Voltage-gated calcium channels. *Cold Spring Harb Perspect Biol* 2011;**3**(8):a003947.
174. Fu Y, Westenbroek RE, Scheuer T, Catterall WA. Phosphorylation sites required for regulation of cardiac calcium channels in the fight-or-flight response. *Proc Natl Acad Sci USA* 2013;**110**(48):19621–6.
175. Mohapatra DP, Park KS, Trimmer JS. Dynamic regulation of the voltage-gated Kv2.1 potassium channel by multisite phosphorylation. *Biochem Soc Trans* 2007;**35**(Pt 5):1064–8.
176. Park KS, Mohapatra DP, Misonou H, Trimmer JS. Graded regulation of the Kv2.1 potassium channel by variable phosphorylation. *Science* 2006;**313**(5789):976–9.
177. Misonou H, Menegola M, Mohapatra DP, Guy LK, Park KS, Trimmer JS. Bidirectional activity-dependent regulation of neuronal ion channel phosphorylation. *J Neurosci* 2006;**26**(52):13505–14.
178. Malinow R, Malenka RC. AMPA receptor trafficking and synaptic plasticity. *Annu Rev Neurosci* 2002;**25**:103–26.
179. Sommer B, Kohler M, Sprengel R, Seeburg PH. RNA editing in brain controls a determinant of ion flow in glutamate-gated channels. *Cell* 1991;**67**(1):11–19.
180. Mosbacher J, Schoepfer R, Monyer H, Burnashev N, Seeburg PH, Ruppersberg JP. A molecular determinant for submillisecond desensitization in glutamate receptors. *Science* 1994;**266**(5187):1059–62.
181. Partin KM, Fleck MW, Mayer ML. AMPA receptor flip/flop mutants affecting deactivation, desensitization, and modulation by cyclothiazide, aniracetam, and thiocyanate. *J Neurosci* 1996;**16**(21):6634–47.
182. Lipscombe D, Allen SE, Toro CP. Control of neuronal voltage-gated calcium ion channels from RNA to protein. *Trends Neurosci* 2013;**36**(10):598–609.
183. Soejima M, Noma A. Mode of regulation of the ACh-sensitive K-channel by the muscarinic receptor in rabbit atrial cells. *Pflugers Arch* 1984;**400**(4):424–31.
184. Wickman KD, Iniguez-Lluhl JA, Davenport PA, Taussig R, Krapivinsky GB, Linder ME, et al. Recombinant G-protein beta gamma-subunits activate the muscarinic-gated atrial potassium channel. *Nature* 1994;**368**(6468):255–7.
185. Whorton MR, MacKinnon R. X-ray structure of the mammalian GIRK2-betagamma G-protein complex. *Nature* 2013;**498**(7453):190–7.
186. Ikeda SR. Voltage-dependent modulation of N-type calcium channels by G-protein beta gamma subunits. *Nature* 1996;**380**(6571):255–8.
187. Herlitze S, Garcia DE, Mackie K, Hille B, Scheuer T, Catterall WA. Modulation of Ca2+ channels by G-protein beta gamma subunits. *Nature* 1996;**380**(6571):258–62.
188. Bean BP. Neurotransmitter inhibition of neuronal calcium currents by changes in channel voltage dependence. *Nature* 1989;**340**(6229):153–6.

189. Swartz KJ. Modulation of channels by protein kinase C in rat central and peripheral neurons: disruption of G protein-mediated inhibition. *Neuron* 1993;**11**(2):305–20.
190. Swartz KJ, Merritt A, Bean BP, Lovinger DM. Protein kinase C modulates glutamate receptor inhibition of channels and synaptic transmission. *Nature* 1993;**361**(6408):165–8.
191. Suh BC, Leal K, Hille B. Modulation of high-voltage activated Ca(2+) channels by membrane phosphatidylinositol 4,5-bisphosphate. *Neuron* 2010;**67**(2):224–38.
192. Suh BC, Inoue T, Meyer T, Hille B. Rapid chemically induced changes of PtdIns(4,5)P2 gate KCNQ ion channels. *Science* 2006;**314**(5804):1454–7.
193. MacKinnon R. Potassium channels. *FEBS Lett* 2003;**555**(1):62–5.

CHAPTER

7

Protein Conformational Dynamics and Signaling in Evolution and Pathophysiology

Liang Schweizer and Luciano Mueller

Leads Discovery and Optimization, Research & Development, Bristol-Myers Squibb Company, Princeton, NJ, USA

OUTLINE

B. Arey (Ed): Biased Signaling in Physiology, Pharmacology and Therapeutics
DOI: http://dx.doi.org/10.1016/B978-0-12-411460-9.00007-0

INTRODUCTION

Protein conformational dynamics plays important roles in normal physiology, evolution, and pathophysiology. Conformational change of a single protein can propagate its functional effects to others proteins such as signaling cascades downstream, and subsequent biased signaling depending on the specific changes in protein conformation. Essentially, these signaling consequences affect the phenotypes and behaviors of a cell, which in turn may result in evolutionary consequences or pathophysiology of an organism.

The signaling proteins contain segments of well-defined structures, known as tertiary folds. These structures empower proteins to communicate with binding partners along signaling cascades via surface to surface interactions, which match in shape as well as complementary polarity and hydrophobicity profiles. The rigidity of protein interaction surfaces has been shown to correlate with interaction specificity. However, most proteins that perform biological functions, such as those participating in signaling pathways or catalyzing biological reactions, must undergo some degree of conformational transformation during the course of exerting their biological activities. Moreover, it appears that evolution imparted proteins with only as much structural rigidity as needed to perform their biological functions. As we will show in the following paragraphs, internal mobility eases the evolution of proteins to adopt conformational flexibility and therefore provides the opportunity to develop new functions. Furthermore, conformational flexibility permits proteins to better cope with deleterious mutations which can result in loss of function or altered function that leads to disease.

Conformational mobility and plasticity represent key inherent features of proteins through evolution. Small peptides tend to exhibit a high degree of conformational mobility. As peptides evolved into larger polypeptides, they began to adapt into tertiary protein folds. Those tertiary structures enable biochemical physiology, e.g., catalyzing biochemical reactions, or trafficking molecules or transmitting signals etc., and polypeptides appear to have incorporated structural rigidity wherever needed. However, a sizeable fraction of the proteome lacks well-defined three-dimensional conformations in the absence of a binding partner aiming to provide versatility for protein functions. Looking closely, even protein folds which adopt well-defined tertiary structures are not entirely rigid entities. The most conformationally restrained amino acid residues which are embedded into well-defined tertiary structures undergo motion over a very wide range of timescales. Within a given amino acid, there is generally a hierarchy of mobility where backbone atoms appear to be most rigid compared with side chains. Moreover, within side chains, the mobility tends to increase with the number of rotatable bonds separating a given side chain atom from the respective α-carbon atom. Those internal conformational dynamics impart proteins with varying degrees of plasticity, i.e. those internal states of mobility position proteins to respond to external perturbations and to adopt altered conformations. Types of external perturbations include changes in temperature, pH, or the binding to a ligand. The level of conformational plasticity correlates with the degree of internal mobility in a given protein domain.

Given the importance of protein mobility, it is necessary to understand the factors contributing to protein fluidity. Within a protein domain, the rigidity of a given amino acid is influenced by its local environment, packing density, and involvement in hydrogen bond formation. Moreover, the mobility of a given amino acid residue within a protein domain is influenced by its distance from the protein surface as well as its secondary structure type. Secondary structural elements exert a considerable impact on the internal dynamics of proteins. Beta-sheets tend to feature the highest degree of conformational rigidity, due to a combination of extended networks of inter-strand hydrogen bonds as well a steric stacking of amino acid side chains. Alpha-helices impart a protein tertiary fold a higher degree of mobility. While the internal hydrogen bonds rigidify helical structural elements, helical cylinders can undergo overall rotational motions relative to adjacent secondary structural elements. This may be the reason why helical structures occur at elevated abundance in proteins which require tertiary structural rearrangement for a specific biological activity, such as transporting an ion through an ion channel or transmitting a signal across a cell membrane. Residues in loops, on average, are most mobile because a good portion of loops tend to reside on protein surfaces lacking the steric restraints with nearby amino acid residues. Overall, surface-exposed amino acid residues with side chains containing polar or charged groups tend to exhibit the highest degree of mobility. Exceptions to the rule include polar or charged side chains which are involved in salt bridges or rigidified by hydrogen bonds.

The majority of known proteins adopt a single, well-defined, three-dimensional structure. The structures of larger proteins consist of multiple spatially compacted domains which include up to a few hundred amino acid residues. The building blocks of protein domains are secondary structural elements such as helices (primarily alpha-helices), beta-sheets, and turns. Multiple computational tools have been developed to sort the universe of three-dimensional protein structures into clusters of varying structural similarities. The two dominant classification methods which are currently used are SCOP (structural classification of protein)[1–3] and CATH (class architecture topology homology).[4] Both SCOP and CATH classify protein structures in a hierarchical fashion. First, the proteins are parsed into domains where feasible. The tertiary structures at the domain level in both SCOP and CATH are classified based on secondary structure content. SCOP divides protein domains into the following five classes:

1. All alpha-helical
2. All beta-sheets
3. Alpha and beta interspersed
4. Alpha and beta largely segregated
5. Different folds of no known homolog.

CATH employs a similar albeit simpler classification:

1. Mostly alpha-helical
2. Mostly beta-sheet
3. Mixed alpha-helical and beta-sheets
4. Few secondary structures.

The next hierarchical level of structure classification in both SCOP and CATH pertains to protein folds, namely how the secondary structural elements pack against each other and with what topology of connectivity. In turn, protein folds are grouped into families and super-families. In SCOP, the process of classifying protein structures is done mostly manually relying on the fact that the human brain is well versed in differentiating between complex structural topologies. CATH, on the other hand, automates some of the steps in protein classification. The RSCB protein databank (http://www.rcsb.org/pdb/home/home.do) keeps statistics on both SCOP- and CATH-based protein folds. Even though SCOP and CATH have different schemes of protein classification, they both divide the entire ensemble of deposited protein structures into approximately 1300 folds. Most noteworthy is the drastic plateau in the discovery of new folds over recent years. Even though the number of deposited protein structures numbers in the thousands each year, no new folds have been reported in the SCOP database since 2009 and no new folds have been reported in the CATH database since 2010.

It is important to note that protein folding and dynamics are strongly affected by water.[5,6] The major contributor of protein-folding energy is the so-called hydrophobic effect. The packing of hydrophobic amino acid residues in the protein interior releases water molecules from the hydration shell which surrounds water-exposed hydrophobic surfaces. The rapid motion of released water molecules accounts for a large portion of free energy gain upon transitioning from an unfolded to a folded state. In a folded protein the packing density of atoms is higher than in the unfolded state.[7] Furthermore, water molecules can also be found in the interior of proteins to compensate for suboptimal packing of mostly polar atoms.[8–14] The stabilizing effect of buried water molecules in the protein interior is due to the formation of inter-molecular hydrogen bonds. The presence of non-hydrogen-bonded polar atoms in the protein interior has a considerable destabilizing effect, which explains why on average about 90% of buried polar groups are hydrogen-bonded.[15] Apart from driving proteins into adopting stable structures, water also enhances protein flexibility by serving as a lubricant. Via the formation of alternate hydrogen bonds, water catalyzes the transition to alternate conformations.[16,17] Moreover, the formation of fluctuating alternate protein conformers is facilitated by the penetration of individual water molecules, which catalyze the temporary breakage of internal hydrogen bonds. Indeed, non-aqueous solvents render proteins highly rigid.[18]

Water molecules might also play a critical role in mediating ligand binding affinities at the protein–ligand interface. Narrow hydrophobic cavities in proteins constitute energetically very unfavorable solvation sites.[19] The displacement of water molecules in these types of cavities by ligand binding produces a considerable release of free energy via the so-called hydrophobic enclosure effect that gives rise to extreme protein–ligand interactions. For example, streptavidin binding to biotin represents an extreme case of enhanced protein binding affinity via the hydrophobic enclosure effect.[20] In contrast, water molecules have only a modest affect on protein–ligand interactions at shallow protein–ligand interfaces. Furthermore, a very recent publication by Breiten et al.[21] reports on water networks in the active site of human carbonic anhydrase (hCA) which contribute to enthalpy/entropy compensation in the binding of hCA to benzothiazole sulfonamide ligands. Variations in the binding affinities of these ligands were shown to be modulated by interactions with water molecules at the protein–ligand interface.

Enthalpy/entropy compensation is a ubiquitous phenomenon which governs ligand–receptor interactions. This binding event is driven by a combination of entropic and enthalpic changes. Driving forces behind the formation of ligand–receptor complexes include the hydrophobic effects caused by the release of water molecules and the gains in binding enthalpies due to the formation of inter-molecular hydrogen-bond and Coulomb interactions, as well as favorable hydrophobic contacts. On the other hand, this leads to enhanced molecular rigidity which gives rise to entropic loss. Typically, a ligand molecule is more flexible in solution than in the protein-bound state. Likewise, the receptor protein also tends to be more rigid at the ligand–receptor interface. Reductions in protein conformational energy as well as breakage in hydrogen bonds distal to the inter-molecular interface may counteract binding. Hence, the knowledge of the precise structure of a ligand–receptor complex does not suffice in computing its interaction energy.[22] Entropy variations during binding events impart nature a great deal of flexibility in fine-tuning inter-molecular binding networks. Unfortunately, entropic contributions to free energy are notoriously difficult to estimate.

It has been recognized for quite some time that receptor binding to its ligand cannot be determined solely based on structures.[22–24] This has led to the development of a plethora of computational tools for quantifying receptor–ligand interactions.[25] Four classes of computational methods have emerged to estimate affinities of receptor–ligand interactions: The computationally fastest and least accurate are the molecular docking programs including DOCK, Glide, Autodock, Flow, FlexX, ICM, PMF, and GOLD. These tools take into account entropic contributions with varying degrees of approximation. The MM-PBSA/GBSA (molecular mechanic poisson-boltzman/generalized born surface area) methods which originated in Kollman's lab[26–28] proved physically more rigorous by providing better estimates of the entropic contributions of water. Tools for estimating relative protein binding of similar ligands proved to be physically even more rigorous in assessing binding affinities albeit at a drastic increase in computational cost.[29,30] The most powerful computational approach aims to calculate absolute free energies of binding.[31] The ABFE approach employs full molecular dynamics simulations with detailed atomic force fields on both the intact protein–ligand complexes and respective protein and ligand molecule separately. Apart from the obvious limitations caused by imperfections in state of the art force-fields ABFE, computation methods are extremely demanding. Prohibitively long molecular dynamics simulation runs may be required to achieve convergence.[32]

Most recent reports begin to note some success in ranking protein binding of ligands and in deciphering their dominant mechanism of protein binding through the computational methods.[33] Recently, two research groups reported results which implicated changes within the protein conformational entropy during ligand binding as important factors that affect affinities of protein–ligand interactions.[34–37] In 2007, Marlow and co-workers[36] found that the binding of calmodulin to various strains of calmodulin binding protein fragments is strongly influenced by variations in conformational entropy. They analyzed the binding of calcium-bound calmodulin to six peptides of different calmodulin binding domains. The calmodulin binding affinities of these peptides exhibited a high degree of similarity even though significant variance within the enthalpic and entropic portions of the free binding energy was observed. They found that variations in conformational entropy within calmodulin play a significant contribution to the overall free energy of target protein binding. Estimates of variations in conformational entropy were derived from deuterium NMR relaxation experiments of protein samples where all methyl group containing amino acids were deuterated. Although time-consuming, these deuterium relaxation experiments allow the most accurate estimate of conformational entropies in proteins.

The protein universe is populated with proteins which exhibit a wide range of flexibility and plasticity. Protein plasticity reflects the ability of the protein to alter its tertiary structure in response to external perturbation. On the other hand, protein flexibility refers to protein structure internal dynamics. It has been shown that protein plasticity relies on protein conformational flexibility. Inflexible proteins may lack protein plasticity. Proteins which exhibit a high degree of rigidity tend to be inhibitors (e.g., trypsin inhibitors or molecular tags [ubiquitin]) and thus these proteins have limited plasticity and correspondingly limited functions. Therefore, internal flexibility and conformational plasticity are of critical importance in empowering proteins to perform diverse biological functions, such as serving as catalysts, molecular transporters, and transducers of signals or serving as molecular motors. Structured proteins which are involved in signaling, such as G protein-coupled of receptors (GPCRs), kinases, and nuclear hormone receptors, feature both flexibility and plasticity. On the extreme end of high flexibility reside the intrinsically unstructured proteins (ISPs) that require complexing with other structured proteins to adopt well-defined structures that play a role in biological processes.

It is well known that protein kinases are critical for normal cellular signaling functions. The activities of kinases can be linked with different conformational states.[38] Upon activation, a kinase binds to its two substrates—ATP and its cellular physiological substrate such as a polypeptide—and catalyzes phosphorylation. Based on crystal structure analysis, activated kinases have three distinct conformations: open, intermediate, and closed (Figure 7.1). The open structure refers to the apo-kinase, which shows the free-enzyme

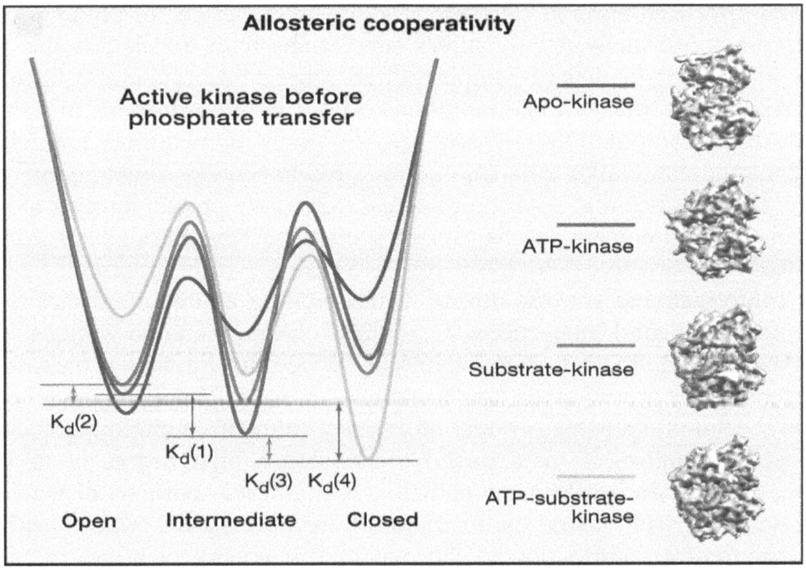

FIGURE 7.1 Three distinct conformations of kinases: open, intermediate, and closed.[38] Allosteric cooperativity is illustrated through the catalytic reaction of protein kinase A (PKA) with three conformations (open, intermediate, and closed) and four kinase binding complexes with combinations of co-substrate ATP and peptide substrate. The structures were derived from PDB entries 1j3h, 1bkx, and 1jlu. *Reproduced from Ref. 38 with permission.*

state. The intermediate conformation describes the kinase bound to either the ATP or the peptide substrate. Binding to both substrates transitions the kinase to the closed conformation. The activated conformations of kinases enable not only the binding of both ATP and substrate, but also the proper orientation of the γ-phosphate of the bound ATP and the hydroxyl group of the phosphate receptor, and the appropriate catalytic residues in the vicinity to facilitate the phosphate transfer reaction. As a result, activated conformations of the different kinase families generally share a strikingly similar three-dimensional structure based on crystallization analysis[39,40] in contrast to a variety of kinase inactivated states. Therefore, different conformations determine various levels of kinase activities which eventually lead to different cellular signaling states and phenotypic behaviors.

GPCRs represent another class of proteins in which the conformational state determines downstream signaling events. GPCRs offer striking illustrations of the interplay between protein structure, internal conformational mobility, and plasticity. Basal receptor activity, for example, is a manifestation of transient adoption of active states. Moreover, agonists tend to increase the conformational dynamics of GPCRs by triggering increases in transient populations of agonist conformers. Upon the binding of different ligands, at either orthosteric or allosteric sites, distinct GPCR-active conformations may be obtained. The conformational differences of these GPCRs enable biased signaling.[38] For example, two different ligands, an agonist and a biased agonist, may result in different conformations for a specific GPCR (Figure 7.2).[38] These two activated and stabilized conformations lead to activation of different signaling pathways. In the agonist (G protein-dependent) pathway, the activated GPCR engages a heterotrimeric G protein, which then signals through a second messenger, such as cyclic AMP or calcium, to activate downstream signaling events. In the meantime, it also recruits the GPCR kinases (GRKs) to phosphorylate Ser/Thr in the cytoplasmic loops and tail of the GPCR. These phosphorylation events initiate the recruitment of β-arrestin to mediate receptor desensitization and internalization. On the other hand, the biased agonist stabilizes another distinct, active GPCR conformation that activates a unique set of G proteins and recruits a different set of GRKs. These different GRKs lead to distinct phosphorylation patterns on the GPCR, which then result in the recruitment of β-arrestin either through orthosteric or allosteric interactions. Due to the dynamic conformation of β-arrestin, the biased signaling pathways activate distinct signaling events, such as ERK 1/2 activation.

To understand protein conformational dynamics, efforts have been made to dissect factors contributing to protein conformational changes. Over the last several years, numerous studies have been performed and, to date, the collected evidence reveals that those conformational changes can arise from both covalent and non-covalent events. The covalent events include reactions with small molecules,[41,42] phosphorylation,[43,44] ubiquitination,[45] and sumoylation, etc.[46] In addition, point mutations play important roles in protein confirmation changes and are well reported in literature.[47] A small number of point mutations may transform the function of a protein. Moreover, internal mobility renders proteins more robust to deleterious point mutants, which occur with great frequency during the course of evolution. A flexible peptide chain possesses a remarkable ability to accommodate a considerable number of amino acid mutations. In contrast, in a highly rigid molecular structure even a conservative mutation can lead to severe steric clashes, which can produce serious disruptions in the tertiary structure according to the Vander Waals law of

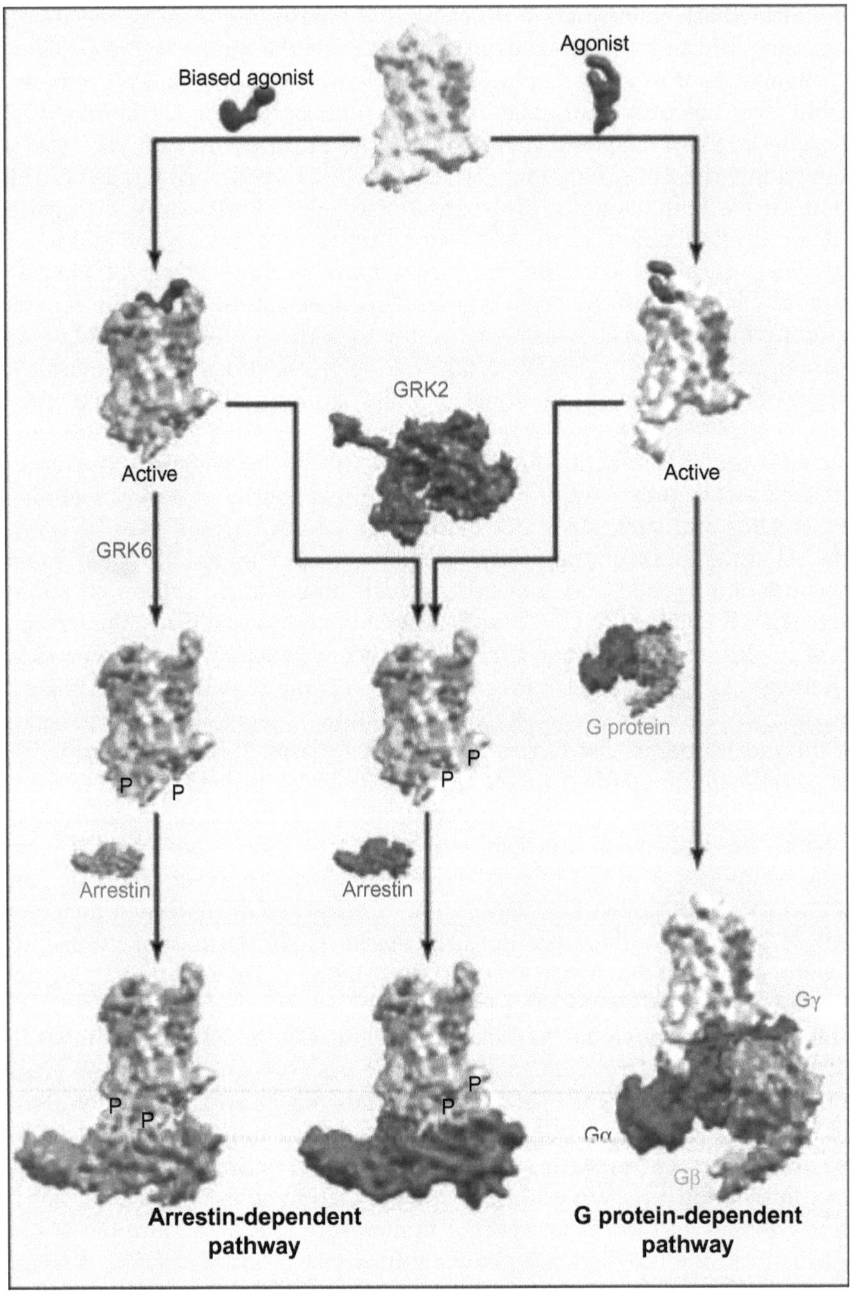

FIGURE 7.2 Conformational changes of GPCR activated by two different agonists lead to different downstream signaling.[38] Allostery can diversify cellular signaling pathways through a single receptor. G protein-coupled receptors use conformational selection to shape signaling. In the agonist (G protein-dependent) pathway, the activated GPCR either activates the heterotrimeric G proteins, which then promote the consequent signaling

repulsion. The non-covalent events, on the other hand, include binding of ions, lipids, cAMP, drugs, proteins, RNA, or DNA.[48–50] Light absorption[51] and other environmental influences such as changes in pH,[52] temperature,[53,54] or irradiation[55] can also contribute to protein conformational dynamics through non-covalent mechanisms. Overall, the factors leading to protein conformational changes play important roles in subsequent protein signaling events, which either are required for physiological functions or have major consequences in evolution and pathophysiology.

In this chapter, we will focus our efforts on detailing the roles of protein conformational changes in evolution and diseases. Finally, we will review therapeutic efforts to target protein conformational aberrations.

THE ROLE OF PROTEIN CONFORMATIONAL DYNAMICS IN EVOLUTION

The evolution of proteins involves mutations that may result in proteins adapting new functions and, in rare cases, novel folds. The rate of amino acid mutation appears strongly influenced by multiple factors, including the local structural environment, overall stability of the protein, and local conformational flexibility. However, our understanding of how protein and species evolve is still rudimentary. There is a lack of detailed understanding of how proteins have evolved. Through examining protein evolution, proteins in single cell organisms tend to be simpler at the earlier stage when the majority of proteins are comprised of single domains. Gradually, proteins have evolved into multi-domains with complex tertiary structures. Recently, the revolutionary advances in genomics, cell biology, and structural biology have contributed significantly to our understanding of how proteins evolve. The following section aims to shed light on our current understanding of this topic. In particular, the role of protein flexibility and stability in evolution will be explored.

The Neutral Theory of Molecular Evolution

Sequencing in the 1960s of the globular proteins hemoglobin and insulin from different species led researchers to conclude that most mutations are neutral with regards to evolutionary fitness.[56,57] These researchers found sizeable variations in amino acid sequences of proteins among various species without any apparent variation in stability and activity.

through a second messenger, such as cyclic AMP, or recruits the GPCR kinases (GRKs) to phosphorylate Ser/Thr in the cytoplasmic loops and tail of the GPCR. In turn, the phosphorylation enables the recruitment of β-arrestins to mediate receptor desensitization and internalization. In the biased agonist (arrestin-dependent) pathway, distinct active GPCR conformations recruit a different set of GRKs. These kinases create distinct phosphorylation patterns on the GPCR. These patterns impart distinct conformations. Via conformational selection, each pattern of modifications recruits a specific conformation of the arrestin either through orthosteric or allosteric interactions. Because the resulting conformation is different, each complex mediates different signaling pathways such as the ERK 1/2 activation. It should be noted that signaling bias can also occur via differential activation of select G proteins. The illustration is adapted from Figure 5 in Ref. 38, with permission. The illustrated structures are at the following PDB codes: GPCR, 3ny8, 4amj, 3sn6; ligands, 3qak and 4amj; GRK, 3nyn; arrestins, 3gd1 and 3p2d; G protein, 3sn6.

These findings led to the formulation of the neutral theory of molecular evolution.[58,59] The neutral theory of molecular evolution states that, at the molecular level, evolutionary changes and polymorphisms are mainly due to mutations that do not alter protein function but have evolutionary consequences. Evolution is driven by multiple factors such as mutations and random genetic drift caused by reproduction. The theory emphasizes that Darwinian selection may not be the dominant force in driving evolution. Especially in small populations, genetic drift appears to have a strong influence on the evolution of proteins. This theory of course only relates to those mutations which render the protein sufficiently fit to be passed on to subsequent generations.

In an attempt to strengthen the neutral theory of molecular evolution, Nei[60] recently relaxed the criterion of neutral mutation by only requiring that a mutant protein still retain *some* of its native activity. Corresponding proteins in different species may possess different thermodynamic stability and flexibility. An example of species-specific fold stability is provided by cytochrome C where the bovine and equine proteins exhibit differences in stability.[61] Early work in molecular evolution also led scientists to hypothesize the existence of a molecular clock. This model suggests that the rate of amino acid substitution is approximately constant in time.[62–64] A recent review article by Bromham and Penny[65] provides an overview on the accuracy and utility of the concept of the molecular clock.

The neutral theory of evolution takes into account the fact that many random mutations appear deleterious with regard to protein stability and activity,[66,67] and, therefore, may not have evolutionary consequences. Truly deleterious mutations which produce very low protein stability folds or inactive proteins may not survive cycles of evolution. Soon after the emergence of the neutral theory of evolution, Dickerson[68] noted that the rate of amino acid mutations is inversely correlated with loss of function due to mutations. Residues which are critical to the stability of a protein fold feature a low rate of mutation. Over the past 40 years numerous studies shed additional light on the mutability of amino acid residues, thereby providing evidence that the rate of mutation is strongly dependent on the local environment of a given residue and its role in contributing to protein activity and stability; for example, Lin et al.[69] observed by analyzing the yeast genome that buried residues which impact structural fold stability exhibit more stringent restraints than surface-exposed residues.

However, a number of reports have noted a lack of correlation between the importance of a protein for survival and the rate of residue mutations (reviewed by Camps et al.[70]). The high rate of mutation in essential proteins has been attributed to a surprisingly high frequency of compensatory mutations.[71–75] Mutations which produce charged residues in the protein interior invariably prove highly destabilizing. A compensating mutation which permits the formation of a salt bridge between two buried charge residues, albeit unstable when mutated individually, would produce a considerable enhancement of protein fold stability.[76,77] However, for the compensating mutations to occur, the initial deleterious mutation needs to preserve adequate protein stability. Therefore, the more stable the protein is, the better it can tolerate mutations which might adversely affect the stability. This was shown by Frances Arnold and co-workers recently when studying cytochrome P450 mutants.[78] They demonstrated that in the cytochrome P450 family of proteins stability promotes evolvability. A reduced requirement for stability may apply to proteins which feature elevated internal mobility such as proteins whose fold is stabilized by sizeable conformational entropy.[79,80] In addition, metamorphic proteins, which are able to interconvert between different folding topologies with various functions, may be well positioned to adapt new activities via mutations.[81] In a simulated evolution experiment, Tawfik and co-workers showed that a metamorphic protein—a fragment from tachylectin-2—can be readily evolved into adopting an alternate fold.[82]

The Link between Protein Evolution and Protein Mobility as well as Stability

To explore the relationship between protein evolution and its mobility and stability, multiple methods have been developed to analyze the correlation between protein evolution and protein mobility as well as its stability. Recently, Liu and Bahar[83] modeled the internal dynamics for 34 enzymes, which represent a diverse set of protein families, functional classes, and sizes, by utilizing the Gaussian network model (GNM).[84] This algorithm represents a protein as an elastic mass-spring network. Analyzing normal modes of vibration within this mass-spring network revealed details of the internal motions of a protein. Liu and Bahar found a significant correlation between structural dynamics and sequence variability whereby flexible residues exhibit a high degree of mutability. They also ranked amino acid types based on their mobility and their co-evolution propensity. As expected, a striking inverse correlation between mobility and conservancy was found. Rigid residues such as cysteine or tryptophan tend to reside in inflexible portions of a protein and are highly conserved, while flexible residues such as lysine and glutamic acids appear poorly conserved. The paper also shows that there is no apparent correlation between the propensity of a residue type in participating in co-evolution and its intrinsic mobility.

Gerek et al.[85] probed the role of structural dynamics in protein evolution using an alternate computational tool—the perturbation response scanning (PRS) technique. This method combines the elastic network model which underlies the GNM with the linear response theory.[86,87] In this method, the structural responses to external perturbations were examined. This is expected to provide more detailed correlations between mobility and mutability of amino acid residues. Amino acid residues are ranked by the flexibility score called the dynamic flexibility index (dfi). The dfi scores were correlated with the absolute evolutionary rate at each residue position in the studied proteins.[88,89] First, they explored whether there is a correlation between the dfi score and disease-associated variants. In order to reduce complexity, they used only Mendelian disease-associated variants where individual amino acid mutations were strongly linked with a disease. They found a strong correlation whereby mutations of rigid residues exhibit a much stronger disease association than flexible residues. They also found that residues which are critical for catalytic function and residues in ligand binding sites feature a high degree of rigidity. Finally, they assessed whether there is a correlation between the average rate of evolution of an amino acid type and its position in the dfi bin range. As illustrated in Figure 7.3, there is a striking correlation whereby rigid residues exhibit a low rate of mutation while flexible residues exhibit a high rate of evolution.

Lockless and Ranganathan[90] employed a statistical method, the statistical coupling analysis technique (SCA), to probe interactions between amino acid positions using

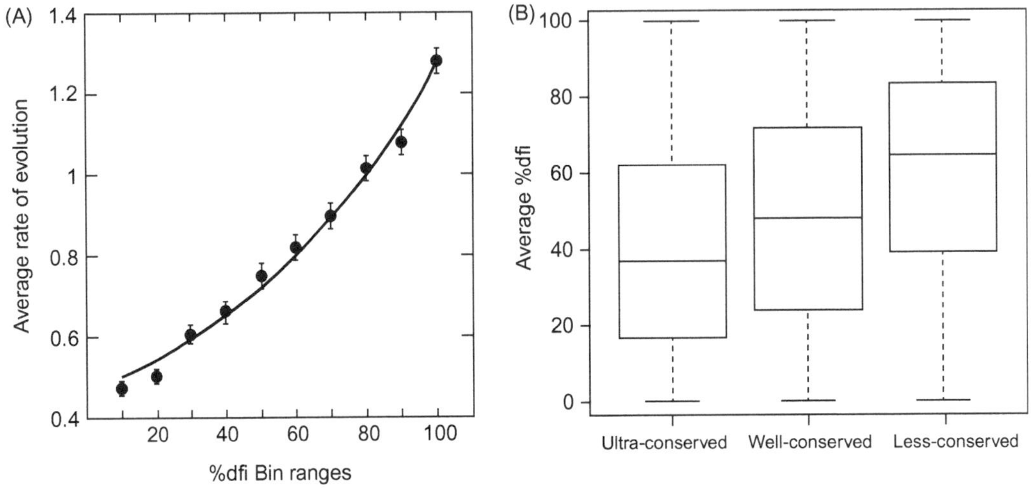

FIGURE 7.3 Relationships of residue evolutionary rates and dynamic flexibilities.[85] (A) Average evolutionary rate of change of residues with increasing dynamic flexibility (%dfi) in a sliding window. The correlation between the average evolutionary rate and the average %dfi is 0.85. (B) Box plot of the average %dfi distributions on ultra-conserved, well-conserved, and less-conserved residues. The amino acid substitution rates (r) for these categories are $r = 0$, $0 < r < = 1$, $r > 1$, respectively. Box plots show median, upper, and lower quartiles, and whiskers show maximum and minimum values. *Reproduced from Ref. 85 with permission.*

evolutionary data of a protein family. This method uncovers correlations of mutation frequencies among pairs of amino acid residues. Applied to the PDZ domain family, they discovered a network of energetic couplings among residue positions which contributes to the binding of its binding partners. As expected, they identified an energetic coupling along the groove which comprises the protein–protein interaction site. In addition, they identified a spatially contiguous network of coupled residues from the protein interaction site to residues on the opposite side of the protein domain. This represents a network of evolutionary coupled residues. This result also underscores the complexity in forces that determine protein–protein interactions, where residues that are distant to a protein binding site may impact the binding properties.

More recently, Ranganathan and co-workers[91] employed the SCA technique further to explore evolutionary coupled units within the S1A family of serine proteases. These enzymes catalyze bonds through a conserved chemical mechanism while its members exhibit a wide range of substrate specificities. The authors took special care to weed out correlation of evolutionary conservation, which may arise from limited sampling of amino acid sequences ("statistical noise") or the phylogenetic relationship between sequences ("historical noise"). As a result, they identified three statistically independent sectors within the proteins. While these three sectors involve only a subset of amino acid residues with an enzyme, each sector comprises a coupled network of residues. Even though this analysis was strictly confined to the primary structure space of the serine protease family, the three sectors feature clearly interpretable tertiary structural properties as illustrated in Figure 7.4. The sector on the upper right of Figure 7.4B comprises a contiguous network of

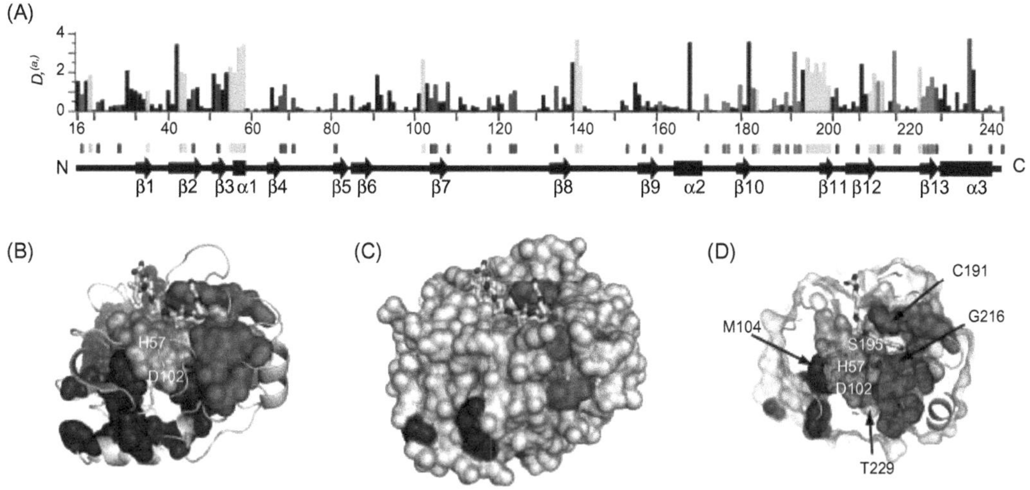

FIGURE 7.4 Results of statistical coupling analysis of correlated conservation of amino acids in the S1A family of proteins during evolution.[91] (A) Frequency of an amino acid participating in correlated conservation shaded by sector identity, positions shaded by sector identity on the primary and secondary structure of a member of the S1A family (rat trypsin); the bar graph shows the global conservation of each position. (B) The shaded sectors are shown together on the three-dimensional structure of rat trypsin (PDB 3TGI); sectors occupy regions but make contact with each other at a few positions. (C) A space-filling representation in the same view as (B), showing that all sectors are similarly buried in the protein core. (D) A slice through the core of rat trypsin at the level of the catalytic triad residues, with sector positions in shaded spheres across the molecular surface of the protein. Two sector positions (M104 and T229) and two sector positions (C191 and G216) that are similarly buried and proximal to the catalytic triad residues are highlighted. *Reproduced from Ref. 91 with permission.*

amino acids around the S1 pocket, which is the primary determinant of substrate specificity.[92] The sector on the bottom left of Figure 7.4B is comprised of a spatially contiguous ring of resides which runs through the protein core. There is no obvious connection of this sector to the catalytic function of this enzyme. Finally, the middle sector of Figure 7.4B forms a spatially contiguous group of residues between the two barrels. Residues within this sector include the catalytic triad (H57, D102, and S195) and surrounding residues which are known to be important for the basic chemical mechanism of this enzyme family.[92,93]

In order to illustrate the independence between protein sectors, Ranganathan and co-workers[91] conducted alanine scanning mutagenesis of residues spanning the range of correlation strength in the upper and lower sectors. In the alanine mutant proteins they measured catalytic activity and thermal stability. Interestingly, mutations within the upper sector significantly impacted catalytic activity while having minimal effect on thermal stability. However, mutations within the lower sector only affected thermal stability. The authors concluded that the finding of independent sectors in serine proteases had important implications on the phylogenetic analysis in this protein family whereby no single measure of the divergence of protein sequences can correctly represent differences in functional properties. In order to illustrate this point, they calculated sequence similarities among residues in individual sectors only. This produced the following interesting results: a principle component analysis of sequence differences involving residues within the upper sector classified proteases effectively by primary catalytic specificity; the PCA analysis of sequence similarities among residues in lower sector neatly separated the enzymes into organism type, i.e., vertebrates and invertebrates; and finally, the PCA analysis in middle sector residues allowed a clean separation of proteins based on enzymatic activity.

Preliminary SCA analysis of four additional protein classes provided additional evidence in support of the reported evolutionary linked sectors of amino acids in proteins. Further validation of the SCA approach was shown in 2012 by McLaughin et al.[94] who systematically tested the effects of mutants within a PDZ domain on protein activities and sensitivity to specificities of binding partners. They found that all mutants that altered protein specificity resided with an SCA determined sector. They concluded that the main functional constraints in a protein are loaded in the SCA identified sector, which contains only a small subset of amino acid residues.

Understanding Protein Evolution from a Structural Biology Perspective

Comparisons of three-dimensional structures of proteins among different species facilitates a deeper understanding of protein evolution. Knowing structures also sheds light on how cascades of mutations can lead to protein evolution.

How mutations induce structural changes in nuclear hormone receptors that alter ligand binding affinity and specificity are illustrated by Thornton and co-workers.[95,96] The authors studied the evolution of the ligand binding domain (LBD) of the glucocorticoid receptor (GR) by determining the crystal structure of the resurrected AncCR (a $\approx$470 million-year-old precursor of the vertebrate GR) and AncGR1 (a $\approx$450 million-year-old precursor of the vertebrate GR). AncCR and GR exhibit distinct differences in ligand binding specificity to 11-Deoxycorticosterone (DOC) and cortisol respectively. In order to elucidate the impact of critical mutations that led to the shift in ligand binding specificity, the authors constructed two homologs of AncCR: the $\approx$450 million AncGR1 (the common ancestor of all jawed vertebrates) and AncGR2 (the 420 million-year-old ancestor of bony vertebrates). The structures of AncGR1 and AncGR2 were determined by homology modeling and crystallization. The structural studies were complemented by functional activity measurements which were performed in expressed protein mutants.

Even though AncGR1 differs from its precursor, AncCR, by 25 amino acid mutations, the two receptors exhibit remarkably similar ligand binding specificities. AncCR exhibits a comparable affinity to aldosterone, the native ligand of AncCG1. The main difference in ligand binding affinity between AncCR and its 20 million-year-younger cousin is a reduction in specificity between DOC, aldosterone, and cortisol. The major shift in ligand specificity occurred during the evolution from AncGR1 to AncGR2, which involved 36 point mutations and one amino acid deletion. The critical mutations that triggered a shift in ligand specificity were found to be S106P and L111Q (Figure 7.5A). Mutating S106 to P triggered a repositioning of helix 7, which caused repositioning of the side chain of Q111 such that its side chain carbonyl was then capable of forming a hydrogen bond with the C17 hydroxyl of cortisol (Figure 7.5B). Substitutions of S106P and L111Q into AncGR1 radically attenuated affinity to aldosterone and DOC while retaining moderate sensitivity to

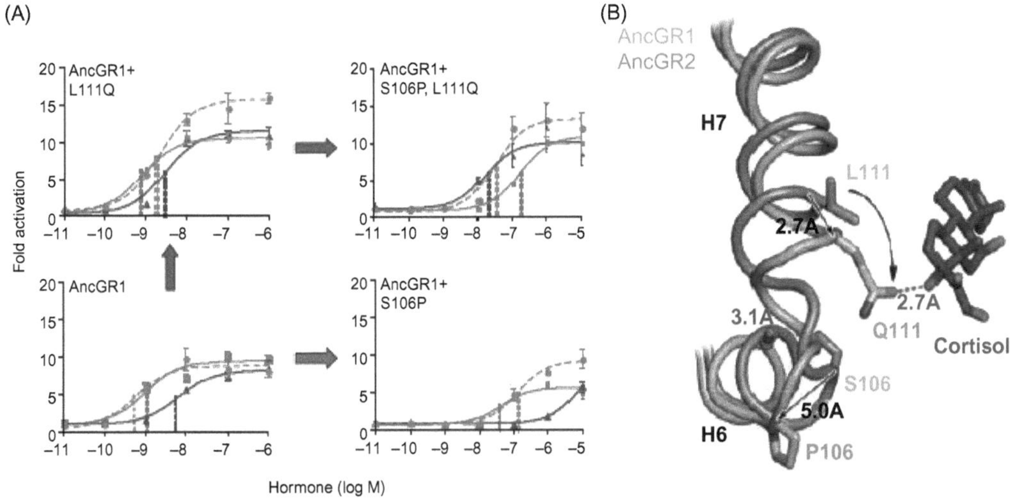

FIGURE 7.5 Mechanism for switching AncGR1's ligand preference from aldosterone to cortisol.[95] (A) Effect of substitutions S106P and L111Q on the resurrected AncGR1's response to hormones. Dashed lines indicate sensitivity to aldosterone, cortisol, and DOC as the EC50 for reporter gene activation. Arrows between top left and top right graphs and between bottom left and top left graphs show probable pathways through a functional intermediate; the arrow between the bottom left and bottom right graphs denotes an intermediate with radically reduced sensitivity to all hormones. (B) Structural changes confer new ligand specificity. Backbones of helices 6 and 7 from AncGR1 and AncGR2 in complex with cortisol are superimposed. Substitution S106P induces a kink in the interhelical loop of AncGR2, repositioning sites 106 and 111 (arrows). In this background, L111Q forms a new hydrogen bond with cortisol's unique C17-hydroxyl (dotted line). *Reproduced from Ref. 95 with permission.*

cortisol. Moreover, a strong epistatic effect was detected between these two mutations whereby L111Q had little impact on affinity to any of the hormones while S106P dramatically reduced binding to all ligands. An inspection of the protein structures revealed that the hydroxyl of S106 forms a hydrogen bond to the backbone carbonyl of Met103 in AncCR and AncGR1.

The authors also identified a second group of mutations which appear to be required to complete the evolution of ligand specificity of AncGR2, i.e., L29M, F98I, and the deletion of S121 (group Y mutations). Surprisingly, the introduction of these mutations into AncGR1 and AncGR1 as well as S106P and L111Q produced a complete non-functional receptor. The authors surmised that some of the additional mutations which differentiate AncGR2 from AncCR must modulate the effect of the group Y mutations. Inspection of the crystal structure of AncCR shed light on the disruptive effect of the Y mutation. Insertion of the Y mutations into AncCR revealed that they destabilize a network of interactions which stabilize helix 3 and the preceding AF-H loop. They hypothesized that some of the mutations which separate AncCR from AncGR1 may be responsible for establishing a tolerance of the group Y mutations. They identified an additional set of mutations, N26T and Q105L, that may be responsible for establishing a tolerance for the group Y mutations. The N26T mutation produces a new hydrogen bond between helix 3 and the AF-H loop, and the Q105L mutation enables helix 7 to pack more tightly against helix 3. Generating a mutant AncGR1 that includes both group X, Y, and Z mutations produced an active receptor which resembles AncGR2 activity. Insertion of group Z mutations in the absence of group Y mutations had little effect on receptor activity. These findings suggest that the permissive mutations Z, plus the X mutations, must have been present when the Y mutations occurred during the course of evolution.

The complex interplay of mutations to achieve a modest change in hormone specificity in the LBD of a nuclear hormone receptor underscores the complexity of inducing a new protein function via an evolutionary string of mutations. Conservative mutations may position proteins to successfully incorporate disruptive mutations which in isolation may seriously destabilize a protein and cause a loss in function. These results also intuitively illustrate why there are many more amino acid sequences than tertiary protein structures.[97] The drastic transition of a protein to a novel fold topology is expected to require a much more extensive cascade of both conservative and disruptive mutations. This may help to explain why the astronomically large number of known amino acid sequences produces structures which adopt a very small number of only about 1500 distinct fold topologies. The remarkable preservation of stable protein fold over a vast time frame has been illustrated in a recent publication[98] which reports the three-dimensional structure of a four billion-year-old protein. Ingles-Prieto et al.[98] have determined the crystal structure of a resurrected Precambrian thioredoxin. In spite of very large sequence variations, over the past four billion years of evolution there appears to be a striking preservation of the structural fold of several proteins among different species (Figure 7.6). In spite of the remarkable structural similarity presented by mesophilic thioredoxins, the ancient protein was found to possess a 30°C increase in its thermal melting temperature.

Evolution of Intracellular Signaling

Even though many signaling proteins contain domains that may have no detectable homologs in single cell organisms,[99] receptor proteins which are involved in intracellular signaling pathways revealed some striking similarities in single celled eukaryotes, plants, and metazoa.[100–102] Further, many components of signaling pathways are conserved from prokaryotic to eukaryotes.[103] This suggests that intracellular communication pathways are more ancient than those involved in extracellular communication. Through evolution, the diversity of extracellular signaling molecules leads to different consequences of downstream signaling. We will use GPCR signaling as an example to demonstrate how biased signaling evolves with the diversity of ligands.

Clearly, GPCRs participate in a sophisticated network of signaling employing multiple signaling pathways including interactions with G proteins, kinases, and arrestins. The evolutionary origin of GPCRs is still a mystery. Seven helix transmembrane (7TM) receptors have already been found in prokaryote genomes, including archaea and bacteria which utilized rhodopsins for energy harvesting.[104] It has also been noted that viral genomes encode GPCR-like proteins, e.g., herpesviruses encode chemokine receptors.[105] Provided that viruses predate cellular life forms, the possibility that 7TM receptors are of viral origin cannot be excluded.[104] An alternative possible origin of 7TM receptors has been proposed by Taylor and Agarwal.[106] They reasoned that the relatively high sequence homology

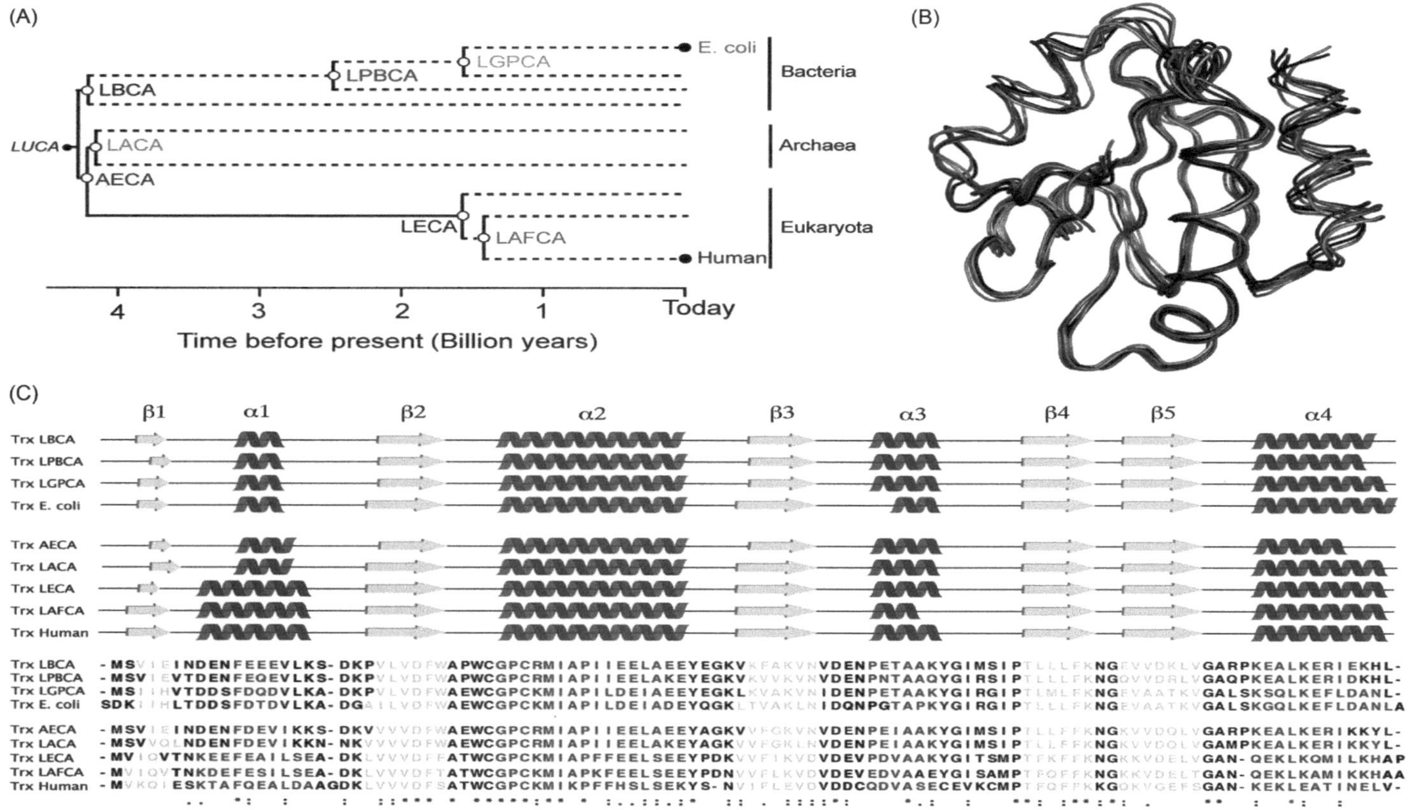

FIGURE 7.6 The remarkable preservation of thioredoxin proton fold during the past four billion years of evolution.[98] (A) Schematic phylogenetic tree showing the geological time[98] and the phylogenetic nodes targeted in this work. (B) Spatial course of the polypeptide chain for the human and E. coli thioredoxins, as well as for the several laboratory resurrections of Precambrian thioredoxins studied in this work. (C) Sequences[98] and secondary structure assignments for the extant thioredoxins and the laboratory resurrections of Precambrian thioredoxins studied in this work. *Reproduced from Ref. 98 with permission.*

between helices 1 and 3 and 5 and 7 could have its origin in gene duplication of an ancestral three-helix trans-membrane protein.[106] Finally, Fredriksson and Schiotn[107] suggested that the original GPCR could be a member of the adhesion/secretin family.

In addition to understanding the complexity of GPCR evolution, the mechanisms of how downstream signaling pathways of GPCRs have evolved attracted the attention of researchers and some progress has been made on this front. GPCRs are best known for their role in transmitting signals into downstream intracellular signaling components via couplings to membrane-anchored intracellular G proteins. Most likely, the signaling mediated by 7TM receptors (GPCR precursors) did not involve G protein coupling in prokaryotes since heterotrimeric G proteins are only present in eukaryotes.[108] Moreover, the role of heterotrimeric G proteins involving GPCR signaling is different in animals and plants. While in animals GPCRs recruit G proteins upon activation and signal downstream to secondary messages, such as cAMP or calcium influx, in plants G proteins are constitutively active and do not require activation by GPCRs. Instead, it was reported that heterotrimeric G proteins can be deactivated by a putative 7TM receptor-regulator of G protein signaling (7TM-RGS) in plants.[109,110]

In contrast to G proteins, arrestins have been implicated in the interactions of 7TM receptors in earlier life forms.[108] Arrestins interact with 7TM receptors in multiple ways, regulating GPCR inactivation,[111] internalization,[112,113] trafficking,[114] and signaling further downstream.[115–117] It is interesting to note that arrestins may mediate GPCR signaling even before the involvement of heterotrimeric G proteins. Alvarez[108] speculated that the emergence of arrestins could have facilitated GPCR precursors signaling in archaea, bacteria, and single cell eukaryotes. Alvarez also established sequence homology between arrestins and SpoOM, a protein which has been implicated in inhibiting sporulation in archaea. Therefore, we hypothesize here that arrestin-mediated GPCR activities may predate heterotrimeric G protein signaling by GPCR from an evolutionary point of view. Biased signaling pathways through GPCRs are thus products of evolution and consequences of ligand diversity.

Evolution Pressure and Consequences

Thermophilic proteins constitute an example of adaption to extreme evolutionary pressure to thrive at temperatures up to the boiling point of water. The questions arise as to how proteins cope with extreme heat and what structures and internal mobility profiles thermophilic proteins adopt at the optimal temperature of activity relative to their mesophilic cousins. Thermophilic organisms appear to have been prevalent at the onset of life as the earth's temperature was high while mesophilic organisms emerged as the earth cooled down over the past four billion years.[118] In order to function at elevated temperature, thermophilic proteins must preserve their tertiary folds in order to maintain their biological function.[119] Hence, thermophilic proteins need to adapt to high temperature environments by means of mutations which enhance conformational stability.[120–125] Detailed comparisons of thermophiles and mesophiles of the related proteins, in particular their tertiary structures and conformational motilities, have helped to shed light on the role of protein dynamics.

It is interesting to understand why thermophilic proteins, especially hyper-thermophilic proteins which function around the boiling point of water, should feature comparable structures and mobility as their mesophilic cousins. Altered properties of water at high temperatures include a distinct lowering of its dielectric constant, plus altered mobility and stability of putative substrates might in principle allow proteins to embrace altered rigidity profiles. However, a growing body of research suggests a remarkable degree of preservation of key tertiary structural features in thermophilic proteins. In order to maintain their structure in their native environment, thermophilic proteins feature enhanced thermal stability with upward-shifted thermal melting points. Proteins resort primarily to three mechanisms to achieve enhanced thermal stability, i.e., tighter packing in their hydrophic core, increased number of Coulomb interactions by replacing polar residues with charged residues, and a systematic shortening of surface-exposed loops.[121] Thermophiles are found in both archaea, which originated in a hot environment, and bacteria that adapted to hot environments. Proteins from archaea tend to be more compact due to tighter packing[126] while thermophile bacteria tend to be stabilized by salt bridges.[127] Salt bridges between pairs of charged residues produce bigger enhancements in protein stability at elevated temperatures due to a reduction in the dielectric constant of water.[128]

In addition to tertiary structural similarities, thermophilic enzymes have been reported to feature plasticity profiles—at least in segments that are responsible for activity—which are comparable to their corresponding mesophile cousins.[129] An early articulation of the idea that proteins may adopt comparable flexibility to perform the same catalytic activity was first formulated by M. Vihinen[130] who named it the "corresponding state hypothesis". Vihinen reached this tentative conclusion from examining B-factors, which estimate the rigidity of an atom or a group of atoms in crystal structures, for a small number of proteins with available thermal stability. In a recent study, Radestock and Gohlke[123] did an extensive comparison of local flexibility in 19 pairs of homologous proteins from mesophilic and thermophilic organisms in order to further probe the corresponding state hypothesis. These researchers analyzed protein mobility using constraint network analysis (CNA). The CNA technique probes the degree of conformational freedom within a protein structure which is compatible with the network of boundary constraints of chemical bonds, hydrogen bonds, and salt bridges. Radestock and Gohlke found that adaptive mutations of thermophilic enzymes maintain a balance between overall rigidity, which is important for thermostability, and local flexibility, which is important for activity at the respective temperature at which the protein functions.

From the dynamic behaviors of thermophilic proteins at the temperatures of their optimal activities, it appears that the internal mobility of existing proteins which have evolved over the course of the past four billion years could be reaching the optimal balance between structural rigidity and protein flexibility. Various researchers have addressed the question of what the internal mobility of thermophiles at ambient temperature is. The current predominant view is that thermophilic proteins tend to feature enhanced rigidity and reduced enzymatic activity at ambient temperature. Papers cited in the review article by C. Vieille[131] indicate that thermophilic proteins appear to feature low activities at temperatures where mesophylic proteins exert their optimal activities.

Interestingly, some studies indicate enhanced internal dynamics in thermophilic proteins, particularly in the picosecond to nanosecond timescale across a range of temperatures.[132–138] The emerging picture from these studies suggests that thermophilic proteins may possess elevated internal mobility in the picosecond to nanosecond timescale compared with the corresponding mesophiles. Moreover, thermophiles show remarkably lower dependence on temperature for the amplitudes of internal motions. Finally, thermophilic proteins tend to exhibit more uniformed motion throughout the protein structure compared with mesophiles.

The uniformity of internal motion within a thermophilic protein domain reduces the onset of local thermal unfolding. One likely mechanism for the reduced temperature sensitivity of internal motion in thermophilic proteins is the increased abundance of charged residues, which help to stabilize the protein fold by the formation of salt bridges. For example, Tehei et al.[138] demonstrated by neutron scattering that protein motions measured by mean square amplitude fluctuations have reduced temperature dependence for hyperthermophilic malate dehydrogenase in *Methanococus jannaschii*. They explained the observation by the increased abundance of charged residues in this protein. The stabilizing effect of salt bridges between charged residues enhances with increasing temperature due to the previously noted temperature dependence of the dielectric constant of water. The amplitudes of high frequency motions in thermophilic proteins increase conformational entropy compared to mesophiles and therefore improve thermodynamic stability.[139] In contrast to rapid motion dynamics, the plasticity of enzymes, i.e., their ability to undergo large-scale structural rearrangements to enable catalysis, does not contribute much to a protein's conformational entropy.[129]

ABERRANT PROTEIN CONFORMATION AND ASSOCIATED DISEASES

It has been well documented that aberrant protein conformation can lead to diseases such as cancer, type 2 diabetes, Parkinson's disease (PD), Alzheimer's disease (AD), prion disease, etc.[38] Loss of controlled protein conformational dynamics can result in activation of aberrant cellular signaling pathways leading to disease. While the mechanisms of altered protein conformation can be diverse, one of the major causes of these alterations is genetic mutation within the protein sequence. As we have discussed earlier, evolution selects mutations that provide novel functions and signaling activities. However, many mutations demonstrate deleterious effects. Many mutations go undetected as they lead to still birth or altered development of the organism that ends in terminated pregnancy *in utero*. But those deleterious mutations which occur in human adults, or which can be passed on to subsequent generations, will eventually lead to diseases. We will focus here on the link between mutation, protein conformational changes, and associated diseases.

With the development of recombinant DNA technology in the 1980s, systematic efforts to identify genes linked with human diseases were initiated. In recent years, numerous human disease genes have been mapped. Corresponding with those genes, unregulated proteins and their related signaling pathways were reported to be involved in the etiology of those human diseases. Most recently, following the optimization of next generation sequencing techniques, rapid progress has been made regarding the identification of human disease gene mutations. Further research has been conducted to understand the functional significance of those mutations at the protein levels. Aberrant protein conformations lead to dysregulated cellular signaling events, and consequently affect cellular functions and phenotypes. Here we describe how those unregulated protein conformational changes lead to various diseases, using examples covering different classes of proteins and their associated mutations.

Conformational Changes in CFTR and Cystic Fibrosis

One of the first well-characterized genetic disorders is cystic fibrosis (CF). CF is a common genetic disorder among Caucasians, with a frequency of about 1 in 2500 live births. This disease affects multiple organs, ranging from chronic pulmonary obstruction, elevated sweat chloride levels, and reduced fertility in male patients. It was first described in 1938 by Dorothy Andersen.[140] It was not until 1985 that studies of sibling pairs affected by CF mapped the gene to chromosome 7.[141–144] Subsequently, through laborious chromosome walking and chromosome jumping techniques, the protein cystic fibrosis transmembrane conductance regulator (CFTR) was identified to be linked to CF in 1988 and the first mutation for CF, ΔF508, was discovered on the 7th chromosome.[145–147]

CFTR is a cAMP-regulated plasma membrane chloride channel that is a 1480 amino acid membrane-bound glycoprotein. Being a member of the ATP binding cassette (ABC) superfamily of proteins, CFTR is organized into two transmembrane domains (TMD1 and TMD2), two nucleotide binding domains (NBD1 and NBD2), and a regulatory (R) domain. Over the years, research has found over 1000 different mutations in CFTR that cause CF. The most common mutation, deletion of phenylalanine 508 (ΔF508), which accounts for approximately 70% of the disease alleles, impairs CFTR folding. This primary CF defect in CFTR, which was localized to the NBD1 domain, causes misfolding and biosynthetic arrest of CFTR. Through structural analysis, it was demonstrated that delta F508 mutation provides a distinct elevation in conformational flexibility. The enhanced flexibility may reduce protein stability compared to the wild-type CFTR. Consequently, this mutation affects CFTR biosynthetic and endocytic processing as well as chloride channel function.[148–150] This example illustrates that mutations leading to aberrant protein conformation subsequently results in functional dysregulation and a diseased state.

Furthermore, continuous research worldwide has led to an understanding that CF is not just one disease, but indeed a disorder of varying severity linked to almost 2000 mutations. In the CF Mutation Database (http://www.genet.sickkids.on.ca/cftr/), 1943 mutations are currently listed. While 269 sequence variations reported on this website are due to polymorphisms, indicating that they are non-disease causing, most of the mutations are from missense, frame shift, and nonsense mutations and may lead to various levels of CF phenotypes due to the different molecular mechanisms of these mutations. As mentioned above, the ΔF508 mutation is a protein-folding mutant affecting the trafficking of CFTR, and thus resulting in ER retention. In contrast, another fairly common CF-associated mutation, G551D, which is also located in NBD1 as ΔF508 affects primarily the function of CFTR but not its trafficking,[151] and the mutant G1349D, at an equivalent position in NBD2 as G551in NBD1, similarly affects the function of CFTR.[152] However, G551D-CFTR was characterized as having a lower open probability than wild-type channels, and patients who carry the G551D mutation are associated with a more severe clinical phenotype. On the other hand, G1349D, also a mutant with gating dysfunction, presents a milder clinical phenotype in patients.[152] Further experiments reveal that the G551D mutation completely eliminates ATP-dependent gating, while G1349D-CFTR maintains some ATP dependence, but with a lower open probability than wild-type channels due to a lower opening rate. These results suggest that different mutations could lead to a differential alteration in protein conformation, and, therefore, varied levels of functional effects. Eventually those differences can be translated into diverse clinical outcomes.

Protein Kinase Mutations and Diverse Diseases

Protein kinases play an important role in cellular signaling and a single mutation can often affect the function of these enzymes. This can lead to altered signaling and ultimately to certain diseases. For example, mutations resulting in activation of some tyrosine kinases have been reported to cause cancer.[153] The mechanisms of action for those mutants that can shift a kinase from an inactive state to an active one either stabilize an active conformation or disrupt critical interactions in an inactive conformation, or a combination of both.

The epidermal growth factor receptor (EGFR) belongs to the HER family of receptor tyrosine kinases which consists of at least four members: EGFR (ErbB1, Her1), Her2 (EGFR2, Her2/neu), Her3 (EGFR3, ErbB3), and Her4 (EGFR4 or ErbB4). Upon binding of epidermal growth factor (EGF) to its receptor, the receptor transitions from an inactive monomer to an active homodimer or heterodimer with another family member. This activation leads to tyrosine phosphorylation of the cytoplasmatic tail that activates downstream signaling pathways, principally the MAPK, Akt, and JNK pathways.[154] These pathways result in subsequent DNA synthesis and cell proliferation, and play important roles in cell migration and adhesion. Therefore, it is not surprising that the EGFR was reported to be linked to cell proliferation and survival pathways,[155,156] and that mutations that lead to EGFR constitutive activity or overexpression have been associated with a number of cancers, including lung cancer, anal cancers, and glioblastoma multiforme.[157,158]

A considerable amount of work on EGFR crystal structures in the presence or absence of ligands has identified two conformations of the EGFR kinase domain: an active conformation that is similar to other active protein kinases and an inactive state.[159] Kinase conformational switching from the inactive conformations to the active ones due to mutations could be a major driver for tumorigenesis. Indeed, oncogenic mutation of EGFR L834R (also referred to as L858R), which accounts for close to half of the EGFR mutations in lung cancer patients,[160] illustrates the importance of correct protein conformation in physiology and pathophysiology. This mutation disrupts interactions critical for maintaining the inactive conformation, and therefore shifts the kinase conformation toward an active state, resulting in constitutively active kinase.

On the other hand, a kinase gate-keeper mutation may allosterically enhance protein mobility in the inactive state and then restore structural integrity of the activated form. It has been reported that T790M mutation in EGFR, T315I in BCR-ABL, T334I in c-ABL, T341I in Src, T670I in KIT, and T674I in PDGFRA promote the assembly of an enzymatically active kinase conformation.[38] These examples illustrate how kinase mutations can exploit different mechanisms, even within a single protein family. Regardless of the mechanisms these kinase mutations utilize, oncogenic mutations often bypass the autoinhibited kinase inactivity and result in a constitutive activation state.

In addition to EGFR, other receptor tyrosine kinases activate cell proliferation upon ligand binding, such as the closely related platelet-derived growth factor receptors, c-Kit, Flt3, and c-Fms. Mutations in these proteins, which can lead to ligand-independent activation, were found in human cancers as well. Those mutants tend to cluster in the juxtamembrane (JM) and catalytic tyrosine kinase domain (TKD) regions. Interestingly, these different mutations could result in different downstream signaling consequences and therefore different pathophsyiology. For example, in acute myeloid leukemia (AML), the JM and TKD mutations for Flt3 could differ in their clinical outcomes with a range of spectra. The differences in disease pathologies cannot be simply attributed to a change in substrate specificity or signaling strength of the catalytic domain, even though the mechanisms of aberrant activation could be elucidated by biochemical and structural analyses of mutant kinases. A working model of differential Flt3 signaling based on mislocalized juxtamembrane autophosphorylation was proposed to account for the disease variations.[161] This model suggests diverse mutations of receptor tyrosine kinases could lead to different signaling activation and therefore those biased signaling events could associate with a variety of pathologies.

Another receptor tyrosine kinase, KIT, was reported to play a crucial role in the pathogenesis of systemic mastocytosis (SM). It was shown that aberrant activation of the KIT receptor can lead to increased production of mast cells in extracutaneous organs, which results in organ failure and even early death of the patient. In SM patients, mutations within various domains of the KIT receptor were identified to constitutively activate the KIT receptor kinase. The frequently detected mutations were found in the kinase activation loop of the KIT receptor in patients with mastocytosis and, therefore, the altered conformation of the KIT kinase.[162] Different mutations cause various levels of kinase activation and therefore a diverse range of SM can be observed.

Kinase mutations were also identified in other diseases beyond cancer. The insulin receptor (IR) is another receptor tyrosin kinase which is activated upon the binding of its ligand insulin, IGF-1 and IGF-II. Insulin plays an important role in regulating glucose homeostasis and is secreted by pancreatic β-cells. The main activity of the IR activation by insulin is to induce glucose uptake. Therefore, "insulin insensitivity," or a decrease in IR signaling, could lead to type 2 diabetes. In this situation, cells are unable to take up glucose resulting in hyperglycemia. Some mutations associated with IR, therefore, can lead to insulin-resistant diabetes.[163] Furthermore, several other forms of insulin mutations can cause rare and severe forms of the disease such as syndrome of type A insulin resistance, Donohue syndrome, and syndrome of Rabson-Mendenhall.[164] Detailed analyses have been performed to understand the nature of these IR mutations. For example, Donohue syndrome, which is also known as Leprechaunism, can be caused by a nonsense mutation of IR that results in a frame shift, a single missense mutation, and in the milder form of a single codon change that alters isoleucine to methionine in the receptor protein.[165] In sum, mutations with the loss of function of IR can be categorized into five classes:

1. Defected IR biosynthesis
2. Impaired receptor transport to the cell surface
3. Decreased affinity of insulin binding
4. Reduced tyrosine kinase activity
5. Accelerated receptor degradation.[166]

All of these classes may result in different receptor conformation and thus varied levels of signaling activities for the receptor that result in different disease states.

The link between kinase mutations with protein conformation aberration and a variety of diseases can be observed beyond receptor tyrosine kinases. In 2004, it was reported that mutations in leucine-rich repeat kinase 2 (LRRK2) could lead to the most common inherited form, and some sporadic forms, of PD.[167,168] Subsequently, it was suggested that the clinical phenotypes from LRRK2 mutations resemble idiopathic, late-onset PD[169,170] and therefore LRRK2 has attracted a lot of attention for scientific research, as well as being explored as a therapeutic target for the treatment of PD.

The most common PD-linked mutation for LRRK2 has been identified as G2019S.[170,171] Applying a metadynamics method, it has been suggested that the G2019S mutation stabilizes the DYG motif, since G2019S mutation lies in the DXG motif (DYG in LRRK2 but DFG in most other kinases) of the activation loop. This results in an increase in the conformational barrier between the active and inactive forms of LRRK2, and biases the enzyme to the stabilization of the active form. Thus, the mutant enzyme is "locked" into the active state and becomes hyperactive and contributes to the Parkinsonian phenotype.[172] It has been suggested that LRRK2 may phoshorylate alpha-synuclein, which is another major cause of autosomal dominant Parkinson disease when mutated. Furthermore, it has been shown that the G2019S mutation in LRRK2 has a significantly greater capacity to phosphorylate alpha-synuclein than the wild-type enzyme. Therefore, the G2019S mutant protein may lead to PD by generating pathological levels of phosphorylated alpha-synuclein.[173] More recently, it has been suggested that changes in LRRK2 activities (due to its mutations) are also linked to signaling alterations in mitogen-activated protein kinase, tumor necrosis factor alpha/Fas ligand, and Wnt signaling pathways, resulting in unregulated translational controls and vesicle trafficking. Therefore, LRRK2 acts as an upstream regulator in events leading to neurodegeneration.[174]

Nuclear Hormone Receptor Mutations and Associated Diseases

Nuclear hormone receptors (NHRs) regulate diverse physiological functions, such as homeostasis, reproduction, development, and metabolism.[175] They function as ligand-activated transcription factors, which regulate gene expression by interacting with specific DNA sequences.[176] NHRs are activated by hormonal ligands, such as glucocorticoids, mineralocorticoids, the sex steroids (estrogen, progesterone, and androgen), thyroid hormones, and vitamin D3. The role of the hormone ligand in the transcriptional process is to modulate and change the functionality of the NHR by inducing conformational changes in the receptor. NHRs share common structural features, which include a variable N-terminal region containing at least one constitutionally active transactivation region (AF-1) and several autonomous transactivation domains (AD), a conserved DNA binding domain (DBD) responsible for targeting the receptor to highly specific DNA sequences with a response element, and finally the ligand binding domain (LBD) at the C-terminal half of the receptor that recognizes a specific hormonal ligand for the corresponding biologic response.

Naturally occurring mutations of NHRs have been identified in increasing numbers in recent years in patients with abnormalities in hormonal responses. For example, there are more than 200 different inactivating mutations in the androgen receptor (AR) reported in patients with various forms of the X-linked androgen insensitivity syndrome (AIS). The severe cases of complete AIS (CAIS) are often caused by truncation mutants in which genotypic males (46XY) have a female phenotype, but with a failure of menstruation (no uterus is present) or inguinal masses (testes). The less severe phenotypes can result from missense mutations in the AR which are associated with both complete and partial forms of AIS (PAIS). Finally, a milder loss of AR function due to a subset of missense mutations has been identified in men with oligospermic infertility.

One specific type of AR gene mutations consists of increased size of a polymorphic tandem CAG repeat in the coding region. These mutations lead to expansion of the polyglutamine tract at the amino-terminal region of the receptor. This expansion was reported to associate with X-linked spinal and bulbar muscular atrophy (Kennedy disease).[177,178] As expected, expansions with less repeats showed less severe phenotypes, which exhibit as an increased risk of impaired spermatogenesis and moderate undermasculinization of males.[179] Interestingly, shorter polyglutamine repeats were reported to associate with more aggressive forms of prostate cancer tumors.[180] In addition, several somatic AR mutations were found in metastatic prostate tumors. The best-characterized mutant, T877A, alters the structure of the ligand binding pocket and confers inappropriate responsiveness to progesterone, glucocorticoids, and other C17, C19, and C21 circulating steroids at concentrations found *in vivo*.[181,182]

Recently, AR mutants are further reclassified based on their transcriptional activities. A novel prevalent class of AR mutation was reported that showed loss of function at low levels of androgen yet transformed to a gain of function at physiological levels.[183] All of these data point to the fact that different mutations in AR lead to different consequences of protein structural and functional changes that impact downstream transcriptional activities. The various transcriptional events associated with AR mutations could result in diverse disease types as well as different levels of severity. They provide a useful example of the impact of mutations on conformationally-dependent activities in this class of receptors.

Protein Misfolding and Diseases

The original concept of "conformational diseases" referred to disorders with accumulation of protein aggregates intracellularly or extracellularly. Quite a few diseases were reported to associate with protein aggregation due to protein misfolding, particularly when these aggregates occur in the highly organized form known as amyloid fibrils. Amyloids were found to be associated with more than 20 serious human diseases due to protein misfolding, including tau protein and amyloid-beta (Aβ) in AD, alpha-synuclein (α-syn) in PD, islet amyloid polypeptide (IAPP) in type 2 diabetes mellitus (T2DM), and polyQ in a group of neurodegenerative diseases including Huntington disease (HD). In these misfolded states, the proteins can either gain toxic function or lose their normal physiological function.

Currently, the number of known diseases associated with misfolding is high, and constantly increasing. Since various control and regulation strategies have evolved in biological systems to control the folding process, these diseases related to misfolding are often associated not just with genetic mutations, which are familial forms of diseases, but also with aging phenomena, which include environmentally induced alterations in protein structure that interfere with the correct folding, assembly, and trafficking of proteins. For example, as discussed earlier, CF is a clinically relevant pulmonary disorder associated with protein misfolding. This misfolding may be due to the mutation ΔF508 CFTR or a non-genetic mechanism which can also damage protein structure and induce protein misfolding in the lung, such as cigarette smoking.[184] While we appreciate that environmental factors can contribute to protein misfolding, here we have highlighted this genetic aspect, and the relationship between protein misfolding and related diseases is discussed in depth.

Aβ and AD

AD was first described in 1906 by a German psychiatrist and neuropathologist, Alois Alzheimer, as a neuropsychiatric condition affecting the elderly. Today, AD is the most prevalent neurodegenerative disorder, with an estimated 35 million cases worldwide. Based on information from the National Institute on Aging, around 5.1 million Americans have AD, and the number is increasing every year due to the increased aging population. At the initial stage of AD, the symptoms are mild, consisting of some changes in behaviors and personality accompanied by memory loss. But symptoms can become severe at later stages, imparting speech deprivation and an inability to perform daily tasks. The progressive loss of cognitive function followed by total incapacitation eventually results in death.

The leading cause of AD has been identified as neuritic plaques with extracellular Aβ peptide accumulation and neurofibrillary tangles with intracellular aggregates of hyperphosphorylated tau protein. It has been suggested that neuritic plaques (also called senile plaques) might be critical for initiating AD pathogenesis, and tau-related tangles could be involved in toxicity and impairment of neuronal function.[43]

Aβ is by far the major constituent of senile plaques, accounting for ≈90% of their mass. Within Aβ forms, Aβ40 is the most abundant form (60–70%), with the remainder consisting of Aβ42 (≈15%) and minor amounts of other peptide forms such as Aβ1-28, Aβ3-34, and Aβ1-39.[185] These Aβ forms are derived from the proteolytic processing of the amyloid precursor protein (APP). APP is a transmembrane glycoprotein which can be found in several cell types, such as glial and neuronal cells. In addition to the single transmembrane region, it has a large extracellular amino-terminus and a short cytoplasmic carboxy-terminal tail (47 residues). Cleavage within the transmembrane domain of APP by γ-secretase produces the C-terminus of Aβ, a proteolytic product prone to aggregation that has been reported to be strongly linked to AD. Functioning as part of the gamma-secretase intramembrane protease complex, presenilins are a family of related multi-pass transmembrane proteins that can facilitate the cleavage processes. Thus mutations in these three genes, namely APP, presenilin-1 (PS1), and presenilin-2 (PS2), were identified to cause familial forms of AD (FAD).[186–188] It has been shown that mutations in each of the genes leads to elevated levels of Aβ production, and possibly promotes its aggregation.

The complex Aβ fibrillogenesis process consists of changes in Aβ conformation, and the self-association of Aβ to form cross-β pleated sheets, subprotofibrils, protofibrils, and fibrils. The fibrillization of Aβ appears to initiate a cascade of events that lead to neuronal cell death and the cognitive and behavioral decline which are characteristics of AD. It has been reported that phosphorylation of Aβ promotes conformational transition and protein misfolding that facilitates the formation of toxic aggregates.[43] It has been suggested that functioning as endogenous seeds, phosphorylated Aβ could trigger soluble, extracellular Aβ to aggregate further into senile plaques. The phosphorylated Aβ aggregates are stabilized against degradation by various proteases.[43]

Alpha-synuclein Misfolding and PD

PD is a chronic neurodegenerative disorder accompanied by progressive and selective loss of nigrostriatal dopaminergic neurons. It is named after the English doctor, James Parkinson, who published *An Essay on the Shaking Palsy* in 1817, providing the first detailed description of this disease. (This essay has been recently republished in the *Journal of Neuropsychiatry and Clinical Neurosciences.*[189]) PD is characterized by the accumulation of a neuronal protein, alpha-synuclein, in Lewy bodies, which are inclusions found within neurons. PD affects about seven to ten million people worldwide and approximately one million people in the United States, ranking as the second most common neurodegenerative disorder after AD.[190]

The aggregation of alpha-synuclein has been implicated in both sporadic and familial forms of PD. Alpha-synuclein is a presynaptic protein which is involved in controlling plasticity of dopamine (DA) overflow in presynaptic terminals, synaptic vesicle recycling, storage, and compartmentalization of neurotransmitters.[191–194] Structurally, alpha-synuclein is a natively unfolded protein consisting of 140 amino acids with an N-terminal amphipathic region (1–60aa), a hydrophobic middle region (61–95aa), and an acidic C-terminal region (96–140aa). Alpha-synuclein has an increased propensity to aggregate due to its hydrophobic middle region, which is particularly influenced by multiple factors including oxidative stressors (such as environmental toxins), altered pH, and mutations that promote misfolding of alpha-synuclein.

The mutations of alpha-synuclein have been found to be associated with autosomal dominant PD.[195] Three missense mutations were identified in the alpha-synuclein gene (A53T, A30P, and E46K).[196–198] In addition, genomic triplications of an alpha-synuclein region were reported to associate with autosomal dominant PD.[199,200] It has been reported that overexpression of wild-type alpha-synuclein, or mutation forms of A30P/A53T alpha-synuclein in animal models, could lead to motor deficits and neuronal inclusions reminiscent of PD.[201–203] Alpha-synuclein has been shown to physically interact with at least 30 proteins, underlying its important role in cell signaling.[204,205] In cells, alpha-synuclein normally adopts an alpha-helical conformation. However, under pathological circumstances, the protein can undergo a profound conformational transition to a beta-sheet-rich structure that polymerizes to form toxic oligomers and amyloid plaques.[206] Therefore, the conformational changes for alpha-synuclein result in different signaling consequences in the cells and thus biased signaling for the formation of PD. Efforts to study the role of alpha-synuclein, both physiologically and pathologically, in pivotal pathways such as apoptosis and oxidative stress, mitochondrial function and trafficking, cell–cell communication at gap junctions, and protein degradation pathways have been intensively explored.[207–210]

It has been suggested that alpha-synuclein is a prion-like protein.[211] In normal neuronal cells, alpha-synuclein adopts an alpha-helical conformation. However, under the circumstances we have discussed above, the protein can undergo a conformational transition to a beta-sheet-rich structure that polymerizes to form toxic oligomers and amyloid plaques. Evidence from autopsy studies of patients with advanced PD implies alpha-synuclein is an aberrantly folded, beta-sheet-rich form that might be able to migrate from affected neurons to unaffected ones. Using animal models, recent studies demonstrate that a single intracerebral inoculation of misfolded alpha-synuclein can induce a Lewy-like pathology in cells that can spread from affected to unaffected nerve cells, as well as neurodegeneration in both transgenic and normal mice. It appears that the misfolded protein can act as a template to promote misfolding of host alpha-synuclein. This results in the formation of large aggregates, neuronal dysfunction, and neurodegeneration. In sum, these findings indicate that alpha-synuclein is a prion-like protein that can adopt a self-propagating conformation and thus leads to neurodegeneration.[212,206]

Polyglutamine (PolyQ) in a Group of Neurodegenerative Diseases

Poly Q diseases consist of a group of pathological disorders that lead to dysfunction and atrophy of certain neural cells and affect different parts of the brain. These diseases share the common characteristics of extended polyQ tracts due to mutations in various cellular genes, specifically huntingtin, ataxins, and androgen receptor. These mutant proteins subsequently form oligomers, aggregates, and, finally, aggregsomes with distinct functions and different degrees of cytotoxicity.[213]

Huntington's disease (HD) is a neurodegenerative disease accompanied by abnormal motor movements, personality changes, and early death. It has been reported that an extended polyQ stretch in the huntingtin protein can lead to HD.[214] This expansion of polyQ is prone to protein misfolding/aggregation, and the huntingtin protein with this expansion is considered to be a mutant form featuring intracellular aggregates called inclusion bodies. The normal physiological function of huntingtin was thought to involve transcription, cell signaling, and intracellular transporting.[215] In addition, huntingtin also facilitates vesicular transport and synaptic transmission.[216] While increased expression of huntingtin could counter balance the expression of mutated huntingtin, the disruption of the normal gene does not cause HD, suggesting that there are different signaling consequences of wild-type and mutant huntingtin protein. HD is thought not to be caused by inadequate production of huntingtin, but by a gain of toxic function of mutant huntingtin.[214] This highlights another example of how aberrant protein conformational changes lead to different cellular signaling events and therefore result in pathological consequences.

THERAPEUTIC STRATEGIES AGAINST PROTEIN CONFORMATIONAL ABERRATION

Since protein conformational aberration has been associated with different diseases, numerous efforts have been made to correct the conformational changes. Using examples which have been discussed above, we demonstrate the therapeutic strategies developed to target altered protein conformation.

Targeting CF Mutations

Over the years, therapies targeting CF have focused on a pharmacological mutation-specific approaches aiming to correct the mechanisms by which mutations lead to impairment of chloride conductance due to conformational changes. From these efforts, seven candidate drugs (CPX, 4PBA, gentamicin, PTC124, Ivacaftor, Lumacaftor, and Miglustat) have been investigated in CF patients.[217] One approach to address gating mutations associated with defective conductance, such as the G551D-CFTR mutation, is through modulation by CFTR potentiators. Ivacaftor is an approved CFTR potentiator for the treatment of CF patients with at least one copy of the G551D-CFTR mutation.[218,219] It is known that this mutated CFTR is trafficked correctly to the epithelial cell surface, but once there the altered channel cannot transport chloride efficiently. Ivacaftor improves the transport of chloride through the ion channel by binding to the channels directly to induce a non-conventional mode of gating. As a result, it increases the probability that the channel is open, and there are sustained improvements in sweat chloride concentrations and lung function, as well as a reduction in pulmonary exacerbations over a long period.[220]

As ΔF508 is the most common mutation for CF, attempts to promote normal processing and function of F508del-CFTR have been made as potential therapeutic strategies for the majority of CF patients. Using proteomic approaches, such as two-dimensional electrophoresis, mass spectrometry (MS), and bioinformatics tools, efforts have been made to identify proteins which could potentially rescue F508del-CFTR folding and trafficking progression. Part of the proteostasis network, such as the unfolded protein response (UPR) signaling pathways, was identified as a possible candidate to rescue the process of F508del-CFTR. Hopefully the complete characterization of these signaling pathways and their regulators in CF will contribute to the development of novel therapeutic strategies against CF, especially those with defined mutations where altered protein conformation exists.[221]

Treatments for Diseases Associated with Kinase Mutations

Currently, there are more than 20 drugs which have been approved for clinical use and many more in clinical trials targeting kinases. However, most of the approved compounds target the ATP binding site on these enzymes. Potent and selective drugs have been identified targeting both active and inactive conformations of protein kinases. Since kinase inactive conformations are diverse, targeting the inactive conformation can provide high specificity. However, active conformations tend to reflect the disease states which have arisen from activating mutations, and therefore drugs targeting the active conformation are favorable. However, it is very challenging for such inhibitors to achieve specificity.

Imatinib (Glivec), the first approved small molecular drug for protein kinase inhibitors, targets the inactive conformation of ABL tyrosine kinase. It functions by stabilizing the inactive conformations of its kinase target, rather than by directly inhibiting the active form. This provides a valuable proof-of-concept for targeting the inactive conformation of kinases as anti-cancer drug design.[222] It was observed that mutations in the transformed kinases often exist in an equilibrium between the "on" and "off" states. Stabilizing the inactive conformations can tip the balance and reduce the "on" state and thus achieve therapeutic effects. But some activating mutations for kinases can be drug-resistant due to the inhibitors targeting inactive conformations. Dasatinib (Sprycel), which has increased potency and targets the ABL active state, is effective for some Glivec-resistant mutations.[223] In addition, inhibitors targeting the active form of the kinase, such as gefitinib (Iressa) and erlotinib (Tarceva), are advantageous for cancer patients with constitutively active EGFR mutations (delL747−P753insS and L858R) compared with wild-type EGFR.[224,225]

Interestingly, unlike the ligand-activated EGFR signaling which stimulates the extracellular signal-regulated kinase and results in proliferation, these mutant EGFRs selectively activate Akt and signal transduction and activator of transcription (STAT) pathways. Therefore, these drugs affect survival and induce apoptosis in NSCLC cells with mutant EGFRs through specifically inhibiting Akt and STAT signaling.[226] These drugs are thus acting upon alternative pathways for EGFR signaling. It has been reported that a second-site of EGFR mutation, T790M, could confer resistance to gefitinib (Iressa) and erlotinib (Tarceva) in patients with L858R mutation.[227] Further analysis indicated double mutations, L858R and T790M, could dramatically reduce binding affinity relative to a prevalent cancer causing mutation L858R and leading to molecular interaction and conformational changes.[228] Thus, drug resistance can be caused by the conformational changes due to mutations that might lead to alternate pathways activation.

In certain situations, kinase inhibitors can work in both active and inactive conformations. For example, type II inhibitors for LRRK2, which bind kinases in their inactive conformation, were studied against wild-type LRRK2 and the most common PD-linked mutation, G2019S. It was demonstrated that those inhibitors exhibit different inhibition mechanisms between the wild-type and the G2019S mutant of LRRK2. While they function as ATP competitors against the G2019S mutant, they appear to follow the expected non-competitive mechanism against wild-type LRRK2.[172] Since the G2019S mutation "locked" the enzyme into the activated state, the inhibitors work by binding to a novel allosteric pocket of the mutated enzyme and further functioning in an ATP-competitive manner.[172]

Targeting Protein Misfolding and Associated Diseases

Since protein misfolding and aggregation are central in the pathogenesis of protein conformational disorders, therapeutic efforts for those disorders were directed to inhibit and/or reverse the conformational changes that result in the formation of the pathological aggregates. Over time, different strategies have been developed to address protein misfolding.

One strategy to address protein misfolding is to stabilize the native state. If protein folding is at the native state, globular proteins are not likely to be converted to amyloid fibrils, or other aggregates. This strategy is well illustrated by the drug Tafamidis (Vyndaqel)[229] which was developed to treat transthyretin-related hereditary amyloidosis (also familial amyloid polyneuropathy, or FAP). FAP is a rare but deadly autosomal-dominant neurodegenerative disease[230] which is caused by a mutation of the transthyretin (TTR) gene. The TTR protein consists of a tetramer and mutated TTR protein which may dissociate into misfolded monomers and then aggregate into structures, including amyloid fibrils.[231,232] Tafamidis functions by kinetically stabilizing the native tetrameric form of the TTR protein.[233] As a result, treatment with tafamidis slows the process of amyloid fibril formation significantly and therefore induces a therapeutic effect.

Reduction of the concentration of aggregation-prone species is another therapeutic strategic approach. Autophagy is a major clearance pathway for the removal of disease-causing aggregate-prone proteins associated with neurodegenerative disorders such as mutant huntingtin,[234,235] the A53T or A30P point mutants of alpha-synuclein,[236] and mutant forms of tau.[237] The autophagy pathway is negatively regulated by the mammalian target of rapamycin (mTOR), and mTOR inhibitor rapamycin can induce autophagy in all mammalian cell types.[235,238] It has been shown that rapamycin may enhance the degradation of several aggregate-prone mutant proteins, such as huntingtin, alpha-synucleins, and tau, and reduce the number of aggregates, as well as protect cells from mutant protein-associated toxicity.[234–237] This strategy could be highly effective due to the fact that the rates of aggregation are often dependent on the concentrations of aggregation-prone proteins.

An alternative therapeutic strategy involves blockade of the nucleation or growth of aggregates. Since aggregation is often a nucleated process, prevention of nuclei development or their ability to spread can provide effective therapy; for example, particular attention has been focused on developing agents that can bind to the end of a developing fibril and "cap" the growth process in amyloid diseases. Apomorphine has been reported to inhibit amyloid-beta fibril formation, and may serve as a potential therapeutic for AD.[239] Another drug that has been reported to interact with amyloid-beta and to inhibit its aggregation is melatonin,[240,241] but it does not reverse fibril formation or oligomers of amyloid-beta after their formation. Applying an ultrafiltration LC-MS screening assay, or an analytical method for amyloid-beta fibrils using CE-laser- induced fluorescence, has revealed more potential therapeutic agents for the treatment of AD based on the prevention of Aβ aggregation,[242,243] including daunomycin, 3-indolepropionic acid, melatonin, and also methysticin.

Finally, enhancement of the natural "housekeeping" mechanisms in the body could be another effective therapeutic strategy. One such approach is through immunization, which could stimulate the immune system to remove aggregated or otherwise misfolded proteins. Another viable approach is to increase the production of chaperones (or, more generally, heat shock proteins). This could potentially neutralize misfolded species, e.g., by binding to exposed hydrophobic surfaces.[54]

CONCLUSION

In this chapter, we provide some basic concepts for protein dynamics, which include the role of rapid motion in protein stability and slow motion for protein plasticity. Furthermore, we illustrate the importance of protein dynamics in normal physiological processes, evolution, and diseases. Protein dynamics are critical in biological processes such as catalyzing chemical reactions and initiating responses for signaling events. For example, the binding of a ligand at the extracellular domain of a cell surface receptor is known to trigger structural changes which activate a downstream signaling cascade in cells. Protein dynamics also facilitate the acquisition of new protein function by molecular evolution. From an evolutionary perspective, protein structural mobility carries considerable benefits. Therefore, evolutionary decision making requires the involvement of protein dynamics.

Finally, dysregulation of proteins during pathological orthosteric and allosteric events can trap the proteins in either active or inactive conformations. The single active or inactive state of the dysregulated protein can then affect the signaling proteins downstream, which render the signal transmitted in a permanently switched "on" (or "off") state. This leads to the related signaling pathway being aberrantly activated (or inactivated) and may result in certain diseases. Furthermore, if a mis-regulated protein conformation is involved in crosstalk between pathways, multiple disease consequences might occur. Therapeutic strategies can focus on modulating aberrant activities in mutant proteins; or targeting compensatory inhibition/activation of other proteins within the signaling pathways; or restoring normal protein dynamics by reversing conformational changes which have occurred due to mutations or other deleterious events. Therefore, understanding dynamic properties of proteins also plays an important role in disease pathology, etiology, and treatment.

References

1. Murzin AG, Brenner SE, Hubbard T, Chothia C. SCOP: a structural classification of proteins database for the investigation of sequences and structures. *J Mol Biol* 1995;**247**(4):536–40.
2. Andreeva A, Howorth D, Brenner SE, Hubbard TJP, Chothia C, Murzin AG. SCOP database in 2004: refinements integrate structure and sequence family data. *Nucleic Acids Res* 2004;**32**(Database):D226–9.
3. Andreeva A, Howorth D, Chandonia J-M, Brenner SE, Hubbard TJP, Chothia C, et al. Data growth and its impact on the SCOP database: new developments. *Nucleic Acids Res* 2008;**36**(Database Iss):D419–25.
4. Orengo CA, Michie AD, Jones S, Jones DT, Swindells MB, Thornton JM. CATH—a hierarchic classification of protein domain structures. *Structure* 1997;**5**(8):1093–108.
5. Mattos C, Clark AC. Minimizing frustration by folding in an aqueous environment. *Arch Biochem Biophys* 2008;**469**(1):118–31.
6. Levy Y, Onuchic JN. Water mediation in protein folding and molecular recognition. *Annu Rev Biophys Biomol Struct* 2006;**35**:389–415.
7. Harpaz Y, Gerstein M, Chothia C. Volume changes on protein folding. *Structure* 1994;**2**(7):641–9.
8. Griffin S, Vitello A, Wittung-Stafshede P. Buried water molecules contribute to cytochrome f stability. *Arch Biochem Biophys* 2002;**404**(2):335–7.
9. Nguyen DM, Reynald RL, Gittis AG, Lattman EE. X-ray and thermodynamic studies of staphylococcal nuclease variants I92E and I92K: Insights into polarity of the protein interior. *J Mol Biol* 2004;**341**(2):565–74.

10. Park S, Saven JG. Statistical and molecular dynamics studies of buried waters in globular proteins. *Proteins: Struct Funct Bioinformatics* 2005;**60**(3):450–63.
11. Parthasarathy S, Altuve A, Terzyan S, Zhang X, Kuczera K, Rivera M, et al. Accommodating a nonconservative internal mutation by water-mediated hydrogen bonding between β-sheet strands: a comparison of human and rat type B (mitochondrial) cytochrome b5. *Biochemistry* 2011;**50**(24):5544–54.
12. Prakash P, Sayyed-Ahmad A, Gorfe AA. The role of conserved waters in conformational transitions of Q61H K-ras. *PLoS Comput Biol* 2012;**8**(2):e1002394.
13. Reichmann D, Phillip Y, Carmi A, Schreiber G. On the contribution of water-mediated interactions to protein-complex stability. *Biochemistry* 2008;**47**(3):1051–60.
14. Takano K, Yamagata Y, Yutani K. Buried water molecules contribute to the conformational stability of a protein. *Protein Eng* 2003;**16**(1):5–9.
15. Fleming PJ, Rose GD. Do all backbone polar groups in proteins form hydrogen bonds? *Protein Sci* 2005;**14** (7):1911–7.
16. Barron LD, Hecht L, Wilson G. The lubricant of life: a proposal that solvent water promotes extremely fast conformational fluctuations in mobile heteropolypeptide structure. *Biochemistry* 1997;**36**(43):13143–7.
17. Rupley JA, Careri G. Protein hydration and function. *Adv Protein Chem* 1991;**41**:37–172.
18. Klibanov AM. Improving enzymes by using them in organic solvents. *Nature* 2001;**409**(6817):241–6.
19. Young T, Abel R, Kim B, Berne BJ, Friesner RA. Motifs for molecular recognition exploiting hydrophobic enclosure in protein–ligand binding. *Proc Natl Acad Sci USA* 2007;**104**(3):808–13.
20. Weber PC, Ohlendorf DH, Wendoloski JJ, Salemme FR. Structural origins of high-affinity biotin binding to streptavidin. *Science* 1989;**243**(4887):85–8.
21. Breiten B, Lockett MR, Sherman W, Fujita S, Al-Sayah M, Lange H, et al. Water networks contribute to enthalpy/entropy compensation in protein-ligand binding. *J Am Chem Soc* 2013;**135**(41):15579–84.
22. Gilson MK, Given JA, Bush BL, McCammon JA. The statistical-thermodynamic basis for computation of binding affinities: a critical review. *Biophys J* 1997;**72**(3):1047–69.
23. Homans SW. Water, water everywhere—except where it matters? *Drug Discov Today* 2007;**12**(13–14):534–9.
24. Homans SW. Dynamics and thermodynamics of ligand–protein interactions. *Top Curr Chem* 2007;**272** (Bioactive Conformation I):51–82.
25. Mobley DL, Dill KA. Binding of small-molecule ligands to proteins: "what you see" is not always "what you get". *Structure* 2009;**17**(4):489–98.
26. Cheatham III TE, Srinivasan J, Case DA, Kollman PA. Molecular dynamics and continuum solvent studies of the stability of polyG-polyC and polyA-polyT DNA duplexes in solution. *J Biomol Struct Dyn* 1998;**16** (2):265–80.
27. Srinivasan J, Miller J, Kollman PA, Case DA. Continuum solvent studies of the stability of RNA hairpin loops and helices. *J Biomol Struct Dyn* 1998;**16**(3):671–82.
28. Kollman PA, Massova I, Reyes C, Kuhn B, Huo S, Chong L, et al. Calculating structures and free energies of complex molecules: combining molecular mechanics and continuum models. *Acc Chem Res* 2000;**33**(12):889–97.
29. Shirts MR, Mobley DL, Chodera JD. Alchemical free energy calculations: ready for prime time? *Annu Rep Comput Chem* 2007;**3**:41–59.
30. Chodera JD, Mobley DL, Shirts MR, Dixon RW, Branson K, Pande VS. Alchemical free energy methods for drug discovery: progress and challenges. *Curr Opin Struct Biol* 2011;**21**(2):150–60.
31. Boresch S, Tettinger F, Leitgeb M, Karplus M. Absolute binding free energies: a quantitative approach for their calculation. *J Phys Chem B* 2003;**107**(35):9535–51.
32. Genheden S, Ryde U. Will molecular dynamics simulations of proteins ever reach equilibrium? *Phys Chem Chem Phys* 2012;**14**(24):8662–77.
33. Deng N-J, Zhang P, Cieplak P, Lai L. Elucidating the energetics of entropically driven protein-ligand association: calculations of absolute binding free energy and entropy. *J Phys Chem B* 2011;**115**(41):11902–10.
34. Tzeng S-R, Kalodimos CG. Protein activity regulation by conformational entropy. *Nature* 2012;**488** (7410):236–40.
35. Frederick KK, Marlow MS, Valentine KG, Wand AJ. Conformational entropy in molecular recognition by proteins. *Nature* 2007;**448**(7151):325–9.
36. Marlow MS, Dogan J, Frederick KK, Valentine KG, Wand AJ. The role of conformational entropy in molecular recognition by calmodulin. *Nat Chem Biol* 2010;**6**(5):352–8.

37. Kasinath V, Sharp KA, Wand AJ. Microscopic insights into the NMR relaxation-based protein conformational entropy meter. *J Am Chem Soc* 2013.
38. Nussinov R, Tsai CJ. Allostery in disease and in drug discovery. *Cell* 2013;**153**(2):293–305.
39. Huse M, Kuriyan J. The conformational plasticity of protein kinases. *Cell* 2002;**109**(3):275–82.
40. Nolen B, Taylor S, Ghosh G. Regulation of protein kinases; controlling activity through activation segment conformation. *Mol Cell* 2004;**15**(5):661–75.
41. Nicklaus MC, Wang S, Driscoll JS, Milne GW. Conformational changes of small molecules binding to proteins. *Bioorg Med Chem* 1995;**3**(4):411–28.
42. Morgan HP, Walsh MJ, Blackburn EA, Wear MA, Boxer MB, Shen M, et al. A new family of covalent inhibitors block nucleotide binding to the active site of pyruvate kinase. *Biochem J* 2012;**448**(1):67–72.
43. Kumar S, Walter J. Phosphorylation of amyloid beta (Aβ) peptides—a trigger for formation of toxic aggregates in Alzheimer's disease. *Aging* 2011;**3**(8):803–12.
44. Barefield D, Sadayappan S. Phosphorylation and function of cardiac myosin binding protein-C in health and disease. *J Mol Cell Cardiol* 2010;**48**(5):866–75.
45. Mohapatra B, Ahmad G, Nadeau S, Zutshi N, An W, Scheffe S, et al. Protein tyrosine kinase regulation by ubiquitination: critical roles of Cbl-family ubiquitin ligases. *Biochim Biophys Acta, Mol Cell Res* 2013;**1833**(1):122–39.
46. Liu J, Nussinov R. The role of allostery in the ubiquitin-proteasome system. *Crit Rev Biochem Mol Biol* 2013;**48**(2):89–97.
47. Studer RA, Dessailly BH, Orengo CA. Residue mutations and their impact on protein structure and function: detecting beneficial and pathogenic changes. *Biochem J* 2013;**449**(3):581–94.
48. Venkatakrishnan AJ, Deupi X, Lebon G, Tate CG, Schertler GF, Babu MM. Molecular signatures of G protein-coupled receptors. *Nature* 2013;**494**(7436):185–94.
49. Morgner N, Robinson CV. Linking structural change with functional regulation-insights from mass spectrometry. *Curr Opin Struct Biol* 2012;**22**(1):44–51.
50. Janin J, Bahadur RP, Chakrabarti P. Protein–protein interaction and quaternary structure. *Q Rev Biophys* 2008;**41**(2):133–80.
51. Strickland D, Moffat K, Sosnick TR. Light-activated DNA binding in a designed allosteric protein. *Proc Natl Acad Sci USA* 2008;**105**(31):10709–14.
52. Grimm S. Respiratory chain complex II as general sensor for apoptosis. *Biochim Biophys Acta, Bioenerg* 2013;**1827**(5):565–72.
53. Schrank TP, Bolen DW, Hilser VJ. Rational modulation of conformational fluctuations in adenylate kinase reveals a local unfolding mechanism for allostery and functional adaptation in proteins. *Proc Natl Acad Sci USA* 2009;**106**(40):16984–9.
54. Csermely P, Palotai R, Nussinov R. Induced fit, conformational selection and independent dynamic segments: an extended view of binding events. *Trends Biochem Sci* 2010;**35**(10):539–46.
55. Rabotyagova OS, Cebe P, Kaplan DL. Collagen structural hierarchy and susceptibility to degradation by ultraviolet radiation. *Mater Sci Eng C Mater Biol Appl* 2008;**28**(8):1420–9.
56. Kimura M. Evolutionary rate at the molecular level. *Nature* 1968;**217**(5129):624–6.
57. King JL, Jukes TH. Non-Darwinian evolution. *Science* 1969;**164**(3881):788–98.
58. Kimura M. The neutral theory of molecular evolution and polymorphism. *Scientia* 1977;**112**(9–12):687–721.
59. Kimura M. *The neutral theory of molecular evolution*. London: Cambridge University Press; 1983.
60. Nei M. Selectionism and neutralism in molecular evolution. *Mol Biol Evol* 2005;**22**(12):2318–42.
61. Thielges MC, Zimmermann J, Dawson PE, Romesberg FE. The determinants of stability and folding in evolutionarily diverged cytochromes c. *J Mol Biol* 2009;**388**(1):159–67.
62. Margoliash E. Primary structure and evolution of cytochrome C. *Proc Natl Acad Sci USA* 1963;**50**:672–9.
63. Doolittle RF, Blombaeck B. Amino-acid sequence investigations of fibrinopeptides from various mammals: evolutionary implications. *Nature* 1964;**202**:147–52.
64. Zuckerkandl E, Pauling L. Evolutionary divergence and convergence in proteins. *Evolving Genes Proteins, Symp Rutgers, State Univ* 1965;97–166.
65. Bromham L, Penn D. The modern molecular clock. *Nat Rev Genet* 2003;**4**(3):216–24.
66. Thorn KS, Bogan AA. ASEdb: a database for alanine mutations and their effects on the free energy of binding in protein interactions. *Bioinformatics* 2001;**17**(3):284–5.

67. Hecht M, Bromberg Y, Rost B. News from the protein mutability landscape. *J Mol Biol* 2013;**425**(21):3937–48.
68. Dickerson RE. The structures of cytochrome c and the rates of molecular evolution. *J Mol Evol* 1971;**1** (1):26–45.
69. Lin Y-S, Hsu W-L, Hwang J-K, Li W-H. Proportion of solvent-exposed amino acids in a protein and rate of protein evolution. *Mol Biol Evol* 2007;**24**(4):1005–11.
70. Camps M, Herman A, Loh E, Loeb LA. Genetic constraints on protein evolution. *Crit Rev Biochem Mol Biol* 2007;**42**(5):313–26.
71. Poon A, Chao L. The rate of compensatory mutation in the DNA bacteriophage phiX174. *Genetics* 2005;**170** (3):989–99.
72. Poon A, Davis BH, Chao L. The coupon collector and the suppressor mutation: estimating the number of compensatory mutations by maximum likelihood. *Genetics* 2005;**170**(3):1323–32.
73. Kondrashov AS, Sunyaev S, Kondrashov FA. Dobzhansky-Muller incompatibilities in protein evolution. *Proc Natl Acad Sci USA* 2002;**99**(23):14878–83.
74. Maisnier-Patin S, Berg OG, Liljas L, Andersson DI. Compensatory adaptation to the deleterious effect of antibiotic resistance in Salmonella typhimurium. *Mol Microbiol* 2002;**46**(2):355–66.
75. Kulathinal RJ, Bettencourt BR, Hartl DL. Compensated deleterious mutations in insect genomes. *Science* 2004;**306**(5701):1553–4.
76. Tissot AC, Vuilleumier S, Fersht AR. Importance of two buried salt bridges in the stability and folding pathway of barnase. *Biochemistry* 1996;**35**(21):6786–94.
77. Packer LE, Song B-B, Raleigh DP, McKnight CJ. Competition between intradomain and interdomain interactions: a buried salt bridge is essential for villin headpiece folding and actin binding. *Biochemistry* 2011;**50**(18):3706–12.
78. Bloom JD, Labthavikul ST, Otey CR, Arnold FH. Protein stability promotes evolvability. *Proc Natl Acad Sci USA* 2006;**103**(15):5869–74.
79. Frederick KK, Kranz JK, Wand AJ. Characterization of the backbone and side chain dynamics of the CaM–CaMKIp complex reveals microscopic contributions to protein conformational entropy. *Biochemistry* 2006;**45**(32):9841–8.
80. Wand AJ, Moorman VR, Harpole KW. A surprising role for conformational entropy in protein function. *Top Curr Chem* 2013;**337**:69–94.
81. Goodchild SC, Curmi PMG, Brown LJ. Structural gymnastics of multifunctional metamorphic proteins. *Biophys Rev* 2011;**3**(3):143–53.
82. Yadid I, Kirshenbaum N, Sharon M, Dym O, Tawfik DS. Metamorphic proteins mediate evolutionary transitions of structure. *Proc Natl Acad Sci USA* 2010;**107**(16):7287–92 S/1–S/12
83. Liu Y, Bahar I. Sequence evolution correlates with structural dynamics. *Mol Biol Evol* 2012;**29**(9):2253–63.
84. Yang LW, Rader AJ, Liu X, Jursa CJ, Chen SC, Karimi HA, et al. oGNM: online computation of structural dynamics using the Gaussian Network Model. *Nucleic Acids Res* 2006;**34**(Web Server issue):W24–31.
85. Gerek ZN, Kumar S, Ozkan SB. Structural dynamics flexibility informs function and evolution at a proteome scale. *Evol Appl* 2013;**6**(3):423–33.
86. Atilgan C, Atilgan AR. Perturbation-response scanning reveals ligand entry–exit mechanisms of ferric binding protein. *PLoS Comput Biol* 2009;**5**(10):e1000544.
87. Atilgan C, Gerek ZN, Ozkan SB, Atilgan AR. Manipulation of conformational change in proteins by single-residue perturbations. *Biophys J* 2010;**99**(3):933–43.
88. Kumar S, Suleski MP, Markov GJ, Lawrence S, Marco A, Filipski AJ. Positional conservation and amino acids shape the correct diagnosis and population frequencies of benign and damaging personal amino acid mutations. *Genome Res* 2009;**19**(9):1562–9.
89. Kumar S, Dudley JT, Filipski A, Liu L. Phylomedicine: an evolutionary telescope to explore and diagnose the universe of disease mutations. *Trends Genet: TIG* 2011;**27**(9):377–86.
90. Lockless SW, Ranganathan R. Evolutionarily conserved pathways of energetic connectivity in protein families. *Science* 1999;**286**(5438):295–9.
91. Halabi N, Rivoire O, Leibler S, Ranganathan R. Protein sectors: evolutionary units of three-dimensional structure. *Cell* 2009;**138**(4):774–86.
92. Hedstrom L. Serine protease mechanism and specificity. *Chem Rev* 2002;**102**(12):4501–24.
93. Baird Jr. TT, Wright WD, Craik CS. Conversion of trypsin to a functional threonine protease. *Protein Sci* 2006;**15**(6):1229–38.

94. McLaughlin Jr. RN, Poelwijk FJ, Raman A, Gosal WS, Ranganathan R. The spatial architecture of protein function and adaptation. *Nature* 2012;**491**(7422):138–42.
95. Ortlund EA, Bridgham JT, Redinbo MR, Thornton JW. Crystal structure of an ancient protein: evolution by conformational epistasis. *Science* 2007;**317**(5844):1544–8.
96. Carroll SM, Ortlund EA, Thornton JW. Mechanisms for the evolution of a derived function in the ancestral glucocorticoid receptor. *PLoS Genet* 2011;**7**(6):e1002117.
97. Caetano-Anolles G, Wang M, Caetano-Anolles D, Mittenthal JE. The origin, evolution and structure of the protein world. *Biochem J* 2009;**417**(3):621–37.
98. Ingles-Prieto A, Ibarra-Molero B, Delgado-Delgado A, Perez-Jimenez R, Fernandez JM, Gaucher EA, et al. Conservation of protein structure over four billion years. *Structure* 2013;**21**(9):1690–7.
99. Patthy L. Modular assembly of genes and the evolution of new functions. *Genetica* 2003;**118**(2–3):217–31.
100. Aravind L, Subramanian G. Origin of multicellular eukaryotes—insights from proteome comparisons. *Curr Opin Genet Dev* 1999;**9**(6):688–**94**.
101. Copley RR, Schultz J, Ponting CP, Bork P. Protein families in multicellular organisms. *Curr Opin Struct Biol* 1999;**9**(3):408–15.
102. King N. The unicellular ancestry of animal development. *Dev Cell* 2004;**7**(3):313–25.
103. Bakal CJ, Davies JE. No longer an exclusive club: eukaryotic signalling domains in bacteria. *Trends Cell Biol* 2000;**10**(1):32–8.
104. Stevens CW. Deciphering the evolution of G protein-coupled receptors in vertebrates. *Neuromethods* 2011;**60**:71–102.
105. Maussang D, Vischer HF, Leurs R, Smit MJ. Herpesvirus-encoded G protein-coupled receptors as modulators of cellular function. *Mol Pharmacol* 2009;**76**(4):692–701.
106. Taylor EW, Agarwal A. Sequence homology between bacteriorhodopsin and G protein-coupled receptors: exon shuffling or evolution by duplication? *FEBS Lett* 1993;**325**(3):161–6.
107. Fredriksson R, Schioth HB. The repertoire of G protein-coupled receptors in fully sequenced genomes. *Mol Pharmacol* 2005;**67**(5):1414–25.
108. Alvarez CE. On the origins of arrestin and rhodopsin. *BMC Evol Biol* 2008;**8**:222.
109. Chen J-G, Willard FS, Huang J, Liang J, Chasse SA, Jones AM, et al. A seven-transmembrane RGS protein that modulates plant cell proliferation. *Science (Washington, DC, United States)* 2003;**301**(5640):1728–31.
110. Bradford W, Buckholz A, Morton J, Price C, Jones AM, Urano D. Eukaryotic G protein signaling evolved to require G protein-coupled receptors for activation. *Sci Signal* 2013;**6**(276):ra37.
111. Benovic JL, Kuhn H, Weyand I, Codina J, Caron MG, Lefkowitz RJ. Functional desensitization of the isolated beta-adrenergic receptor by the beta-adrenergic receptor kinase: potential role of an analog of the retinal protein arrestin (48-kDa protein). *Proc Natl Acad Sci USA* 1987;**84**(24):8879–82.
112. Goodman Jr. OB, Krupnick JG, Santini F, Gurevich VV, Penn RB, Gagnon AW, et al. Beta-arrestin acts as a clathrin adaptor in endocytosis of the beta2-adrenergic receptor. *Nature* 1996;**383**(6599):447–50.
113. Laporte SA, Oakley RH, Zhang J, Holt JA, Ferguson SS, Caron MG, et al. The beta2-adrenergic receptor/betaarrestin complex recruits the clathrin adaptor AP-2 during endocytosis. *Proc Natl Acad Sci USA* 1999;**96**(7):3712–7.
114. Shenoy SK, Lefkowitz RJ. Trafficking patterns of beta-arrestin and G protein-coupled receptors determined by the kinetics of beta-arrestin deubiquitination. *J Biol Chem* 2003;**278**(16):14498–506.
115. Luttrell LM, Ferguson SS, Daaka Y, Miller WE, Maudsley S, Della Rocca GJ, et al. Beta-arrestin-dependent formation of beta2 adrenergic receptor-Src protein kinase complexes. *Science* 1999;**283**(5402):655–61.
116. McDonald PH, Chow CW, Miller WE, Laporte SA, Field ME, Lin FT, et al. Beta-arrestin 2: a receptor-regulated MAPK scaffold for the activation of JNK3. *Science* 2000;**290**(5496):1574–7.
117. Wisler JW, DeWire SM, Whalen EJ, Violin JD, Drake MT, Ahn S, et al. A unique mechanism of beta-blocker action: carvedilol stimulates beta-arrestin signaling. *Proc Natl Acad Sci USA* 2007;**104**(42):16657–62.
118. Gerday C. Life at the extremes of temperature. In: Storz G, Hengge R, editors. *Bacterial stress responses*. 2nd ed John Wiley & Sons; 2011. p. 425–44.
119. Sterner R, Liebl W. Thermophilic adaptation of proteins. *Crit Rev Biochem Mol Biol* 2001;**36**(1):39–106.
120. Vetriani C, Maeder DL, Tolliday N, Yip KS, Stillman TJ, Britton KL, et al. Protein thermostability above 100 degreesC: a key role for ionic interactions. *Proc Natl Acad Sci USA* 1998;**95**(21):12300–5.

121. Kumar S, Nussinov R. How do thermophilic proteins deal with heat? *Cell Mol Life Sci: CMLS* 2001;**58**(9):1216–33.
122. Zhou XX, Wang YB, Pan YJ, Li WF. Differences in amino acids composition and coupling patterns between mesophilic and thermophilic proteins. *Amino Acids* 2008;**34**(1):25–33.
123. Radestock S, Gohlke H. Protein rigidity and thermophilic adaptation. *Proteins* 2011;**79**(4):1089–108.
124. Sawle L, Ghosh K. How do thermophilic proteins and proteomes withstand high temperature? *Biophys J* 2011;**101**(1):217–27.
125. Balasco N, Esposito L, DeSimone A, Vitagliano L. Role of loops connecting secondary structure elements in the stabilization of proteins isolated from thermophilic organisms. *Protein Sci* 2013;**22**(7):1016–23.
126. Berezovsky IN, Shakhnovich EI. Physics and evolution of thermophilic adaptation. *Proc Natl Acad Sci USA* 2005;**102**(36):12742–7.
127. Mizuguchi K, Sele M, Cubellis MV. Environment specific substitution tables for thermophilic proteins. *BMC Bioinformatics* 2007;**8**(Suppl. 1):S15.
128. Malmberg CG, Maryott AA. Dielectric constant of water from 0° to 100°. *J Res Natl Bur Stand (United States)* 1956;**56**:1–8.
129. Wolf-Watz M, Thai V, Henzler-Wildman K, Hadjipavlou G, Eisenmesser EZ, Kern D. Linkage between dynamics and catalysis in a thermophilic-mesophilic enzyme pair. *Nat Struct Mol Biol* 2004;**11**(10):945–9.
130. Vihinen M. Relationship of protein flexibility to thermostability. *Protein Eng* 1987;**1**(6):477–80.
131. Vieille C, Burdette DS, Zeikus JG. Thermozymes. *Biotechnol Annu Rev* 1996;**2**:1–83.
132. Fitter J, Heberle J. Structural equilibrium fluctuations in mesophilic and thermophilic alpha-amylase. *Biophys J* 2000;**79**(3):1629–36.
133. Hernandez G, Jenney Jr. FE, Adams MW, LeMaster DM. Millisecond time scale conformational flexibility in a hyperthermophile protein at ambient temperature. *Proc Natl Acad Sci USA* 2000;**97**(7):3166–70.
134. Tehei M, Zaccai G. Adaptation to extreme environments: macromolecular dynamics in complex systems. *Biochim Biophys Acta* 2005;**1724**(3):404–10.
135. Grottesi A, Ceruso MA, Colosimo A, Di Nola A. Molecular dynamics study of a hyperthermophilic and a mesophilic rubredoxin. *Proteins* 2002;**46**(3):287–94.
136. Wintrode PL, Zhang D, Vaidehi N, Arnold FH, Goddard III WA. Protein dynamics in a family of laboratory evolved thermophilic enzymes. *J Mol Biol* 2003;**327**(3):745–57.
137. Vinther JM, Kristensen SM, Led JJ. Enhanced stability of a protein with increasing temperature. *J Am Chem Soc* 2011;**133**(2):271–8.
138. Tehei M, Madern D, Franzetti B, Zaccai G. Neutron scattering reveals the dynamic basis of protein adaptation to extreme temperature. *J Biol Chem* 2005;**280**(49):40974–9.
139. Wand AJ. The dark energy of proteins comes to light: conformational entropy and its role in protein function revealed by NMR relaxation. *Curr Opin Struct Biol* 2013;**23**(1):75–81.
140. Andersen D. Cystic fibrosis of the pancreas and its relation to celiac disease: a clinical and pathological study. *Am J Dis Child* 1938;**56**:344–99.
141. White R, Woodward S, Leppert M, O'Connell P, Hoff M, Herbst J, et al. A closely linked genetic marker for cystic fibrosis. *Nature* 1985;**318**(6044):382–4.
142. Wainwright BJ, Scambler PJ, Schmidtke J, Watson EA, Law HY, Farrall M, et al. Localization of cystic fibrosis locus to human chromosome 7cen-q22. *Nature* 1985;**318**(6044):384–5.
143. Knowlton RG, Cohen-Haguenauer O, Van Cong N, Frezal J, Brown VA, Barker D, et al. A polymorphic DNA marker linked to cystic fibrosis is located on chromosome 7. *Nature* 1985;**318**(6044):380–2.
144. van Heyningen V, Hayward C, Fletcher J, McAuley C. Tissue localization and chromosomal assignment of a serum protein that tracks the cystic fibrosis gene. *Nature* 1985;**315**(6019):513–5.
145. Rommens JM, Iannuzzi MC, Kerem B, Drumm ML, Melmer G, Dean M, et al. Identification of the cystic fibrosis gene: chromosome walking and jumping. *Science* 1989;**245**(4922):1059–65.
146. Kerem B, Rommens JM, Buchanan JA, Markiewicz D, Cox TK, Chakravarti A, et al. Identification of the cystic fibrosis gene: genetic analysis. *Science* 1989;**245**(4922):1073–80.
147. Riordan JR, Rommens JM, Kerem B, Alon N, Rozmahel R, Grzelczak Z, et al. Identification of the cystic fibrosis gene: cloning and characterization of complementary DNA. *Science* 1989;**245**(4922):1066–73.
148. Thomas PJ, Pedersen PL. Effects of the delta F508 mutation on the structure, function, and folding of the first nucleotide-binding domain of CFTR. *J Bioenerg Biomembr* 1993;**25**(1):11–9.

149. Welsh MJ, Denning GM, Ostedgaard LS, Anderson MP. Dysfunction of CFTR bearing the delta F508 mutation. *J Cell Sci Suppl* 1993;**17**:235–9.
150. Lukacs GL, Verkman AS. CFTR: folding, misfolding and correcting the ΔF508 conformational defect. *Trends Mol Med* 2012;**18**(2):81–91.
151. Drumm ML, Wilkinson DJ, Smit LS, Worrell RT, Strong TV, Frizzell RA, et al. Chloride conductance expressed by delta F508 and other mutant CFTRs in Xenopus oocytes. *Science* 1991;**254**(5039):1797–9.
152. Bompadre SG, Sohma Y, Li M, Hwang TC. G551D and G1349D, two CF-associated mutations in the signature sequences of CFTR, exhibit distinct gating defects. *J Gen Physiol* 2007;**129**(4):285–98.
153. Lengyel E, Sawada K, Salgia R. Tyrosine kinase mutations in human cancer. *Curr Mol Med* 2007;**7**(1):77–84.
154. Oda K, Matsuoka Y, Funahashi A, Kitano H. A comprehensive pathway map of epidermal growth factor receptor signaling. *Mol Syst Biol* 2005;**1** 2005 0010
155. Yarden Y, Sliwkowski MX. Untangling the ErbB signalling network. *Nat Rev Mol Cell Biol* 2001;**2**(2):127–37.
156. Jorissen RN, Walker F, Pouliot N, Garrett TP, Ward CW, Burgess AW. Epidermal growth factor receptor: mechanisms of activation and signalling. *Exp Cell Res* 2003;**284**(1):31–53.
157. Zhang H, Berezov A, Wang Q, Zhang G, Drebin J, Murali R, et al. ErbB receptors: from oncogenes to targeted cancer therapies. *J Clin Invest* 2007;**117**(8):2051–8.
158. Sharma SV, Settleman J. ErbBs in lung cancer. *Exp Cell Res* 2009;**315**(4):557–71.
159. Kumar A, Petri ET, Halmos B, Boggon TJ. Structure and clinical relevance of the epidermal growth factor receptor in human cancer. *J Clin Oncol* 2008;**26**(10):1742–51.
160. Gu D, Scaringe WA, Li K, Saldivar JS, Hill KA, Chen Z, et al. Database of somatic mutations in EGFR with analyses revealing indel hotspots but no smoking-associated signature. *Hum Mutat* 2007;**28**(8):760–70.
161. Chan Perry M. Differential signaling of Flt3 activating mutations in acute myeloid leukemia: a working model. *Protein Cell* 2011;**2**(2):108–15.
162. Verstovsek S. Advanced systemic mastocytosis: the impact of KIT mutations in diagnosis, treatment, and progression. *Eur J Haematol* 2013;**90**(2):89–98.
163. Taylor SI, Kadowaki T, Kadowaki H, Accili D, Cama A, McKeon C. Mutations in insulin-receptor gene in insulin-resistant patients. *Diabetes Care* 1990;**13**(3):257–79.
164. Musso C, Cochran E, Moran SA, Skarulis MC, Oral EA, Taylor S, et al. Clinical course of genetic diseases of the insulin receptor (type A and Rabson-Mendenhall syndromes): a 30-year prospective. *Medicine (Baltimore)* 2004;**83**(4):209–22.
165. Psiachou H, Mitton S, Alaghband-Zadeh J, Hone J, Taylor SI, Sinclair L. Leprechaunism and homozygous nonsense mutation in the insulin receptor gene. *Lancet* 1993;**342**(8876):924.
166. Taylor SI, Cama A, Accili D, Barbetti F, Imano E, Kadowaki H, et al. Genetic basis of endocrine disease. 1. Molecular genetics of insulin resistant diabetes mellitus. *J Clin Endocrinol Metab* 1991;**73**(6):1158–63.
167. Paisan-Ruiz C, Jain S, Evans EW, Gilks WP, Simon J, van der Brug M, et al. Cloning of the gene containing mutations that cause PARK8-linked Parkinson's disease. *Neuron* 2004;**44**(4):595–600.
168. Zimprich A, Muller-Myhsok B, Farrer M, Leitner P, Sharma M, Hulihan M, et al. The PARK8 locus in autosomal dominant parkinsonism: confirmation of linkage and further delineation of the disease-containing interval. *Am J Hum Genet* 2004;**74**(1):11–9.
169. Aasly JO, Toft M, Fernandez-Mata I, Kachergus J, Hulihan M, White LR, et al. Clinical features of LRRK2-associated Parkinson's disease in central Norway. *Ann Neurol* 2005;**57**(5):762–5.
170. Khan NL, Jain S, Lynch JM, Pavese N, Abou-Sleiman P, Holton JL, et al. Mutations in the gene LRRK2 encoding dardarin (PARK8) cause familial Parkinson's disease: clinical, pathological, olfactory and functional imaging and genetic data. *Brain* 2005;**128**(Pt 12):2786–96.
171. Goldwurm S, Di Fonzo A, Simons EJ, Rohe CF, Zini M, Canesi M, et al. The G6055A (G2019S) mutation in LRRK2 is frequent in both early and late onset Parkinson's disease and originates from a common ancestor. *J Med Genet* 2005;**42**(11):e65.
172. Liu M, Bender SA, Cuny GD, Sherman W, Glicksman M, Ray SS. Type II kinase inhibitors show an unexpected inhibition mode against Parkinson's disease-linked LRRK2 mutant G2019S. *Biochemistry* 2013;**52**(10):1725–36.
173. Qing H, Wong W, McGeer EG, McGeer PL. LRRK2 phosphorylates alpha synuclein at serine 129: Parkinson's disease implications. *Biochem Biophys Res Commun* 2009;**387**(1):149–52.

174. Berwick DC, Harvey K. LRRK2 signaling pathways: the key to unlocking neurodegeneration? *Trends Cell Biol* 2011;**21**(5):257–65.
175. Gronemeyer H, Gustafsson JA, Laudet V. Principles for modulation of the nuclear receptor superfamily. *Nat Rev Drug Discov* 2004;**3**(11):950–64.
176. Olefsky JM. Nuclear receptor minireview series. *J Biol Chem* 2001;**276**(40):36863–4.
177. La Spada AR, Wilson EM, Lubahn DB, Harding AE, Fischbeck KH. Androgen receptor gene mutations in X-linked spinal and bulbar muscular atrophy. *Nature* 1991;**352**(6330):77–9.
178. Kumar R. Role of androgen receptor polyQ chain elongation in Kennedy's disease and use of natural osmolytes as potential therapeutic targets. *IUBMB Life* 2012;**64**(11):879–84.
179. Zhang L, Leeflang EP, Yu J, Arnheim N. Studying human mutations by sperm typing: instability of CAG trinucleotide repeats in the human androgen receptor gene. *Nat Genet* 1994;**7**(4):531–5.
180. Culig Z, Klocker H, Bartsch G, Hobisch A. Androgen receptors in prostate cancer. *Endocr Relat Cancer* 2002;**9**(3):155–70.
181. Sack JS, Kish KF, Wang C, Attar RM, Kiefer SE, An Y, et al. Crystallographic structures of the ligand-binding domains of the androgen receptor and its T877A mutant complexed with the natural agonist dihydrotestosterone. *Proc Natl Acad Sci USA* 2001;**98**(9):4904–9.
182. Krishnan AV, Zhao XY, Swami S, Brive L, Peehl DM, Ely KR, et al. A glucocorticoid-responsive mutant androgen receptor exhibits unique ligand specificity: therapeutic implications for androgen-independent prostate cancer. *Endocrinology* 2002;**143**(5):1889–900.
183. Hay CW, McEwan IJ. The impact of point mutations in the human androgen receptor: classification of mutations on the basis of transcriptional activity. *PLoS One* 2012;**7**(3):e32514.
184. McElvaney NG, Greene CM. Mechanisms of protein misfolding in conformational lung diseases. *Curr Mol Med* 2012;**12**(7):850–9.
185. Golde TE, Eckman CB, Younkin SG. Biochemical detection of Abeta isoforms: implications for pathogenesis, diagnosis, and treatment of Alzheimer's disease. *Biochim Biophys Acta* 2000;**1502**(1):172–87.
186. Sherrington R, Rogaev EI, Liang Y, Rogaeva EA, Levesque G, Ikeda M, et al. Cloning of a gene bearing missense mutations in early-onset familial Alzheimer's disease. *Nature* 1995;**375**(6534):754–60.
187. Goate AM. Molecular genetics of Alzheimer's disease. *Geriatrics* 1997;**52(Suppl. 2)**:S9–12.
188. Lopera F, Ardilla A, Martinez A, Madrigal L, Arango-Viana JC, Lemere CA, et al. Clinical features of early-onset Alzheimer disease in a large kindred with an E280A presenilin-1 mutation. *JAMA* 1997;**277**(10):793–9.
189. Parkinson J. An essay on the shaking palsy. 1817. *J Neuropsychiatry Clin Neurosci* 2002;**14**(2):223–36.
190. de Lau LM, Breteler MM. Epidemiology of Parkinson's disease. *Lancet Neurol* 2006;**5**(6):525–35.
191. Yavich L, Oksman M, Tanila H, Kerokoski P, Hiltunen M, van Groen T, et al. Locomotor activity and evoked dopamine release are reduced in mice overexpressing A30P-mutated human alpha-synuclein. *Neurobiol Dis* 2005;**20**(2):303–13.
192. Yavich L, Tanila H, Vepsalainen S, Jakala P. Role of alpha-synuclein in presynaptic dopamine recruitment. *J Neurosci* 2004;**24**(49):11165–70.
193. Pelkonen A, Yavich L. Neuromuscular pathology in mice lacking alpha-synuclein. *Neurosci Lett* 2011;**487**(3):350–3.
194. Pelkonen A, Kallunki P, Yavich L. Effects of exogenous alpha-synuclein on stimulated dopamine overflow in dorsal striatum. *Neurosci Lett* 2013;**554**:141–5.
195. Lavedan C. The synuclein family. *Genome Res* 1998;**8**(9):871–80.
196. Polymeropoulos MH, Lavedan C, Leroy E, Ide SE, Dehejia A, Dutra A, et al. Mutation in the alpha-synuclein gene identified in families with Parkinson's disease. *Science* 1997;**276**(5321):2045–7.
197. Kruger R, Kuhn W, Muller T, Woitalla D, Graeber M, Kosel S, et al. Ala30Pro mutation in the gene encoding alpha-synuclein in Parkinson's disease. *Nat Genet* 1998;**18**(2):106–8.
198. Zarranz JJ, Alegre J, Gomez-Esteban JC, Lezcano E, Ros R, Ampuero I, et al. The new mutation, E46K, of alpha-synuclein causes Parkinson and Lewy body dementia. *Ann Neurol* 2004;**55**(2):164–73.
199. Singleton AB, Farrer M, Johnson J, Singleton A, Hague S, Kachergus J, et al. alpha-Synuclein locus triplication causes Parkinson's disease. *Science* 2003;**302**(5646):841.
200. Farrer M, Kachergus J, Forno L, Lincoln S, Wang DS, Hulihan M, et al. Comparison of kindreds with parkinsonism and alpha-synuclein genomic multiplications. *Ann Neurol* 2004;**55**(2):174–9.

201. Masliah E, Rockenstein E, Veinbergs I, Mallory M, Hashimoto M, Takeda A, et al. Dopaminergic loss and inclusion body formation in alpha-synuclein mice: implications for neurodegenerative disorders. *Science* 2000;**287**(5456):1265–9.
202. Feany MB, Bender WW. A Drosophila model of Parkinson's disease. *Nature* 2000;**404**(6776):394–8.
203. Kuwahara T, Koyama A, Gengyo-Ando K, Masuda M, Kowa H, Tsunoda M, et al. Familial Parkinson mutant alpha-synuclein causes dopamine neuron dysfunction in transgenic Caenorhabditis elegans. *J Biol Chem* 2006;**281**(1):334–40.
204. Payton JE, Perrin RJ, Clayton DF, George JM. Protein-protein interactions of alpha-synuclein in brain homogenates and transfected cells. *Brain Res Mol Brain Res* 2001;**95**(1–2):138–45.
205. Uversky Vladimir N. Alpha-synuclein misfolding and neurodegenerative diseases. *Curr Protein Pept Sci* 2008;**9**(5):507–40.
206. Olanow CW, Brundin P. Parkinson's disease and alpha synuclein: is Parkinson's disease a prion-like disorder? *Mov Disord* 2013;**28**(1):31–40.
207. Eliezer D. Synuclein structure and function in Parkinson's disease. *Focus Struct Biol* 2009;**7**:159–74.
208. Alberio T, Lopiano L, Fasano M. Cellular models to investigate biochemical pathways in Parkinson's disease. *FEBS J* 2012;**279**(7):1146–55.
209. Dikiy I, Eliezer D. Folding and misfolding of alpha-synuclein on membranes. *Biochim Biophys Acta* 2012;**1818**(4):1013–8.
210. Ozansoy M, Basak AN. The central theme of Parkinson's disease: α-synuclein. *Mol Neurobiol* 2013;**47**(2):460–5.
211. Desplats P, Lee HJ, Bae EJ, Patrick C, Rockenstein E, Crews L, et al. Inclusion formation and neuronal cell death through neuron-to-neuron transmission of alpha-synuclein. *Proc Natl Acad Sci USA* 2009;**106**(31):13010–5.
212. Dunning CJ, George S, Brundin P. What's to like about the prion-like hypothesis for the spreading of aggregated alpha-synuclein in Parkinson disease? *Prion* 2013;**7**(1):92–7.
213. Margulis BA, Vigont V, Lazarev VF, Kaznacheyeva EV, Guzhova IV. Pharmacological protein targets in polyglutamine diseases: mutant polypeptides and their interactors. *FEBS Lett* 2013;**587**(13):1997–2007.
214. Walker FO. Huntington's disease. *Lancet* 2007;**369**(9557):218–28.
215. Harjes P, Wanker EE. The hunt for huntingtin function: interaction partners tell many different stories. *Trends Biochem Sci* 2003;**28**(8):425–33.
216. Cattaneo E, Zuccato C, Tartari M. Normal huntingtin function: an alternative approach to Huntington's disease. *Nat Rev Neurosci* 2005;**6**(12):919–30.
217. Leonard A, Leal T, Lebecque P. Mucoviscidosis: CFTR mutation-specific therapy: a ray of sunshine in a cloudy sky. *Arch Pediatr* 2013;**20**(1):63–73.
218. Jones AM, Helm JM. Emerging treatments in cystic fibrosis. *Drugs* 2009;**69**(14):1903–10.
219. McPhail GL, Clancy JP. Ivacaftor: the first therapy acting on the primary cause of cystic fibrosis. *Drugs Today (Barc)* 2013;**49**(4):253–60.
220. Sermet-Gaudelus I. Ivacaftor treatment in patients with cystic fibrosis and the G551D-CFTR mutation. *Eur Respir Rev* 2013;**22**(127):66–71.
221. Gomes-Alves P, Neves S, Penque D. Signaling pathways of proteostasis network unraveled by proteomic approaches on the understanding of misfolded protein rescue. *Methods Enzymol* 2011;**491**:217–33.
222. Hosfield DJ, Mol CD. Targeting inactive kinases: structure as a foundation for cancer drug discovery. *Cancer Drug Des Discov* 2008;229–52.
223. Lee F, Fandi A, Voi M. Overcoming kinase resistance in chronic myeloid leukemia. *Int J Biochem Cell Biol* 2008;**40**(3):334–43.
224. Pao W, Miller V, Zakowski M, Doherty J, Politi K, Sarkaria I, et al. EGF receptor gene mutations are common in lung cancers from "never smokers" and are associated with sensitivity of tumors to gefitinib and erlotinib. *Proc Natl Acad Sci USA* 2004;**101**(36):13306–11.
225. Greulich H, Chen TH, Feng W, Janne PA, Alvarez JV, Zappaterra M, et al. Oncogenic transformation by inhibitor-sensitive and -resistant EGFR mutants. *PLoS Med* 2005;**2**(11):e313.
226. Sordella R, Bell DW, Haber DA, Settleman J. Gefitinib-sensitizing EGFR mutations in lung cancer activate anti-apoptotic pathways. *Science* 2004;**305**(5687):1163–7.

227. Balak MN, Gong Y, Riely GJ, Somwar R, Li AR, Zakowski MF, et al. Novel D761Y and common secondary T790M mutations in epidermal growth factor receptor-mutant lung adenocarcinomas with acquired resistance to kinase inhibitors. *Clin Cancer Res* 2006;**12**(21):6494–501.
228. Balius TE, Rizzo RC. Quantitative prediction of fold resistance for inhibitors of EGFR. *Biochemistry* 2009;**48** (35):8435–48.
229. Bulawa CE, Connelly S, Devit M, Wang L, Weigel C, Fleming JA, et al. Tafamidis, a potent and selective transthyretin kinetic stabilizer that inhibits the amyloid cascade. *Proc Natl Acad Sci USA* 2012;**109** (24):9629–34.
230. Ando Y, Ueda M. Novel methods for detecting amyloidogenic proteins in transthyretin related amyloidosis. *Front Biosci* 2008;**13**:5548–58.
231. Colon W, Kelly JW. Partial denaturation of transthyretin is sufficient for amyloid fibril formation in vitro. *Biochemistry* 1992;**31**(36):8654–60.
232. Sousa MM, Cardoso I, Fernandes R, Guimaraes A, Saraiva MJ. Deposition of transthyretin in early stages of familial amyloidotic polyneuropathy: evidence for toxicity of nonfibrillar aggregates. *Am J Pathol* 2001;**159** (6):1993–2000.
233. Hammarstrom P, Wiseman RL, Powers ET, Kelly JW. Prevention of transthyretin amyloid disease by changing protein misfolding energetics. *Science* 2003;**299**(5607):713–6.
234. Ravikumar B, Duden R, Rubinsztein DC. Aggregate-prone proteins with polyglutamine and polyalanine expansions are degraded by autophagy. *Hum Mol Genet* 2002;**11**(9):1107–17.
235. Ravikumar B, Vacher C, Berger Z, Davies JE, Luo S, Oroz LG, et al. Inhibition of mTOR induces autophagy and reduces toxicity of polyglutamine expansions in fly and mouse models of Huntington disease. *Nat Genet* 2004;**36**(6):585–95.
236. Webb JL, Ravikumar B, Atkins J, Skepper JN, Rubinsztein DC. Alpha-synuclein is degraded by both autophagy and the proteasome. *J Biol Chem* 2003;**278**(27):25009–13.
237. Berger Z, Ravikumar B, Menzies FM, Oroz LG, Underwood BR, Pangalos MN, et al. Rapamycin alleviates toxicity of different aggregate-prone proteins. *Hum Mol Genet* 2006;**15**(3):433–42.
238. Sarkar S, Ravikumar B, Floto RA, Rubinsztein DC. Rapamycin and mTOR-independent autophagy inducers ameliorate toxicity of polyglutamine-expanded huntingtin and related proteinopathies. *Cell Death Differ* 2009;**16**(1):46–56.
239. Lashuel HA, Hartley DM, Balakhaneh D, Aggarwal A, Teichberg S, Callaway DJ. New class of inhibitors of amyloid-beta fibril formation. Implications for the mechanism of pathogenesis in Alzheimer's disease. *J Biol Chem* 2002;**277**(45):42881–90.
240. Pappolla M, Bozner P, Soto C, Shao H, Robakis NK, Zagorski M, et al. Inhibition of Alzheimer beta-fibrillogenesis by melatonin. *J Biol Chem* 1998;**273**(13):7185–8.
241. Skribanek Z, Balaspiri L, Mak M. Interaction between synthetic amyloid-beta-peptide (1-40) and its aggregation inhibitors studied by electrospray ionization mass spectrometry. *J Mass Spectrom* 2001;**36**(11):1226–9.
242. Cheng X, van Breemen RB. Mass spectrometry-based screening for inhibitors of beta-amyloid protein aggregation. *Anal Chem* 2005;**77**(21):7012–5.
243. Kato M, Kinoshita H, Enokita M, Hori Y, Hashimoto T, Iwatsubo T, et al. Analytical method for beta-amyloid fibrils using CE-laser-induced fluorescence and its application to screening for inhibitors of beta-amyloid protein aggregation. *Anal Chem* 2007;**79**(13):4887–91.

CHAPTER

8

Promises of Biased Signaling in the Development of Improved Therapeutics

Barbara Bosier and Emmanuel Hermans

Institute of Neuroscience, Group of Neuropharmacology, Université Catholique de Louvain, Brussels, Belgium

OUTLINE

B. Arey (Ed): *Biased Signaling in Physiology, Pharmacology and Therapeutics*
DOI: http://dx.doi.org/10.1016/B978-0-12-411460-9.00008-2

INTRODUCTION

As for the vast majority of functional proteins, receptor recognition and activation by endogenous ligands and by drugs are governed by conformational dynamics. Thus, the conformational state adopted by the receptor upon binding of a ligand determines its ability to transduce this interaction into cellular responses. More specifically, the current receptor model postulates that a given ligand qualitatively and quantitatively dictates the signaling functions of the receptor by altering its molecular conformation or, more likely, by stabilizing one of the active or inactive conformations.[1,2] Acting as biochemical switches, the receptor proteins are thought to oscillate between active and inactive conformations.[3–5] The binding of a ligand contributes to change the spontaneous equilibrium by promoting selected conformations that are associated with functional interactions with defined signaling partners. Considering the diversity of these partners, it is indeed assumed that receptors exist in different active conformations that distinctly but specifically favor these functional interactions.[6–8] This model implies that certain ligands could display distinct affinities for the different active states of the receptor, and could thereby orientate the receptor response. At the extreme, a given ligand may show full selectivity toward a particular signaling cascade for which it will be classified as a full agonist. At the same time, as this ligand occupies the receptor, it will impede other possible conformations and the associated signaling, so it will be considered as an antagonist or even as an inverse agonist.

Indeed, supporting the concept proposing that a ligand stabilizes distinct receptor conformations, several examples of receptor agonists offering selectivity amongst a variety of signaling have been identified in unrelated classes of receptors. Raloxifene, marketed as a selective oestrogen receptor agonist, shows beneficial oestrogenic activity in bones and in the cardiovascular system while countering oestrogen proliferative actions in the uterus and breasts.[9,10] A unique conformational change in the oestrogen receptor induced by raloxifene that favors interaction with a specific subset of co-activators was proposed as the molecular mechanism supporting this selectivity.[11–13] Indeed, the conformational changes induced by raloxifene as an oestrogen receptor trigger co-factor recognition/recruitment that results in oestrogenic action in the bones but not in the breasts and the uterus, where it essentially functions as an antagonist.[12]

In addition, the existence of two non-redundant oestrogen receptor subtypes (α and β) with distinct signaling properties should be considered as an additional mechanism possibly contributing to this pharmacological selectivity.[14–16] Indeed, these receptor subtypes act as homodimers or heterodimers of which combinations can be distinctly stabilized upon binding of agonists.[16] While the ability of diverse ligands to stabilize distinct active conformations has been mostly described for metabotropic receptors and mainly G protein-coupled receptors (GPCRs), similar regulations are not excluded for the ionotropic receptors. As an example, different types of modulations have been reported for the transient receptor potential (TRP) superfamily of cation channels, leading to variable regulations of cation-selective permeability.[17,18]

Specific drug-driven changes in receptor conformational equilibrium appear as elegant mechanisms to explain selectivity amongst distinct signaling pathways and/or ion permeability driven by receptors from all classes. This concept has, however, been more extensively described for the family of seven transmembrane domain GPCRs. In regard to this large receptor family, studies established the basis of the concept of signaling-selective agonism,[19] also referred to as agonist-specific trafficking of receptor signaling, largely developed by Kenakin[6] and more recently termed functional selectivity,[20,21] or biased signaling.[22,23] As there is currently no consensus about the adequate nomenclature, in this chapter we will use the term biased signaling.

Given the compelling role of GPCRs in physiology and in the physiopathology of diverse diseases, and considering the increasing number of marketed compounds that target GPCRs, we will focus on the pharmacological relevance of biased signaling at GPCRs and on the development of novel GPCR-related therapies. The different mechanisms supporting biased signaling will be briefly revised. Several examples will be explored, demonstrating that such complexity should be regarded as an opportunity to gain therapeutic specificity by designing drugs that would discriminate between receptor signaling. Finally, the clinical significance will be emphasized, as the properties of some GPCR agonists are now thought to frequently participate in the atypical pharmacological profile of certain drugs.

"MECHANISMS" OF BIASED SIGNALING

Bias in G Protein-Dependent Signaling

GPCRs are a family of seven transmembrane domain receptors, whose name clearly refers to their property to transduce agonist binding into cellular responses through activation of heterotrimeric G proteins. Common pharmacology views postulate that a single GPCR exists either in an active (R′) or an inactive (R) receptor conformation, and oscillates between these two structural states.[3–5] This concept provides a molecular basis to explain the existence of ligands with agonist and inverse agonist profiles that respectively stabilize the active or inactive conformation, and antagonists that bind to the receptor without affecting the oscillating equilibrium.[1,2] In this simplified concept, affinity and efficacy appear as the sole pharmacological parameters allowing the characterization and/or the prediction of drug action. However, data arising from the experimental exploration of the pharmacological characteristics of a drug in a given biological model cannot always be extrapolated to all systems. Rather, the receptor-mediated cellular responses appear to be quantitatively influenced by diverse tissue-intrinsic factors such as the strength of G protein coupling or the existence of receptor reserve.[24,25]

Furthermore, during the decades at the start of the twenty-first century, the exploitation of a large variety of biological systems used as experimental models to determine ligand-mediated responses, including models overexpressing recombinant receptors, has highlighted the capacity of GPCRs to interact with several unrelated G proteins. Thus, examples have accumulated demonstrating the ability of a single receptor subtype to trigger diverse intracellular signals through activation of distinct G proteins. While initially considered as an artifact related to the high expression level of the receptors in recombinant systems, the multiplicity of G protein couplings is now demonstrated for the vast majority of GPCRs. This concept gave rise to the notion of collateral efficacy,[26] or pluridimensional efficacy,[25] meaning that not all receptor functions are activated to the same extent by a ligand. Rather, activation of a single receptor by a ligand may independently activate an array of intracellular responses, opening avenues for qualitative regulation of receptor signaling. In particular, it is now demonstrated that drugs acting on a given GPCR may distinctly influence the different couplings and should therefore be characterized by multiple efficacy values, depending on the signaling pathway which is considered (Figure 8.1).

At the molecular level, the ability of a given receptor to independently activate distinct G proteins is best explained by considering the existence of several active conformations of the receptor (commonly referred to as R′ and R″) that differ in their coupling efficiencies with distinct G proteins.[27,28] Contrasting with the traditional receptor theory in which the rank order for the activation of related signaling pathways is the same for every agonist, the existence of multiple active receptor conformations implies that a ligand may activate the distinct signaling cascades with unrelated potency values, reflecting the differential stabilization of active conformations.[6,8,29] Extending the concept outlined with the two-conformation model mentioned above, the functional responses will here depend on the ability of a drug to modify the equilibrium between multiple conformations. As a consequence, the pharmacodynamic properties of a drug would merely reflect its affinity for selected conformations, which are stabilized at the expense of others (Figure 8.1).

As ligands may differentially influence the coupling with distinct G proteins, examining two independent signaling pathways as receptor activation readouts may generate different intrinsic activity values. At the extreme, a drug could be characterized as an agonist for one signaling and as a competitive antagonist for another signaling. Consolidating the theoretical concept, several experimental approaches including fluorescence resonance energy transfer (FRET), bioluminescence resonance energy transfer (BRET), plasmon waveguide resonance spectroscopy, circular dichroism, and X-ray crystallography have helped to demonstrate the ability of distinct agonists to stabilize different active conformations, ensuing different signaling cascade subsets.[30] Also, the pharmacological characterization of drug properties on GPCRs whose sequences were manipulated by site-directed mutagenesis has provided additional experimental validation of this multiple conformation model. During these last few years, the concept of multiplicity of coupling and biased signaling—initially viewed as a peculiarity of some receptors and ligands—turned into a major pharmacological reality with putative therapeutic relevance for most GPCRs.

β-Arrestin-Dependent Bias

Decades after the first description of G protein coupling as the key mechanism for signal transduction and generation of second messengers (for what is now identified as the largest family of cell surface receptors), it was demonstrated that signaling at these proteins is not restricted to G protein-mediated biochemical cascades. Indeed, descriptions of biased ligands identified as those activating G protein-independent signaling have accumulated during these last few years. Of major importance is the physical and functional recruitment of β-arrestins that essentially occurs after receptor phosphorylation by GPCR kinases (GRK).

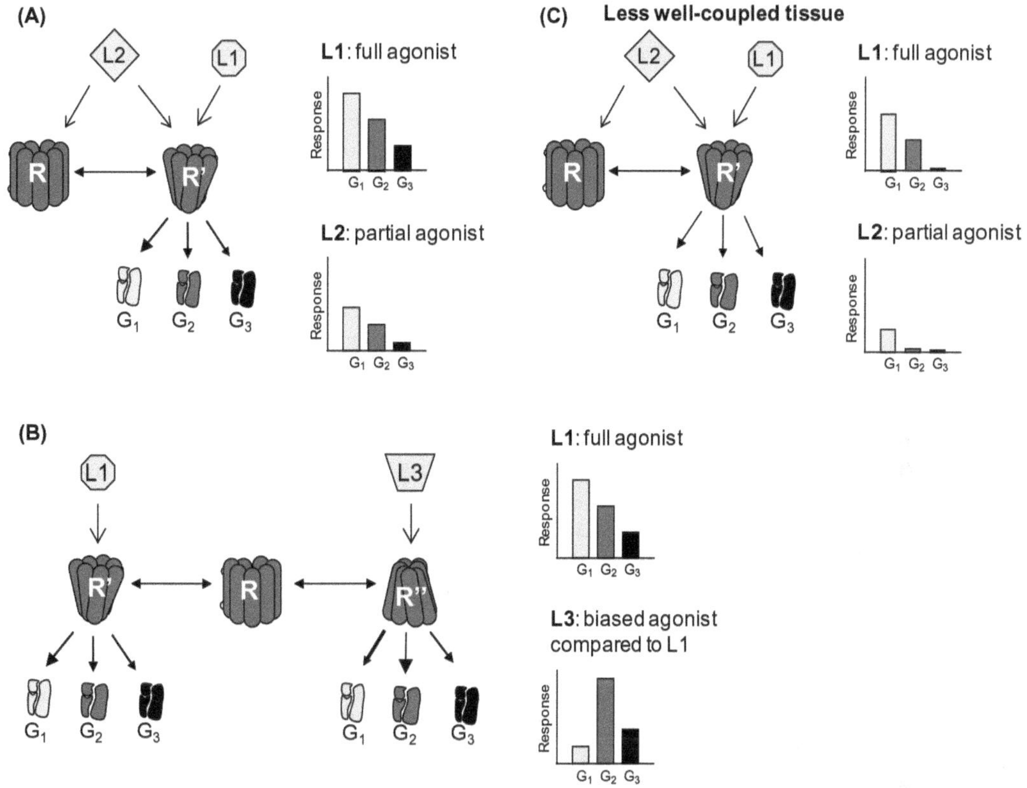

FIGURE 8.1 From classical two-state equilibrium model to multiple active conformation model, and biased ligand properties. (A) Initial receptor theory predicts that in the absence of ligand, an equilibrium spontaneously establishes between the inactive (R) and the active conformation state of the receptor that efficiently couples to a particular pattern of G proteins (R′). Ligand endowed with a full agonist profile (L1) displays high affinity for the R′ conformation, therefore completely displacing equilibrium toward this state and inducing signaling cascades and functional responses of which amplitude partially depends on the coupling efficiencies with different G proteins. Besides, ligands acting as partial agonists (L2) modestly displace the equilibrium towards the R′ conformation, while inverse agonists stabilize the inactive R conformation and antagonists do not modulate spontaneous equilibrium. (B) The complexity of the responses associated with GPCRs is best explained by considering a model in which the receptor can adopt multiple active conformations (R′ and R″) that differ in their coupling efficiencies with the distinct G proteins. Depending on the relative affinity of different interacting ligands for these different conformations, a particular biased ligand (L3) may promote the activation of a subset of G proteins while reducing the activity of others in comparison to a reference ligand. Thus, ligand L1 preferentially induces G1 activation while ligand L3 predominantly activates G3 with only a modest activation of G1. (C) Functional responses observed with particular ligands are intimately tied to the sensitivity of the tissue in which activity is measured. This can be modulated depending on the strength of coupling with a defined pattern of G proteins, the amplification of the response, the existence of receptor reserve, etc. Indeed, in a given environment the functional responses induced by a full agonist (L1) may appear with reduced amplitude. In this case, a partial agonist (L2) may appear as a neutral antagonist. As an example, when investigating G2-mediated response in less well-coupled tissue, L1 may appear as an agonist while L2 would be considered as an antagonist. However, in spite of this apparent functional discrepancy, this does not represent real biased signalling. Real biased signalling may only be considered if distinct potency orders (L1- versus L3-mediated responses, as represented in panel B) for different signaling cascades are observed with different ligands.

Initially viewed as key actors in the desensitization of GPCR, β-arrestins are now clearly identified as major partners in the signaling at these receptors. Upon activation by agonists, phosphorylation of key amino acid residues by GRK promotes the subsequent interaction of the receptor with arrestins. In the early 1990s, GRK-mediated β-arrestin recruitment was primarily associated with G protein interaction impairment and the cessation of G protein signaling (receptor desensitization), and with subsequent receptor internalization and recycling (receptor trafficking).[31–38] However, from the year 2000, it has become increasingly obvious that β-arrestins also act as G protein-independent signal transducers. While many of the β-arrestin-mediated signaling pathways contribute to their well-known functions in receptor desensitization and trafficking (i.e., by modulating endocytosis and second messenger degradation[39]), more recent studies, primarily based on the functional evaluation of β-arrestin knock-out mice, revealed the implication of β-arrestin-dependent cascades in cellular processes unrelated to receptor desensitization and/or trafficking. Indeed, the ability of β-arrestins to activate extracellular signal-regulated kinase (ERK) independently of G protein was demonstrated.

Denoting the importance of the origin of signaling for the functional outcome, β-arrestin-dependent ERK activation is generally referred to as low-level, long-lasting ERK stimulation and is associated with reduced apoptosis and increased activation of survival pathways. In contrast, G protein-dependent activation of ERK is recognized as rapid and of short-term duration and is frequently associated with increased apoptosis.[40] Furthermore, it is now frequently assumed that β-arrestins act as multifunctional scaffold proteins that interact with a variety of protein partners, including protein kinases, thereby leading to the phosphorylation and regulation of a broad set of intracellular targets.[39] Indeed, the activation of β-arrestins by GPCRs was shown to influence major cellular effectors, such as the NF-κB transcription factor.

Accumulating data have also demonstrated that β-arrestins regulate Src family tyrosine kinase activation, cRaf/ERK, Akt/c-Jun N-terminal kinases (JNK), and the small G protein "Ras homolog gene family, member A" (RhoA)/RhoA kinase (ROCK) signaling cascades.[39,40] Importantly, many of these effectors are known to be regulated by G proteins as well, suggesting that GPCRs/β-arrestin/effectors function in parallel with GPCRs/G protein/effectors for ensuring timely and finely-tuned cellular responses. Also, considering that β-arrestin signaling would essentially arise from endocytosed receptors, one may speculate on a distinct subcellular localization of the response. Thus, β-arrestin signaling would develop more deeply inside the cells while G protein-dependent signals are expected to develop close to the inner side of the plasma membrane, before receptor phosphorylation and desensitization.[39]

Due to the multifunctional scaffolding role of β-arrestins to desensitize/internalize receptors, assemble signaling complexes, and trigger intracellular signals,[39,41,42] β-arrestin-dependent bias should be categorized as having consequences either on receptor trafficking and/or on receptor signaling. So far, both types of biases have been reported for several receptors with putatively different pharmacological implications.[41] The first type of bias is thought to influence the efficiency and duration of receptor activation upon agonist binding, which can be exploited therapeutically as a means to modulate tolerance (such as opioid-related tolerance) or cell surface receptor expression (such as in pathological processes involving regulation of receptor expression). In contrast, biased manipulation of signaling cascades could be directly exploited in the regulation of pathophysiological processes. Different examples of this way of selectively influencing signaling at a given receptor, and the possible therapeutic exploitations, are listed in Table 8.1 and discussed in the section "Biased Signaling Holds Promise for Drug Development" below. It is important to note that the duality of β-arrestin functions inevitably imposes a limitation in the use of β-arrestin knock-out animals. Indeed, the loss of all β-arrestin-mediated functions in these models precludes any definitive interpretation of the role of β-arrestin in specific signaling or GPCR desensitization.

While offering a large set of perspectives regarding the pharmacological manipulation of GPCR-mediated signaling, such diversity is conceptually difficult to reconcile with the ubiquitous expression of β-arrestins. In contrast to G proteins, which consist of three different subunits, each existing in different subtypes, there are only two non-visual β-arrestins that are ubiquitously expressed outside the retina and that have the ability to bind to the vast majority of GPCRs. Nevertheless, there is evidence that β-arrestins are subjected to specific modulations of their regulatory or signaling functions.[39,105,106] Indeed, β-arrestins may adopt different conformations depending on which receptor they bind to and which GRK subtype caused their phosphorylation, and these conformations may specify the effector coupling.[107] Accordingly, the molecular mechanisms supporting β-arrestin-dependent desensitization and/or β-arrestin-dependent signaling have been successively dissected. As an example, the phosphorylation of the vasopressin receptor V2 by GRK2 or GRK3 was found to support β-arrestin-dependent desensitization while its phosphorylation by GRK5 and GRK6 induced β-arrestin-dependent ERK activation.[108] One may therefore speculate that selective drugs will be useful to promote different active conformations in order to manipulate the β-arrestin functions.

Referring back to the dual role of β-arrestins in either cellular signaling or receptor internalization, it is worth noting that β-arrestin-dependent desensitization/internalization can be further divided into sustained receptor endocytosis or its rapid recycling to the cell surface. Supporting the hypothesis of ligands that would selectively manipulate β-arrestin functions, abundant examples of agonists that differentially manipulate the receptor turnover have been reported regarding opioid receptors. Thus, while fentanyl and methadone are known to induce consequent and rapid receptor internalization, presumably followed by rapid recycling of the receptor, morphine has been shown to weakly induce receptor internalization possibly leading to sustained desensitization.[109–111] Underlining the importance of discriminating between these diverse β-arrestin-mediated processes, it was initially suggested that such biased agonism influencing the rate of receptor trafficking could support the level of tolerance associated with distinct opiate analgesics.[112,113] A parallelism could be drawn here with the attempts to systematically design specific drugs that discriminate between activation of individual G protein subtypes.

Besides heterotrimeric G proteins, β-arrestins are certainly the best-characterized GPCR interacting proteins involved in cell signaling. However, other intracellular proteins have been shown to interact with GPCRs and to support cell responses. Thus, several GPCRs contain a PDZ recognition motif within their carboxy-terminus, supporting the interaction with proteins possessing complementary PDZ domain.[114,115] Numerous functions related to intracellular trafficking, transcriptional regulation, and cell growth are supposed to be modulated by this type of interaction and activation. The Na^+/H^+ exchange regulatory

TABLE 8.1 Examples of Different Types of Biased Signaling Possibly Presenting Benefits in the Treatment of Diverse Pathologies

Mechanism	Receptor	Pathology	Bias of Therapeutic Value	Bias-Related Effect	References
G protein-biased coupling	Serotonin 2A receptor (5HT2A)	Neuropsychiatric disorders (depression and schizophrenia)	Antagonist at Gi signaling	Gi/o: hallucinogenic properties through Src pathway	43
	β2-adrenergic receptor	Heart failure	Balanced Gi/Gs activation	Gs: heart contractility through PKA-dependent pathway Gi: antiapoptotic, antiarrhythmic, negative inotropic properties	44–48
	Thyroid-stimulating hormone receptor	Hypothyroidism	Gq > Gs (for selective thyroid hormone synthesis)	Gs: thyroid hormone secretion, follicular cell proliferation, iodide uptake Gq: iodination and thyroid hormone synthesis	49–51
	Histamine H1 receptor	Neuropsychiatric disorders	Adenylate cyclase (AC) > Phospholipase C (PLC); suggesting Gs > Gq	AC (putative Gs coupling): respiratory distress, hyperalgesia, allergic response PLC (putative Gq coupling): modulation of catecholamine neurotransmission, potentially useful in neuropsychiatric disorders	52–54
	Somatostatin type 5 receptor	Acromegaly, pituitary tumor	GoA > GoB, Gi1–3	GoA: inhibition of growth hormone release	55
	Vasopressin 1 A receptor	Small cell lung cancer	Gi > Gq (antag gq)	Gi: antiproliferative properties through ERK activation	56
β-arr biased coupling affecting signaling	GPR109A	Hyperlipidemia	Gi > β-arr	Gi/o: decrease in LDL and TG and increase in HDL β-arr: cutaneous flushing	57–60
	Mu opioid receptor	Pain	Gi > β-arr2	Gi: analgesia β-arr: respiratory depression, constipation, analgesic tolerance	22,61
	Kappa opioid receptor	Pain	Gi > β-arr2	β-arr: dysphoria, hallucinations	62–64
	Delta opioid receptor (DOR)	Mood disorder (anxiety, depression)	Gi > β-arr2	β-arr: seizures	65,66

Muscarinic M3 receptor	Cognitive disorders	β-arr > Gq	β-arr: learning and memory in fear conditioning	67
Dopamine D2 receptor	Schizophrenia	Antagonist for D2L/β-arr2-mediated signaling	β-arr2 antagonism is shared by different types of antipsychotics independently of their activity on G proteins	68
β2-adrenergic receptor	Congestive heart failure	β-arr > G proteins	β-arr: survival signal	48,69,70
β1-adrenergic receptor	Congestive heart failure	β-arr (EGFR transactivation) > PKA cascade and CaMKII recruitment	Gs (PKA cascade): heart contractility β-arr (CaMKII recruitment): cardiomyocyte apoptosis and hypertrophy, maladaptive cardiac remodeling β-arr (EGFR transactivation): cardioprotective antiapoptotic signals	71–75
Angiotensin type 1A receptors	Cardiovascular diseases (heart failure, hypertension)	β-arr > Gq	β-arr: cardioprotective antiapoptotic signal, inotropic response, improved cardiac performance, increased renal performance Gq: vasoconstriction, cardiac contractility, cardiac fibrosis	23,76–81
Serotonin 2A receptor (5HT2A)	Neuropsychiatric disorders (depression and schizophrenia)	Antagonist for β-arr2-mediated signaling	β-arr: hallucinogenic properties of serotonin	82
Parathyroid hormone receptor	Osteoporosis	β-arr2 > Gs	β-arr2: anabolic bone formation through control of cell cycle, cell survival, apoptosis, and migration Gs: bone remodeling, osteoblast differentiation, increase in osteoclast number, catabolic bone resorption	83–87
Glucagon-like peptide-1 receptor (Glp-1R)	Diabetes	β-arr1 activation	β-arr1: insulin secretion and receptor signaling (ERK and CREB activation, stimulation of cAMP production)	88
Alpha2-adrenergic receptor	Mood disorders	β-arr2 activation	β-arr2: induction of sedation	89,90

(Continued)

TABLE 8.1 (Continued)

Mechanism	Receptor	Pathology	Bias of Therapeutic Value	Bias-Related Effect	References
	Muscarinic M3 receptor	Diabetes	Activation of β-arr-dependent signaling	Gq: transient, early phase of insulin release β-arr: insulin release during sustained, late phase; glucose homeostasis	91–94
β-arr biased coupling affecting desensitization	Kappa-opioid receptor	Stress-related disorders, depression and addiction	JNK activation (through β-arr)	JNK: long-term receptor inactivation	63,95
	Opioid receptors	pain	No β-arr recruitment and receptor regulation	Suppression of β-arr2 –mediated receptor regulation: prolonged analgesia, reduced tolerance	96,97
	Dopamine D1 receptor	Parkinson	No receptor regulation	Improved efficacy at D1 receptor	98
	C chemokine receptor type 5	HIV	β-arr recruitment	β-arr: rapid receptor internalization and endosomal destruction	99–101
	β2-adrenergic receptor	Asthma	No β-arr recruitment	Suppression of β-arr2-mediated receptor regulation: enhanced airway relaxation	102,103
	CB1 cannabinoid receptor	Pain	No β-arr recruitment	Suppression of β-arr2-mediated receptor regulation: sustained antinociception	104

factor is a well-known example of a protein directly regulated by GPCRs in a ligand-dependent but G protein- and β-arrestin-independent manner.[116–118] While for the moment no evidence of biased signaling directly involving this type of regulation has been reported, one cannot ignore the fact that particular ligands would be able to stabilize receptor conformations that specifically ensure the interaction with functional PDZ domain proteins.

Modulation of Biased Signaling

The ability of GPCRs to trigger a large variety of unrelated signaling raises the issue of the specificity of the response of individual receptors to the endogenous transmitter. Several mechanisms that contribute to control and limit signal scattering at these receptors have been identified. Indeed, diverse extracellular and intracellular molecular partners selectively interact with GPCRs[119] to organize their associated signaling, to modulate receptor trafficking, and, finally, to confine receptor responses in a temporal and spatial fashion (Figure 8.2). Such interactions that eventually impact on the receptor signaling may be constitutive and thus operate in a ligand-independent manner. However, several studies have shown that the interactions of GPCRs with their partner proteins were influenced by agonist stimulation. In this model, such molecular interaction can be viewed as one of the mechanisms that directly drive the biased signaling. Importantly, not only can the physical interaction be regulated by the receptor ligands, but also the expression and intracellular trafficking of the partners can be modulated as well as the receptor transcription (RNA splicing) and translation (post-translational modifications).

Amongst the key mechanisms influencing G protein coupling, compelling evidence supports the concept that phosphorylation of GPCRs not only regulates their desensitization but also modulates their couplings to different G proteins.[120–123] Best documented in simplified models of cell cultures is the switch from G_i to G_s coupling consecutive to receptor phosphorylation,[33,120,124] providing a sophisticated mechanism that permits temporal resolution of cellular signaling elicited during sustained stimulation of a receptor. Further supporting this hypothesis, evidence has been presented indicating that a phosphorylation-mediated switch in G protein coupling also operates *in vivo* for a variety of GPCRs upon long-lasting receptor activation.[125,126] Not only does the coupling to distinct G proteins appear to be influenced by receptor phosphorylation, but the activation of β-arrestin-mediated signaling is directly dependent on such a process (see above). Moreover, altered palmitoylation of the receptors or changes in their distribution at the cell surface also contribute to biased receptor signaling.[127]

In the last few years, there have been several discoveries which have supported the hypothesis that GPCRs interact with each other to form dimers, oligomers, or higher order complexes. Studies have suggested that such heteromer complexes may be involved in mediating functional selectivity.[128] Accumulating evidence indicates that homodimers or heterodimers show altered G protein coupling compared with individual receptor monomers[129–132]: A concept commonly referred to as heterodimer-directed specificity. Importantly, the coupling selectivity of the heteromer appears clearly dependent on the nature of the interacting ligand,[133] further suggesting that the dimerization could be one

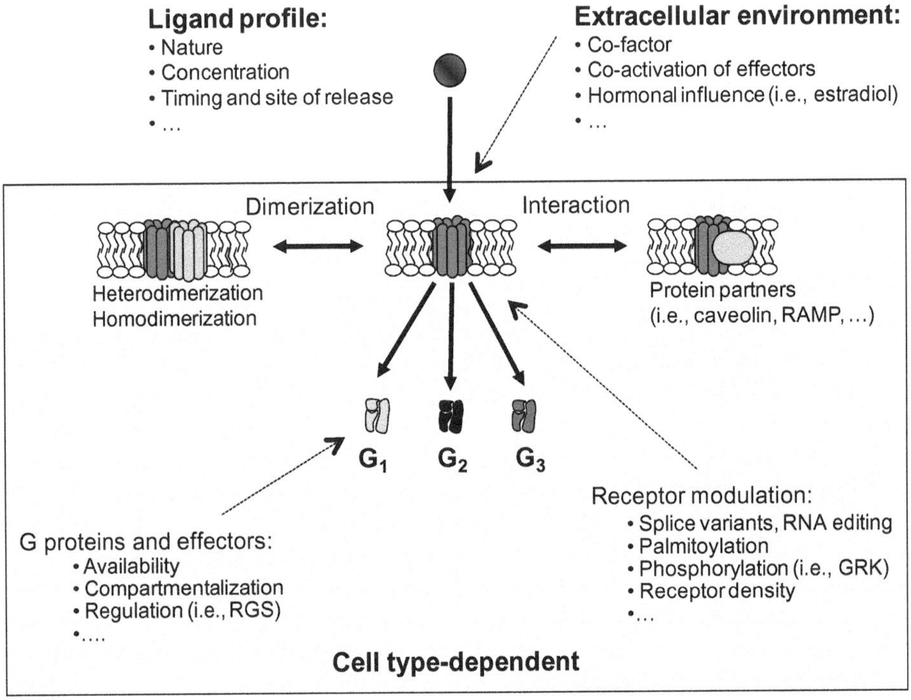

FIGURE 8.2 Different levels GPCR signaling modulations. Numerous mechanisms, interfering at different levels of signaling cascades, are now reported to influence the signaling of GPCRs either through different G proteins or through non G protein transducers (i.e., β-arrestins). While the ligand profile is obviously crucial in dictating receptor-mediated responses, the final outcome of receptor stimulation is additionally finely controlled by both cell type-dependent machinery and extracellular factors. Signaling cascades regulated by GPCRs are pre-organized and dependent of complex machinery that governs their cellular responses. Indeed, several interactions with cellular proteins, which are in general confined to defined membrane microdomains, serve to organize the different partners of signaling cascades, modulate coupling efficiency, and control receptor trafficking. While the expression and the accurate functionality of this machinery is cell type dependent, pharmacologists currently dispose of several tools to modulate this pre-organized signaling. Besides the classically described agonists, inverse agonists and antagonists, ligands possessing more complex profiles, including biased ligands, are shown to independently orchestrate the diverse cellular responses. In addition, it is now well described that different ligands can directly modulate receptors in a ligand-specific manner (i.e., all the agonists do not induce the same extent of receptor phosphorylation) or influence interactions of the receptor with other receptors or other protein partners. Finally, different co-activators or other GPCR ligands may also directly interfere with this pre-organization or with the receptor-mediated signaling.

of the molecular mechanisms supporting biased signaling. While initial studies mainly focused on the role of receptor dimerization on the control of G protein coupling and related switches, there is now also evidence showing the importance of dimerization on the functional interaction with β-arrestin. Indeed, dimerization may influence both the recruitment and the activation of β-arrestins.[134–136] While most of these studies have found distinct profiles of receptor trafficking/recycling for dimers, the pharmacological relevance of this concept remains largely elusive. The data from Gupta and colleagues[137] are quite convincing, showing that repeated administration of morphine induces the dimerization of mu and delta opioid receptors, promoting the preferential activation of β-arrestin2 and supporting both analgesia and clinical tolerance.[138]

Cell Type-Dependent Biased Signaling

Quantitative differences in the response to a given agonist in distinct tissues or cell models are commonly observed and result from differences in the density of the targeted receptor or in the availability of signaling partners. Also, the cell environment in which the receptor is characterized has been shown to influence the level of constitutive activities as well as the intrinsic activity of partial agonists.[21,139–141] Similarly, biased agonism has been shown to arise from intrinsic alterations of the receptor itself (mRNA splicing, receptor glycosylation, phosphorylation, dimerization, etc.) or from the variability of extrinsic factors (interacting proteins, signaling partners, etc.) that qualitatively influence the pharmacodynamic properties of ligands (Figure 8.2). Accordingly, it has been frequently demonstrated that the expression of GRKs, β-arrestins, and G protein subtypes varies between different cell types.[39,142–145] Therefore, the functional response to a given ligand is not solely determined by the ligand occupancy and ligand-mediated conformational stabilization, but also depends on the particular intracellular environment and/or on the specific conditions to which cells/tissues are exposed.

The stoichiometry and sensitivity of the cellular components driving the response can also support cell-dependent selective responses.[25,119] As a consequence, the measured efficacy assigned to a ligand not only depends on the investigated signaling pathway but is also intimately tied to the cell/tissue in which the activity is examined. However, defined responses could be authenticated only if the efficiency of coupling is appropriate. Indeed, leading to possible confusion with the concept of biased agonism, a given ligand could be classified as partial agonist in tissues where the examined coupling is robust while it could behave as an antagonist in tissues in which coupling is weak (Figure 8.1).

A convincing example of how the cell environment can influence the type of observed bias is provided by data obtained during the development of somatostatin analogs. As this hormone inhibits tumor secretion and growth, such drugs have been envisaged for the treatment of neuro-endocrine cancers.[146] However, SOM230 and KE108, two of these analogs that were initially reported as full agonists at somatostatin subtype 2A receptor, were found to antagonize some of the somatostatin actions. Even though both compounds displayed biased signaling in both recombinant models and in a pancreatic cell line, endogenously expressing the somatostatin subtype 2A receptor, different G proteins and signaling pathways were shown to functionally operate in these different models, providing one of the first demonstrations of cell type-dependent ligand-induced biased signaling.[147] Given the variety of phenotypic changes observed in tumor cells, this may be of critical importance for the development of novel therapies with enhanced clinical efficacy and safety.

More recently, it has been demonstrated that the expression level of distinct receptor partners, including G proteins and β-arrestins, directly affects the conformation repertoire of the ghrelin receptor.[148] As these receptor conformations directly support biased signaling and ligand efficacies, this study suggests that cell components are crucial in the determination of the distinct biological responses and biased signaling induced by ligands.

It should also be noted that remarkable cell-specific effects have been reported for other receptor classes including nuclear receptors. Indeed, for several nuclear receptors,[149,150] opposite and cell type-dependent receptor-driven regulations of transcription have been reported. Importantly, these fine-tuned regulations were not satisfactorily justified by anatomical distribution of the receptors but were rather explained by considering cell type-dependent interactions with different co-regulators and/or different post-translational modifications of the receptor.[149–151]

Together, these concepts, which have been validated in several experimental reports, indicate that optimizing therapeutically valuable signaling bias would necessitate the evaluation of new compounds in the most representative model in which the typical pharmacological response is actually sought. Indeed, the action of a drug at a receptor not only depends on the phenotype of the cell but is influenced by the physiological/pathological state of the cell. Therefore, it is important to consider that particular pathological contexts could alter the pre-cited intrinsic and extrinsic factors and consecutively modulate receptor pharmacology and "response." Obviously, this cell-based selectivity renders efforts to guide medicinal chemistry toward the identification of potent novel biased signaling-based therapies much more complex.

BIASED SIGNALING IN PATHOPHYSIOLOGY

Biased Endogenous Ligands

While experimental evidence has accumulated demonstrating the capacity of diverse ligands to induce biased signaling, it is noteworthy that the results have been frequently obtained in recombinant *in vitro* systems where a high density of receptor is artificially achieved in defined cell environments, usually an immortalized clonal cell. Nevertheless, in recent years ligand-induced biased signaling has also been validated in *in vivo* models, opening new avenues for the design of pharmacological tools with enhanced selectivity. Convincingly supporting a role for biased signaling in physiology, and thus probably in pathologies, examples of endogenous ligands endowed with biased signaling profiles have been recently collected.

Amongst the variety of molecular mechanisms driving such specificity of signaling is the glycosylation of numerous secreted proteins, which has been shown to impact how they interact and activate targeted receptors. In particular, the naturally occurring glycosylated form of the follicle-stimulating hormone has been reported to activate both G_s and G_i-dependent signaling while the non-glycosylated hormone is known to only affect G_s signaling.[22,152] Similar glycosylation-induced biased signaling has been reported for interleukin 22 (IL22),[153] bone morphogenetic protein 6 (BMP6),[154] and human chorionic gonadotropin hormone (hCG).[155,156] It has been suggested that this enables the induction or recognition of subtle conformational changes in the receptor, thereby inducing unique signaling profiles and physiological responses in order to accommodate subtle extracellular influences.[22]

Similarly, biased signaling has been documented for different endogenous ligands that bind to the same receptor but trigger distinct responses, suggesting that in physiological conditions such diversity of ligands is endorsed with distinct functions. This has been exemplified for the C-C chemokine receptor type 7, which is expressed on different types of leucocytes and which is recognized by two endogenous chemokines: CCL19 ([C-C motif] ligand 19) and CCL21 ([C-C motif] ligand 21)—both known to induce chemotaxy. However, despite similar affinities and potencies,[157] only CCL19 was shown to induce phosphorylation of the receptor, which in turn is followed by rapid receptor desensitization.[158] Likewise, the binding of neurokinin A to its cognate tachykinin NK2 receptor was shown to elicit modulation of both adenylyl cyclase and phospholipase C, whereas its natural truncated analog neurokinin$^{4-10}$ only mediates phosphoinositide hydrolysis.[159] More recently, it has been demonstrated that endocannabinoids acting at cannabinoid type 2 receptor distinctly bind to the receptor, therefore displaying different rank order of potencies for regulation of several signaling pathways.[160] These studies support the hypothesis that the complexity of physiological tasks regulated by endogenous systems possessing several endogenous agonists not only depends on the receptor-type specificity, the spatial segregation, or the timescale of receptor activation, but is also fine-tuned by the type of effector-mediated responses induced consecutively to receptor activation.

Implications in Diseases

Although more indirect than those supporting an implication in receptor regulation or in discriminating between therapeutic and secondary undesired effects, data are now emerging showing the importance of biased signaling in the pathophysiology of several diseases. Abundant data from the literature have indeed provided functional evidence for the association of several disease conditions with alterations of receptor signaling. Thus, naturally occurring mutations resulting in the specific loss of identified signaling pathways are associated with diverse endocrine diseases such as in forms of acromegaly[161] or hyperthyrotropinemia.[49] In these pathologies, point mutations at critical amino acid residues were reported to cause a loss of some defined couplings of the somatostatin type 5 receptor[55] or the thyrotropin receptor[49] respectively. These mutations were shown to modify the G proteins coupling specificity, thereby influencing particular downstream signaling cascades and altering the secretion of hormones.

Characterizing the signaling cascades regulated by these mutant receptors constitutes an elegant means to decipher the physiological functions specifically associated with specific couplings of the receptor with a given G protein subtype. Such characterization, however, also provides the molecular bases for the identification of signaling cascades to be targeted by alternative therapeutics. Indeed, the specificity in the altered signaling cascades and related functions may explain the therapeutic resistance to conventional receptor agonists that can be observed in such endocrine diseases and justifies the need for biased signaling-based therapies that would specifically address the altered function.

Besides changes in receptor couplings, alterations that specifically affect β-arrestin functions have also been linked to pathologies. In particular, it is thought that β-arrestins are involved in several cancer-promoting signals that involve a large variety of receptor subtypes.[162] Elevated β-arrestin mRNA levels have thus been measured in several cancer cell lines and isolated tumors.[41,163] Other studies have revealed that overexpression of β-arrestin1 in transgenic animals promotes tumor progression and the associated angiogenesis,[164] demonstrating the role of β-arrestins in critical signaling pathways regulating tumor growth and metastasis.[165] Furthermore, tumor proliferation, invasiveness, and metastasis were found to be reduced after interfering with β-arrestin functions,[41,162,166] supporting the relevance of β-arrestins as potential targets for future chemotherapeutic strategies.

Finally, underlining the importance of correctly discriminating between the roles of β-arrestins in the control of receptor internalization and in cell signaling, a truncated form of the C-X-C chemokine receptor type 4 capable of supporting β-arrestin-mediated signaling, but not β-arrestin-mediated receptor regulation, was found in the Warts-hypogammaglubulinemia-infection-myelokatexis syndrome (WHIM syndrome).[167] In this particular form of autoimmune disease, the C-X-C chemokine receptor type 4 displays defective desensitization accompanied by enhanced chemotaxis, which have both been shown to require β-arrestin2 recruitment.[167] This further suggests that β-arrestin may adopt distinct conformations, triggering different regulatory functions, depending on different interactions with the receptor.

BIASED SIGNALING HOLDS PROMISE FOR DRUG DEVELOPMENT

Given their implication in regulating the pathophysiology of diverse human diseases, as well as their potential for therapeutic interventions, GPCRs and associated signaling appear to be ideal molecular targets for drug discovery. Indeed, more than 50% of the drugs in clinical practice act as agonists or antagonists at GPCRs. Yet GPCRs are highly versatile membrane sensors that respond to a large diversity of extracellular stimuli by transducing their signals into a rather low number of intracellular cascades. How do such limited signaling possibilities contribute to specific control of a broad range of physiological and pathological processes? How can a single GPCR manipulate one particular cellular response? These are typical questions pharmacologists have to answer in order to obtain more selective and fine-tuned therapies endowed with maximal effectiveness whilst avoiding side effects.

In most cases, endogenous GPCR ligands show modest affinity and poor selectivity towards related receptor subtypes and therefore trigger complex responses throughout the organism which cannot be assigned to a single target. Hence, early developments in medicinal chemistry were focused on searching for high-affinity ligands with high selectivity for individual receptor subtypes. However, single receptor subtypes often show complex tissue distribution, leading to predictable secondary side effects. Classical examples are those neurotransmitter receptors found in several brain structures and frequently associated with the control of psychiatric behaviors and mood disorders. Optimizing pharmacokinetic properties of the chemical entities may sometimes help to ameliorate tissue targeting. Indeed, tropism for some tissues has been achieved with some antibiotics or proton pump inhibitors. Also, access to the central nervous system is commonly prevented by designing drugs that do not cross the blood–brain barrier.

While these approaches based on targeted drug delivery systems[168–171] may help to decrease unwanted side effects, the clinical relevance remains limited a few examples. Since the early nineties, data have been accumulating suggesting that it might be possible to achieve selectivity at another level by developing drugs that selectively modulate

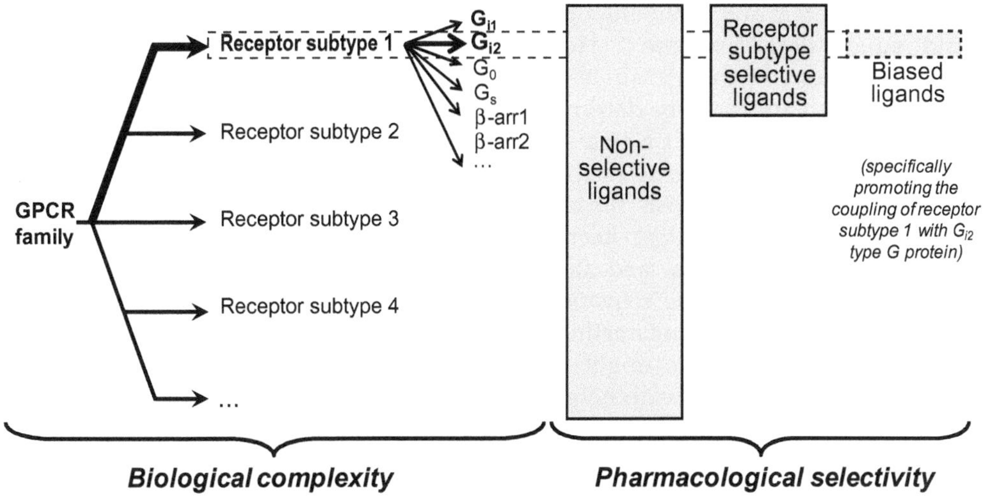

FIGURE 8.3 From physiological complexity to pharmacological selectivity. In a family of GPCRs responding to a given transmitter, the first mechanism governing the diversity of physiological responses triggered by a ligand arises from the multiplicity of closely related receptor subtypes. Even when focusing on the activation of a single receptor subtype, a diversity of cellular responses can be observed, reflecting additional levels of complexity resulting from the activation of multiple G proteins, as well as from the regulation of different unrelated intracellular effectors. Finally, the diversity and complexity of the responses mediated by a single receptor subtype also result from the propensity of the receptor to form multifaceted signaling machinery involving a variety of protein partners. Indeed, a fine-tuned regulation of these complex mechanisms operates to prevent uncontrolled signal scattering during physiological processes. The selective pharmacological manipulation of individual cellular responses involved in the desired therapeutic response could prove to be of clinical benefit while minimizing the side effects. Traditionally, drug design focused on the development of compounds that bind with high affinity and optimized selectively to a receptor of interest. Indeed, considerable efforts in drug discovery are dedicated to the detection of off-target interactions with undesired targets that could lead to undesirable side effects or toxicity. In addition, attention is commonly given to the development of subtype selective drugs. As an alternative, gain in specificity of drug action could be achieved by the development of biased ligands that selectively regulate the signaling cascade of interest. Reaching such selectivity, however, requires considering the environment and pathological conditions in which drugs will act, as well as the potential interacting partners which may interfere with drug action. A better understanding of the complex mechanisms modulating receptor-mediated responses should pave the way for the efficient design of compounds with desirable selectivity profiles.

defined signaling cascades (Figure 8.3). Some recent studies indicate that such approaches could efficiently produce new therapeutic tools with enhanced clinical efficacy and lower risk of secondary effects (Table 8.1). Obviously, further development will largely depend on the thorough understanding of GPCR signaling complexity.

G Protein Bias

Heart failure is a typical disease exemplifying how alteration of GPCR signaling may contribute to pathological dysfunction, and how targeting these signaling biases may give access to safer therapeutic strategies. Congestive heart failure is characterized by tonic activity of adrenergic and renin-angiotensin systems that compensate for cardiac dysfunction and reduces blood pressure.[172] However, in the long term, the high levels of circulating catecholamines and angiotensin lead to progressive myocardial apoptosis, fibrosis, and remodeling, which appear to be detrimental for cardiac function.[173–175]

In particular, the persistent activation of β1-adrenergic receptors, which are primarily coupled to G_s protein in order to mediate contractile response,[176] also triggers cardiomyocyte hypertrophy and apoptosis through calmodulin kinase II (CaMKII)-dependent and PKA-independent mechanisms, thereby inducing maladaptive cardiac remodeling *in vivo*.[71,72,177,178] In addition, and further contributing to impaired cardiac contractile response, sustained adrenaline activation of β2-adrenergic receptors induces a switch from G_s to G_i coupling,[44,120,179] counteracting the G_s-mediated inotropic response; whilst harmful for cardiac function, this might constitute a protective compensatory mechanism. Indeed, G_i coupling of the β2-adrenergic receptor is known to induce survival-supporting signals through phosphatidylinositide 3-kinase (PI3K)/Akt-dependent mechanisms.[45,125] By identifying specific roles of individual signaling pathways in the control of distinct aspects of the cardiac function, these studies underline the necessity to ensure a balanced activation of both G_i- and G_s-type G proteins by distinct adrenergic receptors.

In addition, there is also prevailing evidence that a deregulation of GPCR signaling could be associated with the development of pathological conditions. Thus, altered adrenergic receptor signaling was correlated with enhanced desensitization and down-regulation of β1-adrenergic receptor[180,181] that results in a substantial change in the β1-/β2-adrenergic receptor expression ratio in the failing heart (from 80/20 to 60/40). As a consequence, beside β1-adrenergic receptor, for which blockade is expected to restore expression and signaling, β2-adrenergic receptor could constitute an alternative or complementary therapeutic target. Indeed, given the detrimental effect of β2-adrenergic receptor-mediated G_i activation in the control of cardiac contractile response, it has been suggested that a selective activation of β2-adrenergic receptor-G_s coupling could restore cardiac contractility.[182] While most of the β2-adrenergic receptor agonists were shown to activate both G_i and G_s proteins, the selective β2-adrenergic receptor agonist, fenoterol, was reported to selectively induce G_s coupling.[46] Further supporting the importance of selectively manipulating G_i/G_s activation, it has been shown that prolonged fenoterol treatment improves cardiac function and delays inappropriate remodeling, therefore providing beneficial effects as validated in a rat model of cardiomyopathy.[183,184]

β-Arrestin Biased Signaling

Beside G protein-dependent signaling, studies primarily based on the functional evaluation of β-arrestin knock-out mice revealed the implication of this family of intracellular proteins in diverse G protein-independent signaling cascades and physiological processes. A well-investigated system is the β-arrestin-dependent activation of ERK signaling cascades.[39,40] In the case of cardiac function, it has been demonstrated that both adrenergic receptors (β1- and β2-adrenergic receptors)[73,185] and angiotensin type 1A receptors[76,186] activate ERK signaling pathways through transactivation of the epidermal growth factor receptor (EGFR). Indeed, EGFR is a single transmembrane domain tyrosine kinase receptor that initiates antiopoptotic and proliferative signals through ERK-dependent mechanisms.[187,188] Supporting β-arrestin as a promising pharmacological molecular target to treat cardiovascular diseases such as heart failure, several β1-adrenergic receptor antagonists (β-blockers) and angiotensin type 1A receptor antagonists have been shown to induce cardioprotective responses through biased β-arrestin-dependent EGFR transactivation.[74,77,78,189] It is noteworthy that EGRF also influences nitric oxide (NO) production, which would likely further contribute to the anti-hypertensive properties of these drugs.[190] Indeed, drugs enhancing *no* effects have recognized benefits in a variety of cardiovascular diseases[191,192] and this likely contributes to the therapeutic advantages of β-arrestin-biased GPCR ligands in the treatment of heart failure.

In addition to the modulation of EGFR signaling, it has been reported that activation of β1-adrenergic receptor may induce specific conformational changes in β-arrestin that consecutively promotes the formation of a complex between β-arrestin, CaMKII, and cAMP-dependent guanine nucleotide exchange factor (Epac), which in turn induces CaMKII-dependent apoptosis.[75] Based on the accepted concept that β-arrestin acts as a scaffold protein for other enzymes and signaling partners (see above),[39,41,42] this finding highlights the multifaceted properties of β-arrestin.[193,194] Adding another layer of complexity in the development of biased ligands, studies have demonstrated that distinct ligands may induce specific conformational changes of the β-arrestin proteins[107] with putative implications for cell signaling.

Biased Signaling and Ligand-Induced Receptor Regulation

To restrict or timely control the effects of receptor activation, GPCRs are subjected to a variety of regulatory mechanisms. Balanced molecular and cellular controls of GPCR activation, desensitization, and internalization act in concert to determine the overall degree of biological sensitivity of target cells or tissues. Since the early nineties, a prominent role of β-arrestin has been assigned to the control of GPCR desensitization and downregulation. As a major therapeutic issue, the resulting pharmacodynamic tolerance limits the long-term use of GPCR agonists. Supported by accumulating experimental observations, it appears that G protein-biased ligands that fail to promote β-arrestin recruitment prevent receptor desensitization. These regulation-resistant selective agonists represent a novel strategy for the development of improved therapies with more sustained efficacy.

An archetypal demonstration of this concept has been obtained by the observation of persistent morphine-mediated analgesia in β-arrestin2 knock-out mice.[96,97,195] Indeed, the influence of β-arrestins on opioid-mediated analgesia and tolerance has been examined in several studies.[196] Morphine-induced analgesia is largely impaired in transgenic mice lacking central mu opioid receptor, and it is now accepted that this effect is essentially mediated through interaction of opioid ligands with this receptor subtype.[197,198] Clearly demonstrating the role of β-arrestin in negatively regulating mu opioid receptor homologous desensitization, β-arrestin2 knock-out mice were shown to experience enhanced morphine-induced analgesia with reduced analgesic tolerance.[195]

It is important to note that qualitative differences in the ability of individual ligands to regulate receptor desensitization and internalization do not correlate with differences in G protein activation efficacy, but more closely correlate with the influence on β-arrestin recruitment.[199] Thus, morphine, methadone, etorphine, and fentanyl produce equi-efficacious G protein coupling and similar analgesia in mice; however, these ligands distinctly influence receptor regulation.[196] Indeed, morphine displays a stronger propensity to cause desensitisation and internalisation. This was only observed in wild type animals, while methadone, etorphine, and fentanyl display similar analgesic tolerance in both wild-type and β-arrestin2 knock-out animals.[195,199] Consistent with this it was previously demonstrated that morphine preferentially recruits β-arrestin2, while etorphine, methadone, and fentanyl may recruit both β-arrestin1 and β-arrestin2,[110] allowing for a possible β-arrestin1-dependent compensation of the receptor regulation upon etorphine, methadone, or fentanyl stimulation. In addition, a behavioral study has recently demonstrated that JNK signaling was selectively involved in the morphine- but not in the fentanyl-induced tolerance.[200] Together this indicates that analgesic responses, including analgesic tolerance induced by distinct mu opioid receptor agonists, can be differentially regulated by β-arrestin and downstream signaling cascades *in vivo*.

While this further corroborates differences in receptor internalization, the precise contribution of desensitization and internalization in terms of clinical outcomes remains difficult to predict as it also depends on other regulatory elements.[42] Indeed, morphine, despite the strong influence of β-arrestin2 expression on its pharmacological outcome, weakly promotes receptor internalization while etorphine, methadone, and fentanyl robustly promote it.[109,195,201] This was initially suggested to support differences in opioid tolerance, due to the rapid recycling of the receptor at the cell surface.[112,113,202] Nevertheless, similar analgesic tolerance was observed when these drugs were administered at equi-efficacious doses.[199] To better define the role of β-arrestin recruitment in mu opioid receptor-dependent analgesia and tolerance, the *in vivo* properties of newly developed agonists devoid of β-arrestin activation and receptor internalization properties, such as herkinorin,[203–205] should be further examined. Indeed, encouraging the development of such biased ligands that do not recruit β-arrestin, a recent work revealed a reduced tolerance to herkinorin as compared to morphine following chronic exposure.[206]

It is of importance to note that several morphine-related responses found to be modified in the β-arrestin2 knock-out mice[96] could result from either reinforced receptor activation (lack of receptor desensitization), or from the loss of alternate β-arrestin-dependent signaling. Thus, the consequences of suppressing β-arrestin2 expression are not restricted to an altered receptor desensitization. In a simplistic scenario where β-arrestins would only act as a desensitizing protein partner, all behavioral responses associated with mu opioid receptor activation, including respiratory depression, constipation, and induction of locomotor activity should be enhanced in the β-arrestin knock-out mice. However, experiments showed that all these morphine-induced side effects were actually reduced in the transgenic animals lacking β-arrestin2,[207] indicating that all functional responses to morphine are not controlled by the degree of receptor expression/sensitization but are rather directly mediated by β-arrestin2 acting as an independent intracellular signaling trigger. While alternative explanations may help to elucidate these phenotypic differences, such as involvement of different mu opioid receptor subtypes[208,209] or distinct influences of β-arrestin1 and β-arrestin2 depending on the cellular environment,[42] these studies underline the potential pharmacologic relevance of developing opioid receptor agonists that do not activate β-arrestin.

In contrast, but supported by the same concept, some clinical conditions could benefit from drugs designed to specifically trigger β-arrestin recruitment and associated receptor internalization. Thus, a potential anti-human immunodeficiency virus (HIV) strategy is to block the C-C chemokine receptor type 5 or the C-X-C chemokine receptor type 4, both membrane receptors act as virus entry ports in eukaryote cells. While competitive antagonists of these receptors have been developed in order to interfere with receptor recognition by the virus, the high mutation rate of HIV might quickly lead to the emergence of resistant strains. An alternative approach is to force the internalization of C-C chemokine receptor type 5, which is documented as depending on β-arrestin-mediated signaling.[99] RANTES (regulated on activation, normal T cell expressed and secreted), also known as chemokine 5 (CCL5 or [C-Cmotif] ligand 5) is an endogenous ligand of C-C chemokine receptor type 5[210] that induces receptor internalization followed by rapid recycling. In contrast, the synthetic analog- aminooxypentane-RANTES was shown to induce persistent β-arrestin interaction that supports receptor internalization followed by endosomal sequestration[100,101]—an effect that justifies its potential benefit in the treatment of HIV-mediated acquired immunodeficiency syndrome.

Receptor Heterodimer-Driven Bias in Receptor Signaling

As detailed above, several GPCRs physiologically interact with each other to form dimers, oligomers, or higher order complexes.[211] In addition to the first-characterized homodimers combining identical GPCRs, the existence of heterodimers combining unrelated GPCRs (recognizing distinct endogenous transmitters) has been largely documented, leading to the identification of altered coupling properties. Indeed, heterodimerization can lead to a switch in G protein coupling and/or in β-arrestin recruitment. This phenomenon, termed heterodimer-directed signaling selectivity, was shown to be either ligand dependent (induced upon ligand binding) or ligand independent (constitutive in tissues where receptors are co-expressed)[212] (see above).

Importantly, unique signaling properties arising from defined receptor heterocomplexes have been recently associated with pathophysiological processes calling for possible specificity in therapeutic approaches.[213] As an example, hallucinogenic drugs including lysergic acid diethylamide (LSD), psilocybin, and dimethoxy-iodoamphetamine (DOI) are all high affinity ligands for the serotonin 2A receptor that activate signaling cascades downstream of $G_{q/11}$- and $G_{i/o}$-type G proteins.[43,214] However, not all drugs that activate serotonin 2A receptor show psychodisleptic properties: Lisuride and ergotamine are not psychoactive while they bind to and activate this receptor subtype. Yet these drugs have been shown to only induce $G_{q/11}$ protein-dependent signaling cascades.[43,215]

Demonstrating that this signaling bias is supported by the specific recognition of heteromer receptor complexes, it has recently been shown that DOI activates $G_{q/11}$- and $G_{i/o}$-type G proteins only when the serotonin 2A receptor is molecularly combined with the type 2 metabotropic glutamate receptor.[216] Thus, DOI failed to induce psychotic-related behaviors in type 2 metabotropic glutamate receptor knock-out mice,[217] confirming the involvement of heterocomplexes and, possibly, $G_{q/11}$-biased coupling for the induction of hallucinogenic properties. In addition, highlighting the possible involvement of heterocomplexes in the development of pathological processes, the expression of the heterocomplex formed between serotonin 2A receptor and type 2 metabotropic glutamate receptor was found altered in the brain of post-mortem schizophrenic patients.[216]

Further exemplifying the potential physiopathological significance of heterodimer-directed signaling selectivity, an increased expression of dopamine D1/D2 receptor heterocomplexes was observed in the brains of patients suffering from major depression syndrome.[218] Possibly supporting a switch of G protein coupling as the cause of neuropsychiatric alterations, it has been demonstrated that the dopamine D1/D2 receptor heterodimer triggered G_q protein-dependent signaling cascades while individual receptors activate G_s- and G_i-type G proteins, respectively.[130,132] It is of note that disruption of the dopamine D1/D2 receptor heterocomplex in the mouse prefrontal cortex supports anti-depressant-like effects,[218] suggesting the potential therapeutic benefit of drugs specifically targeting receptor heterodimers.

TRENDS TO UNDERSTAND BIAS AND IMPLICATIONS IN PHARMACEUTICAL DESIGN

As illustrated in Table 8.1, since the start of the twenty-first century biased signaling has emerged as a major issue for pharmacologists as well as a major challenge in various therapeutic areas. While there is growing evidence supporting the concept that by targeting a particular signaling pathway it is possible to achieve therapeutic goals while minimizing or avoiding side effects, this essentially remains a theoretical prediction. Indeed, we still lack definitive evidence of the unique and specific roles played by individual G proteins or other GPCR coupling partners, such as β-arrestins, in mediating defined cellular responses to endogenous transmitters. Predicting how biased signaling will affect drug responses *in vivo* is even more ambiguous, therefore limiting the speculations about which type of signaling bias should be privileged to obtain therapeutic benefits. Nevertheless, a better description of the roles of particular signaling cascades in physiological and pathological conditions has been made possible with the use of numerous transgenic animal models.[39]

Indeed, the generation of β-arrestin1 knock-out mice models in 1997 by the team of J. Seidman and C. Seidman[219] or β-arrestin2 knock-out mice models in 1999 by the team of R.J. Lefkowitz and M.G. Caron[96] has permitted detailed characterization of the behavioural, physiological, and pharmacological alterations associated with the suppression of these proteins,[42] shedding light on their roles in an unexpected number of biological processes. With respect to G proteins, such transgenic approaches, aimed at dissecting the complexity of receptor signaling, appear far more challenging. Indeed, the heterogeneity of α, β, and γ subunits necessitates to envisage hordes of transgenic knock-out animals. Initially, the signaling functions of G proteins have been mainly assigned to the α subunit and several lines of transgenic mice have been generated either deficient for (i.e., $G\alpha_q$, $G\alpha_{11}$, $G\alpha_o$, $G\alpha_z$, ...), or carrying inactivating mutations for (i.e., $G\alpha_{i1}$, $G\alpha_{i2}$, $G\alpha_{i3}$, ...) specific Gα protein subtypes, that allows one to specify their physiological functions.[220,221] Despite similarities between the different Gα subunits, supporting redundancy in mediating conventional intracellular cell signals, individual subtypes were shown to specifically drive alternate functions, unveiling putative processes that could be regulated by appropriate biased ligands.

Besides those studies aimed at interfering with G protein α subunits, several investigations have also shown that the loss of γ subunit also provokes particular phenotypes in transgenic mice. Revealing the requirement of specific γ subunits for the function of particular G protein subtypes in defined biological responses, this adds further complexity to G protein signaling.[222,223] Finally, other strategies have been considered in order to manipulate GPCR-dependent signaling, i.e., the expression of GPCR impaired in either G protein or β-arrestin coupling, the expression of C-terminal peptide fragments of the receptor competing with the activation of G proteins, or the expression of constitutively active mutants of G proteins, etc. All these genetically engineered mice represent valuable tools that have enabled considerable progress in the comprehension of GPCR-selective signaling. Nevertheless, these models also present major limitations due to early embryonic lethality,[187] signaling partner redundancy, lack of complete loss of function of mutant receptors/G protein subunits, compensatory up-regulation of other G protein subtypes/subunits,[39,224,225] etc.

Besides these genetic approaches, pharmacological data suggestive of the therapeutic relevance of biased ligands can be collected through the large-scale screening of ligand libraries or by re-evaluating existing therapies in the light of new pharmacological concepts. Indeed, biased ligands, which display extreme profiles—being strict antagonists for some signaling pathways while activating others, represent unique tools to dissect the specific role played by particular signaling cascades in cell/tissue responses. Nevertheless, the mixed antagonist/agonist profile of these drugs constitutes an additional difficulty for the interpretation of *in vivo* experiments. Thus, by competing with the endogenous transmitter for the receptor, these functionally biased ligands will also inhibit the endogenous effects. The clinical outcome will therefore combine the specific signaling induced by the drug with the blockade of other responses related to the endogenous transmitter.

Despite such limitations, the systematic review aimed at correlating the signaling profile of efficient drugs and their reported clinical efficacy and/or superiority has proven productive. This is exemplified for both anti-psychotics[68] and β-adrenergic receptor blockers,[226] for which the importance of β-arrestin recruitment has been demonstrated. In order to categorize new drugs as biased ligand candidates to be tested in models of diseases, quantification of the biased profile is required. So far, several methods have been described to quantify and compare ligands regarding their influence on defined biased responses.[227–229] To this end, ligands are characterized separately for each signaling pathway. Obviously comparison of the potencies is inadequate for ligands that produce responses with distinct relative efficacies (Emax values). Similarly, determination of individual Emax fails to demonstrate differences in the potency of full agonists. Therefore, selected methods should be based on a single parameter scale including both potency and efficacy. Comparison of these unrelated parameters would allow for the quantification of bias for diverse ligands that ultimately could help medicinal chemists to optimize bias activity.

The identification of molecular features that are positively or negatively correlated with the ability of ligands to trigger individual signaling is essential to optimize the design of new GPCR-biased ligands. Accordingly, first investigations of the structure–biased receptor activation relationship have recently emerged. Exploration of the functional particularities associated with the aripiprazole scaffold resulted in the discovery of the first dopamine, D2, receptor-biased ligand, displaying antagonist properties at G_i proteins and agonist properties at β-arrestin2.[230] Importantly, studies have suggested a contribution of β-arrestin2 regulation in both the anti-psychotic efficacy of aripiprazole and the nature of the side effects.[68,231] In a similar study combining functional investigations with computational docking analyses, the molecular features supporting biased signaling at histamine type 4 receptor were disclosed. Undeniably, these studies constitute the first steps towards a better understanding of the mechanism of biased GRCR activation and of the rational design of biased signaling-based therapeutics.

While the discovery of biased receptor coupling/regulation is revolutionizing GPCR pharmacology by offering possibilities for superior therapeutic profiles, both experimental drug screening and lead optimization processes are affected by this new concept. Indeed, this supplementary level of complexity represents a major pharmacological challenge for drug discovery and development processes. The approaches used for screening and characterization of novel therapeutic agents should be reconsidered to accommodate biased ligands. Indeed, if only the efficacy parameters in a single signaling pathway are collected for screening compound libraries, potentially valuable drugs could be neglected and/or discarded. Therefore, characterization of pluridimentional efficacies for every single given molecule should be considered before directing it toward the clinical process.

The development of GPCR and GPCR ligand screening assays remains a major focus in the field of drug discovery. Indeed, during the last few years, using both G protein-dependent and G protein-independent assays (Table 8.2), it has become possible to systematically screen compound libraries. These different assays have been extensively described.[30,229] Worth mentioning are these elegant, sophisticated approaches of whole cell-based assays that measure integrated cellular responses through the use of optical (resonant waveguide grating) or electrical (impedance-based) biosensors. Using these assays, different response profiles can be observed, revealing the summation of the G protein/signaling pathways activated.[241,243–246] Considering the influence of receptor heterodimerization on receptor signaling, the formation of GPCR heterocomplexes, and the influence of drugs on these complexes, has to be investigated as well.

As described above, while many receptors can function as monomers or homodimers, the heterodimerization (either constitutive or ligand-induced) may lead to distinct and unique signaling properties. Assays that specifically assess the ability of compounds to enhance or disrupt receptor heterodimerization have been developed. These include commonly used energy transfer (FRET-BRET)-based approaches,[247,248] as well as more recently developed assays enabling quantitative analysis of protein–protein interactions (time-resolved FRET and snap-tag technologies)[249,250] and measurement of transactivation effects through the enzyme fragment complementation technology based on the recombination of an active enzyme upon protein–protein interactions. (Importantly, all these assays can be used to detect interactions between two receptors, but also to quantify the influence of a compound on receptor dimerization.)

Finally, in rational drug discovery processes, optimizing the selectivity towards target proteins is often driven by information obtained from protein–ligand structure analyses. For a long time, however, major progress in the GPCR field has been compromised by the lack of receptor crystal structure information. Recently, however, major advances in GPCR

TABLE 8.2 Technologies Used for the Functional Screening of GPCRs

G protein-related assays	G protein activation	Overall guanine nucleotide exchange	$[^{35}S]$-GTPγS binding: quantification of radio-labelled non-hydrolysable GTP analog that binds to G proteins upon receptor activation
		G protein subtype specific guanine nucleotide binding	$[^{35}S]$-GTPγS scintillation proximity assay: quantification of receptor-induced radio-labeled non-hydrolysable GTP analog that binds to a particular G protein subtype through formation of a G protein/$[^{35}S]$-GTPγS/antibody complex and capture of this defined complex by scintillant-containing beads
	Regulation of G protein-associated signaling cascades	Second messenger measurement	Measurement of kinase (ERK, JNK, p38, ...) phorsphorylation
			Determination of cAMP, InsP3, Ca^{++}, ...
		Reporter gene assay	Evaluation of reporter gene expression as indirect readout of second messenger modulating responsive element present on the promoter site of the transfected vector[232]
G protein-unrelated assays	β-arrestin recruitment	β-arrestin redistribution	Recruitment of fluorophore-labeled β-arrestin to cell membranes*
		Proximity-based assay	FRET/BRET: Resonance energy transfer between fluorescent of bioluminescent donor probe linked to β-arrestin and acceptor probe linked to the receptor[233]
			Complementation assay: Enzyme recombination allowing the enzymatic generation of a chemiluminescent product upon GPCR and β-arrestin interaction[234,235]
			Protease activated reporter gene: Release of transcription factor regulating the reporter gene from the C-terminal part of the GPCR consecutively to recruitment of protease tagged β-arrestin to the receptor[236]

	Receptor trafficking (β-arrestin-mediated internalization)	Receptor redistribution	Visualization of fluorophore-labeled receptor on endocytic vesicles*
	G protein independent/β-arrestin-dependent signaling	Activation of signaling cascade	Measurement of kinase (ERK, JNK, p38, ...) phorsphorylation
			Measurement of characterized responses that are mediated by β-arrestins (stress fiber formation, chemotaxis)[237,238]
		β-arrestin conformation	Intramolecular BRET/FRET evaluation of β-arrestin conformational changes related to signalling[107,239]
Whole cell assays	Sum of cellular events including changes in cell density, spreading, morphology, or adhesion...	Impedance-based biosensor	Impedance of a cell layer in close contact with an electrode as a measure of cellular processes[240]
		Resonant waveguide grating or optical biosensor	Angle or wavelength shift of the monochromatic light illuminating cell growth on a grating structure[241]
Receptor dimerization assays	Interaction between receptors	Proximity-based assay	FRET/BRET: Resonance energy transfer between (fluorescence or bioluminescence) donor and acceptor probes linked to distinct receptors, when brought into close proximity
			Tag-lite™ method: FRET measurement between cell-impermeable substrates carrying FRET compatible fluorophore and labeling corresponding N-terminally SNAP or CLIP-tagged receptors
			Complementation assay: enzymatic generation of a chemiluminescent product upon GPCR and β-arrestin interaction by enzyme recombination consecutively to transactivation effect induced by a second untagged GPCR

**Fluorescence can be monitored quantitatively using high-content imaging systems that combine high resolution fluorescence microscopy with automated image analysis*[242] purification, stabilization, and crystallization have resulted in the elucidation of different structures,[251] as well as in the determination of the agonist- and antagonist-bound crystal structures of GPCRs in active and inactive receptor conformations.[252] Modeling the interactions of a given receptor with a variety of drugs that have been shown to display biased receptor activation, as already reported concerning the binding site of the β2-adrenergic receptor,[253] would allow conclusions to be drawn about the capacity of different classes of drugs to promote biased signaling. Furthermore, providing additional insights into the different signaling biased receptor conformations, recent studies have elucidated the crystal structure of the β2-adrenergic receptor combined with its naturally coupled G_s protein,[254] as well as the β1-adrenergic receptor[255] and some serotonin receptors[256] bound to β-arrestin-biased ligands. In the future, identification of structural and molecular features of ligand-induced selective biased receptor conformations could prove crucial for the pharmaceutical design of more potent and selective drugs.

EXISTING THERAPIES WITH BIASED SIGNALING CARRYING PUTATIVE CLINICAL RELEVANCE

For decades, the development of drugs showing selectivity of response was shown to essentially rely on their specific recognition of discrete receptor subtypes. Indeed, an essential issue of modern pharmacology was to design drugs that would selectively interact with closely related receptors, at variance with the endogenous transmitter that potentially activates these receptors with similar potency/efficacy. The discovery of biased signaling has led to new views on the concept of selectivity, which can actually concern a single receptor but distinct associated signaling pathways. While biased signaling has been frequently called upon to explain the clinical peculiarity of some drugs, experimental evidence should be provided in order to confirm the involvement of a single receptor in the complex responses triggered by these drugs. However, as detailed below, there are now several examples of drugs for which biased signaling putatively supports a unique clinical profile.

Carvedilol

For a long time, β-blockers have been used as the major first-line treatment for hypertension. Besides this predominant use, more recent clinical investigations have revealed these drugs as effective therapy in heart failure, as they have been shown to prolong patient survival.[257–259] As mentioned above, heart failure is characterized by high levels of circulating catecholamines that compensate for the cardiac dysfunction.[172] However, in the long term, this is associated with increased risk of myocardial apoptosis, fibrosis, and remodeling, leading to detrimental outcomes for cardiac function.[173–175] The pharmacological basis for the clinical benefits of β-blockers in heart failure is founded on the blockade of catecholamine-induced myocardial adaptations. In addition, these drugs may also normalize the β-adrenergic system through the up-regulation and re-sensitization of β1 adrenergic receptors.[260,261]

However, clinical trials conducted with 16 β-blockers showed that only three of these drugs (and primarily carvedilol) have demonstrated clinical value in this disease,[262–265] suggesting that in addition to β1-adrenergic receptor blockade, carvedilol possesses other properties that distinguish this molecule in the clinical setting. Indeed, carvedilol was the only drug member acting as biased ligand at β1-adrenergic receptor. Thus, in addition to an inverse agonism profile when focusing on G_s protein coupling, this drug also activates ERK1/2 signaling[189] and EGFR[74] through β-arrestin-dependent mechanisms. As both signaling pathways are involved in cardioprotective processes,[73,266] the β-arrestin2 bias may explain the unique beneficial therapeutic properties of carvedilol.

This initial interpretation may however—at least partially—be reconsidered, as a more recent report showed that other β-blockers devoid of ERK1/2 activation ability were endowed with similar clinical efficacies.[267] Adding to this controversy, the long-term administration of metoprolol, a selective β_1-adrenergic receptor blocker used in the treatment of various heart diseases, was shown to induce cardiac fibrosis through a GRK5/β-arrestin2-dependent pathway.[268] While highlighting the need for better characterization of the therapeutic potential of the β-arrestin2 bias in the treatment of cardiovascular diseases, this combination of fundamental and clinical studies at least confirms the physiological importance of biased agonism, and its implication in the diversity of β-blockers-mediated effects.

Aripiprazole

Besides their unique pharmacological complexity that relies on interaction with a large panel of unrelated neurotransmitter receptors, all anti-psychotic drugs share the common property of interacting with the dopamine D2 receptor. Since the discovery of chlorpromazine in the early 1950s, traditional anti-psychotics, being classified as typical or atypical, have all been antagonists at this receptor. Developed in the early 2000s, aripiprazole is the first anti-psychotic showing partial agonism at the dopamine D2 receptor; a peculiarity that is proposed to support a unique clinical profile. Indeed, its high affinity at this receptor, and its rather low intrinsic activity, explain how aripiprazole efficiently inhibits dopamine hyperactivity without fully blocking the receptor. This compound has been described as a dopamine stabilizer and used in clinical settings for the treatment of schizophrenia. Even though aripiprazole has been characterized as a low intrinsic activity partial agonist in experimental models,[269–271] other data suggest that instead aripiprazole acts as a biased ligand.[140,271–274] This compound has indeed been shown to differentially regulate several G protein-dependent signaling cascades, and this balanced regulation of cell signals is now proposed to confer aripiprazole its unique clinical profile.

Several studies have reported on the involvement of the β-arrestin2/Akt/GSK-3 signaling pathway in the development of schizophrenia,[275–277] suggesting that anti-psychotics could act by impeding this signaling cascade. Further supporting this hypothesis, several anti-psychotics, including aripiprazole, were found to prevent dopamine-mediated β-arrestin2 recruitment, independently of their influence on G protein activation.[68] This led to the mechanistic proposal that dopamine D2 receptor-mediated antagonism at β-arrestin2-dependent pathways supports the therapeutic efficacy of anti-psychotics rather than regulating G protein-dependent signaling.

Contradicting this theory, new dopamine receptor ligands with antagonist properties when tested for G protein activation, but inducing β-arrestin recruitment, have been developed and were found to show anti-psychotic-like activity in mice. Importantly, these ligands were without motor side effects,[231] suggesting a protective role of β-arrestin against motor dysfunctions. Further supporting this hypothesis, the genetic deletion of β-arrestin2 restored the propensity of these drugs to induce catalepsy while attenuating their anti-psychotic effects.[231]

Finally, an alternative hypothesis to explain the unique clinical profile of atypical anti-psychotics refers to the complex regulation of serotonin pathways. Indeed, along with other atypical anti-psychotics aripiprazole shares the property to interact with a diversity of other receptors including several serotonin receptor subtypes.[278] By conferring the anti-psychotics an intricate pharmacology, this undeniably makes the discrimination between the dopamine-biased-dependent properties and the off target-dependent (serotonin) properties extremely complex.

Loperamide

Activation of mu opioid receptors is known to decrease gastrointestinal motility; therefore constipation constitutes a major side effect of morphine. Taking advantage of this peripheral influence of mu opioid receptor agonists, loperamide, a peripherally restricted agonist, has been developed to reduce intestinal activity and is commonly used to treat diarrhoea. Like morphine, systemic injection of loperamide delays intestinal transit time in wild-type mice but not in β-arrestin2 knock-out mice, suggesting the role of β-arrestin2-mediated signaling in the control of gastrointestinal motility by mu opioid receptors.[207] Whether β-arrestin2 would constitute a valuable therapeutic target for which activation is privileged or not will therefore directly depend on the therapeutic objective. Thus, while activation of peripheral β-arrestin2 is probably required for the anti-diarrhoeal effect of loperamide, this signaling cascade should probably be avoided in the optimized development of mu opioid receptor-dependent analgesics.

Biased Drugs Under Clinical Development

Based on the observation that β-arrestin2 is involved both in mu opioid receptor desensitization as well as in mu opioid receptor-associated side effects, TRV130 has been developed as an optimized treatment for postoperative pain and is currently undergoing clinical investigations. This biased ligand induces analgesia through the conventional agonism towards G_i protein coupling while it behaves as an antagonist when focusing on β-arrestin2 activation, thereby preventing the development of secondary effects related to classical opioids[279] such as respiratory depression, constipation, and analgesic tolerance.[97,207]

Arising from the numerous studies showing the implication of β-arrestin in the cardio-protective signals mediated by angiotensin type 1 receptor, compounds were intentionally designed as biased ligands targeting this receptor. Both TRV120027 and SH-AngII are biased ligands at the angiotensin type 1 receptor presently undergoing clinical trials to treat acute heart failure. Indeed, as classical angiotensin antagonists, these compounds provoke a reduction of blood pressure. However, offering improved therapeutic efficacy, their bias toward β-arrestin activation leads to improved cardiomyocyte contractility and performance, reduced cardiac fibrosis, and preservation of renal perfusion.[23,78,79]

CONCLUSION

The discovery of an extreme complexity in the cell signaling associated with GPCRs has led to a reconsideration of the diverse pharmacological models classically used to characterize drugs and to the emergence of the concept of biased signaling. Indeed, more and more studies are currently demonstrating the possibility of independently manipulating receptor signaling using appropriate multifaceted ligands. In addition, the number of studies presuming the physiopathological implication of biased receptor regulations is growing, therefore encouraging the re-characterization of several marketed drugs. This has already led to the validation of biased signaling for a number of currently marketed therapies that were not initially developed with such an aim.

Considering some potential therapeutic improvements, it is tempting to speculate about the implication of these biases in the therapeutic efficacies of these particular drugs. However, the complexity of the molecular mechanisms and the poor specificity of these drugs for their respective receptors cloud our understanding of the real benefits of a GPCR bias in therapy. More extended characterization would be required in order to better define these precise mechanisms and their particular involvement in diverse physiological processes. However, given the multiplicity of targets and/or modulators potentially involved, this remains a considerable pharmacological challenge for the future, which becomes even more challenging as GPCR biases are cell type-dependent and influenced by modifications in tissular environments, as happens in a pathological context. Nevertheless, some drugs were recently specifically designed so as to selectively modulate GPCR functions. Complete pharmacological and clinical characterizations of these drugs will probably help to position GPCR bias in future clinical developments.

References

1. Colquhoun D. The quantitative analysis of drug–receptor interactions: a short history. *Trends Pharmacol Sci* 2006;**27**(3):149–57.
2. Park PS, Lodowski DT, Palczewski K. Activation of G protein-coupled receptors: beyond two-state models and tertiary conformational changes. *Annu Rev Pharmacol Toxicol* 2008;**48**:107–41.
3. Leff P. The two-state model of receptor activation. *Trends Pharmacol Sci* 1995;**16**(3):89–97.
4. Black JW, Leff P. Operational models of pharmacological agonism. *Proc R Soc Lond B Biol Sci* 1983;**220**(1219):141–62.
5. Burgen AS. The drug–receptor complex. *J Pharm Pharmacol* 1966;**18**(3):137–49.
6. Kenakin T. Agonist–receptor efficacy. II. Agonist trafficking of receptor signals. *Trends Pharmacol Sci* 1995;**16**(7):232–8.
7. Leff P, Scaramellini C, Law C, McKechnie K. A three-state receptor model of agonist action. *Trends Pharmacol Sci* 1997;**18**(10):355–62.
8. Perez DM, Karnik SS. Multiple signaling states of G protein-coupled receptors. *Pharmacol Rev* 2005;**57**(2):147–61.
9. Ettinger B, Black DM, Mitlak BH, Knickerbocker RK, Nickelsen T, Genant HK, et al. Reduction of vertebral fracture risk in postmenopausal women with osteoporosis treated with raloxifene: results from a 3-year

randomized clinical trial. Multiple Outcomes of Raloxifene Evaluation (MORE) Investigators. *JAMA* 1999;**282**(7):637–45.

10. Delmas PD, Bjarnason NH, Mitlak BH, Ravoux AC, Shah AS, Huster WJ, et al. Effects of raloxifene on bone mineral density, serum cholesterol concentrations, and uterine endometrium in postmenopausal women. *N Engl J Med* 1997;**337**(23):1641–7.
11. Kraichely DM, Sun J, Katzenellenbogen JA, Katzenellenbogen BS. Conformational changes and co-activator recruitment by novel ligands for estrogen receptor-alpha and estrogen receptor-beta: correlations with biological character and distinct differences among SRC co-activator family members. *Endocrinology* 2000;**141**(10):3534–45.
12. Paige LA, Christensen DJ, Gron H, Norris JD, Gottlin EB, Padilla KM, et al. Estrogen receptor (ER) modulators each induce distinct conformational changes in ER alpha and ER beta. *Proc Natl Acad Sci USA* 1999;**96**(7):3999–4004.
13. Brzozowski AM, Pike AC, Dauter Z, Hubbard RE, Bonn T, Engstrom O, et al. Molecular basis of agonism and antagonism in the oestrogen receptor. *Nature* 1997;**389**(6652):753–8.
14. Paech K, Webb P, Kuiper GG, Nilsson S, Gustafsson J, Kushner PJ, et al. Differential ligand activation of estrogen receptors ERalpha and ERbeta at AP1 sites. *Science* 1997;**277**(5331):1508–10.
15. Weihua Z, Saji S, Makinen S, Cheng G, Jensen EV, Warner M, et al. Estrogen receptor (ER) beta, a modulator of ERalpha in the uterus. *Proc Natl Acad Sci USA* 2000;**97**(11):5936–41.
16. Hall JM, McDonnell DP. The estrogen receptor beta-isoform (ERbeta) of the human estrogen receptor modulates ERalpha transcriptional activity and is a key regulator of the cellular response to estrogens and antiestrogens. *Endocrinology* 1999;**140**(12):5566–78.
17. Smani T, Dionisio N, López JJ, Berna-Erro A, Rosado JA. Cytoskeletal and scaffolding proteins as structural and functional determinants of TRP channels. *Biochim Biophys Acta* 2014;**1838**(2):658–64.
18. Cioffi DL, Wu S, Chen H, Alexeyev M, St Croix CM, Pitt BR, et al. Orai1 determines calcium selectivity of an endogenous TRPC heterotetramer channel. *Circ Res* 2012;**110**(11):1435–44.
19. Evans PD, Robb S, Cheek TR, Reale V, Hannan FL, Swales LS, et al. Agonist-specific coupling of G protein-coupled receptors to second-messenger systems. *Prog Brain Res* 1995;**106**:259–68.
20. Urban JD, Clarke WP, von ZM, Nichols DE, Kobilka B, Weinstein H, et al. Functional selectivity and classical concepts of quantitative pharmacology. *J Pharmacol Exp Ther* 2007;**320**(1):1–13.
21. Bosier B, Hermans E. Versatility of GPCR recognition by drugs: from biological implications to therapeutic relevance. *Trends Pharmacol Sci* 2007;**28**(8):438–46.
22. Arey BJ, López FJ. Are circulating gonadotropin isoforms naturally occurring biased agonists? Basic and therapeutic implications. *Rev Endocr Metab Disord* 2011;**12**(4):275–88.
23. Violin JD, DeWire SM, Yamashita D, Rominger DH, Nguyen L, Schiller K, et al. Selectively engaging beta-arrestins at the angiotensin II type 1 receptor reduces blood pressure and increases cardiac performance. *J Pharmacol Exp Ther* 2010;**335**(3):572–9.
24. Furchgott RF. The pharmacology of vascular smooth muscle. *Pharmacol Rev* 1955;**7**(2):183–265.
25. Kenakin T. Functional selectivity and biased receptor signaling. *J Pharmacol Exp Ther* 2011;**336**(2):296–302.
26. Kenakin T. New concepts in drug discovery: collateral efficacy and permissive antagonism. *Nat Rev Drug Discov* 2005;**4**(11):919–27.
27. Kenakin T. New concepts in pharmacological efficacy at 7TM receptors: IUPHAR review 2. *Br J Pharmacol* 2013;**168**(3):554–75.
28. Onaran HO, Costa T. Agonist efficacy and allosteric models of receptor action. *Ann N Y Acad Sci* 1997;**812**:98–115.
29. Hermans E. Biochemical and pharmacological control of the multiplicity of coupling at G protein-coupled receptors. *Pharmacol Ther* 2003;**99**(1):25–44.
30. Zhang R, Xie X. Tools for GPCR drug discovery. *Acta Pharmacol Sin* 2012;**33**(3):372–84.
31. Ferguson SS. Evolving concepts in G protein-coupled receptor endocytosis: the role in receptor desensitization and signaling. *Pharmacol Rev* 2001;**53**(1):1–24.
32. Pierce KL, Lefkowitz RJ. Classical and new roles of beta-arrestins in the regulation of G protein-coupled receptors. *Nat Rev Neurosci* 2001;**2**(10):727–33.
33. Lefkowitz RJ, Pierce KL, Luttrell LM. Dancing with different partners: protein kinase a phosphorylation of seven membrane-spanning receptors regulates their G protein-coupling specificity. *Mol Pharmacol* 2002;**62**(5):971–4.

34. Goodman Jr. OB, Krupnick JG, Santini F, Gurevich VV, Penn RB, Gagnon AW, et al. Beta-arrestin acts as a clathrin adaptor in endocytosis of the beta2-adrenergic receptor. *Nature* 1996;**383**(6599):447–50.
35. Laporte SA, Oakley RH, Holt JA, Barak LS, Caron MG. The interaction of beta-arrestin with the AP-2 adaptor is required for the clustering of beta 2-adrenergic receptor into clathrin-coated pits. *J Biol Chem* 2000;**275**(30):23120–6.
36. Laporte SA, Miller WE, Kim KM, Caron MG. beta-Arrestin/AP-2 interaction in G protein-coupled receptor internalization: identification of a beta-arrestin binging site in beta 2-adaptin. *J Biol Chem* 2002;**277**(11):9247–54.
37. Kim YM, Benovic JL. Differential roles of arrestin-2 interaction with clathrin and adaptor protein 2 in G protein-coupled receptor trafficking. *J Biol Chem* 2002;**277**(34):30760–8.
38. Edeling MA, Smith C, Owen D. Life of a clathrin coat: insights from clathrin and AP structures. *Nat Rev Mol Cell Biol* 2006;**7**(1):32–44.
39. Luttrell LM, Gesty-Palmer D. Beyond desensitization: physiological relevance of arrestin-dependent signaling. *Pharmacol Rev* 2010;**62**(2):305–30.
40. Ibrahim IA, Kurose H. beta-arrestin-mediated signaling improves the efficacy of therapeutics. *J Pharmacol Sci* 2012;**118**(4):408–12.
41. Whalen EJ, Rajagopal S, Lefkowitz RJ. Therapeutic potential of beta-arrestin- and G protein-biased agonists. *Trends Mol Med* 2011;**17**(3):126–39.
42. Schmid CL, Bohn LM. Physiological and pharmacological implications of beta-arrestin regulation. *Pharmacol Ther* 2009;**121**(3):285–93.
43. González-Maeso J, Weisstaub NV, Zhou M, Chan P, Ivic L, Ang R, et al. Hallucinogens recruit specific cortical 5-HT(2A) receptor-mediated signaling pathways to affect behavior. *Neuron* 2007;**53**(3):439–52.
44. Xiao RP, Zhu W, Zheng M, Chakir K, Bond R, Lakatta EG, et al. Subtype-specific beta-adrenoceptor signaling pathways in the heart and their potential clinical implications. *Trends Pharmacol Sci* 2004;**25**(7):358–65.
45. Zhu WZ, Zheng M, Koch WJ, Lefkowitz RJ, Kobilka BK, Xiao RP. Dual modulation of cell survival and cell death by beta(2)-adrenergic signaling in adult mouse cardiac myocytes. *Proc Natl Acad Sci USA* 2001;**98**(4):1607–12.
46. Xiao RP, Zhang SJ, Chakir K, Avdonin P, Zhu W, Bond RA, et al. Enhanced G(i) signaling selectively negates beta2-adrenergic receptor (AR)—but not beta1-AR-mediated positive inotropic effect in myocytes from failing rat hearts. *Circulation* 2003;**108**(13):1633–9.
47. Baker JG, Hall IP, Hill SJ. Agonist and inverse agonist actions of beta-blockers at the human beta 2-adrenoceptor provide evidence for agonist-directed signaling. *Mol Pharmacol* 2003;**64**(6):1357–69.
48. Ponicke K, Groner F, Heinroth-Hoffmann I, Brodde OE. Agonist-specific activation of the beta2-adrenoceptor/Gs-protein and beta2-adrenoceptor/Gi-protein pathway in adult rat ventricular cardiomyocytes. *Br J Pharmacol* 2006;**147**(7):714–9.
49. Grasberger H, Van SJ, Hag-Dahood MA, Tenenbaum-Rakover Y, Refetoff S. A familial thyrotropin (TSH) receptor mutation provides *in vivo* evidence that the inositol phosphates/Ca2 + cascade mediates TSH action on thyroid hormone synthesis. *J Clin Endocrinol Metab* 2007;**92**(7):2816–20.
50. Dumont JE, Maenhaut C, Lamy F. Control of thyroid cell proliferation and goitrogenesis. *Trends Endocrinol Metab* 1992;**3**(1):12–7.
51. Vassart G, Costagliola S. G protein-coupled receptors: mutations and endocrine diseases. *Nat Rev Endocrinol* 2011;**7**(6):362–72.
52. Hill SJ, Ganellin CR, Timmerman H, Schwartz JC, Shankley NP, Young JM, et al. International Union of Pharmacology. XIII. Classification of histamine receptors. *Pharmacol Rev* 1997;**49**(3):253–78.
53. Galeotti N, Malmberg-Aiello P, Bartolini A, Schunack W, Ghelardini C. H1-receptor stimulation induces hyperalgesia through activation of the phospholipase C-PKC pathway. *Neuropharmacology* 2004;**47**(2):295–303.
54. Moniri NH, Covington-Strachan D, Booth RG. Ligand-directed functional heterogeneity of histamine H1 receptors: novel dual-function ligands selectively activate and block H1-mediated phospholipase C and adenylyl cyclase signaling. *J Pharmacol Exp Ther* 2004;**311**(1):274–81.
55. Peverelli E, Busnelli M, Vitali E, Giardino E, Gales C, Lania AG, et al. Specific roles of G(i) protein family members revealed by dissecting SST5 coupling in human pituitary cells. *J Cell Sci* 2013;**126**(Pt 2):638–44.
56. MacKinnon AC, Tufail-Hanif U, Lucas CD, Jodrell D, Haslett C, Sethi T. Expression of V1A and GRP receptors leads to cellular transformation and increased sensitivity to substance-P analogue-induced growth inhibition. *Br J Cancer* 2005;**92**(3):522–31.

57. Benyo Z, Gille A, Bennett CL, Clausen BE, Offermanns S. Nicotinic acid-induced flushing is mediated by activation of epidermal langerhans cells. *Mol Pharmacol* 2006;**70**(6):1844–9.
58. Richman JG, Kanemitsu-Parks M, Gaidarov I, Cameron JS, Griffin P, Zheng H, et al. Nicotinic acid receptor agonists differentially activate downstream effectors. *J Biol Chem* 2007;**282**(25):18028–36.
59. Shen HC, Ding FX, Deng Q, Wilsie LC, Krsmanovic ML, Taggart AK, et al. Discovery of novel tricyclic full agonists for the G protein-coupled niacin receptor 109A with minimized flushing in rats. *J Med Chem* 2009;**52** (8):2587–602.
60. Walters RW, Shukla AK, Kovacs JJ, Violin JD, DeWire SM, Lam CM , et al. beta-Arrestin1 mediates nicotinic acid-induced flushing, but not its antilipolytic effect, in mice. *J Clin Invest* 2009;**119**(5):1312–21.
61. Raehal KM, Bohn LM. Mu opioid receptor regulation and opiate responsiveness. *AAPS J* 2005;**7**(3):E587–91.
62. Bruchas MR, Land BB, Aita M, Xu M, Barot SK, Li S, et al. Stress-induced p38 mitogen-activated protein kinase activation mediates kappa-opioid-dependent dysphoria. *J Neurosci* 2007;**27**(43):11614–23.
63. Bruchas MR, Chavkin C. Kinase cascades and ligand-directed signaling at the kappa opioid receptor. *Psychopharmacology (Berl)* 2010;**210**(2):137–47.
64. Schattauer SS, Miyatake M, Shankar H, Zietz C, Levin JR, Liu-Chen LY, et al. Ligand-directed signaling differences between rodent and human kappa-opioid receptors. *J Biol Chem* 2012;**287**(50):41595–607.
65. Broom DC, Jutkiewicz EM, Folk JE, Traynor JR, Rice KC, Woods JH. Convulsant activity of a non-peptidic delta-opioid receptor agonist is not required for its antidepressant-like effects in Sprague-Dawley rats. *Psychopharmacology (Berl)* 2002;**164**(1):42–8.
66. Pradhan AA, Walwyn W, Nozaki C, Filliol D, Erbs E, Matifas A, et al. Ligand-directed trafficking of the delta-opioid receptor *in vivo*: two paths toward analgesic tolerance. *J Neurosci* 2010;**30**(49):16459–68.
67. Poulin B, Butcher A, McWilliams P, Bourgognon JM, Pawlak R, Kong KC, et al. The M3-muscarinic receptor regulates learning and memory in a receptor phosphorylation/arrestin-dependent manner. *Proc Natl Acad Sci USA* 2010;**107**(20):9440–5.
68. Masri B, Salahpour A, Didriksen M, Ghisi V, Beaulieu JM, Gainetdinov RR, et al. Antagonism of dopamine D2 receptor/beta-arrestin 2 interaction is a common property of clinically effective antipsychotics. *Proc Natl Acad Sci USA* 2008;**105**(36):13656–61.
69. Xiao RP. Beta-adrenergic signaling in the heart: dual coupling of the beta2-adrenergic receptor to G(s) and G (i) proteins. *Sci STKE* 2001;**2001**(104):re15.
70. Tilley DG, Rockman HA. Role of beta-adrenergic receptor signaling and desensitization in heart failure: new concepts and prospects for treatment. *Expert Rev Cardiovasc Ther* 2006;**4**(3):417–32.
71. Zhu WZ, Wang SQ, Chakir K, Yang D, Zhang T, Brown JH, et al. Linkage of beta1-adrenergic stimulation to apoptotic heart cell death through protein kinase A-independent activation of Ca2 + /calmodulin kinase II. *J Clin Invest* 2003;**111**(5):617–25.
72. Sucharov CC, Mariner PD, Nunley KR, Long C, Leinwand L, Bristow MR. A beta1-adrenergic receptor CaM kinase II-dependent pathway mediates cardiac myocyte fetal gene induction. *Am J Physiol Heart Circ Physiol* 2006;**291**(3):H1299–308.
73. Noma T, Lemaire A, Naga Prasad SV, Barki-Harrington L, Tilley DG, Chen J, et al. Beta-arrestin-mediated beta1-adrenergic receptor transactivation of the EGFR confers cardioprotection. *J Clin Invest* 2007;**117**(9):2445–58.
74. Kim IM, Tilley DG, Chen J, Salazar NC, Whalen EJ, Violin JD, et al. Beta-blockers alprenolol and carvedilol stimulate beta-arrestin-mediated EGFR transactivation. *Proc Natl Acad Sci USA* 2008;**105**(38):14555–60.
75. Mangmool S, Shukla AK, Rockman HA. beta-Arrestin-dependent activation of Ca(2 +)/calmodulin kinase II after beta(1)-adrenergic receptor stimulation. *J Cell Biol* 2010;**189**(3):573–87.
76. Rakesh K, Yoo B, Kim IM, Salazar N, Kim KS, Rockman HA. beta-Arrestin-biased agonism of the angiotensin receptor induced by mechanical stress. *Sci Signal* 2010;**3**(125):ra46.
77. Rajagopal R, Chao MV. A role for Fyn in TRK receptor transactivation by G protein-coupled receptor signaling. *Mol Cell Neurosci* 2006;**33**(1):36–46.
78. Wei H, Ahn S, Shenoy SK, Karnik SS, Hunyady L, Luttrell LM, et al. Independent beta-arrestin 2 and G protein-mediated pathways for angiotensin II activation of extracellular signal-regulated kinases 1 and 2. *Proc Natl Acad Sci USA* 2003;**100**(19):10782–7.
79. Boerrigter G, Lark MW, Whalen EJ, Soergel DG, Violin JD, Burnett Jr. JC. Cardiorenal actions of TRV120027, a novel ss-arrestin-biased ligand at the angiotensin II type I receptor, in healthy and heart failure canines: a novel therapeutic strategy for acute heart failure. *Circ Heart Fail* 2011;**4**(6):770–8.

80. Zhai P, Yamamoto M, Galeotti J, Liu J, Masurekar M, Thaisz J, et al. Cardiac-specific overexpression of AT1 receptor mutant lacking G alpha q/G alpha i coupling causes hypertrophy and bradycardia in transgenic mice. *J Clin Invest* 2005;**115**(11):3045–56.
81. Violin JD, Lefkowitz RJ. Beta-arrestin-biased ligands at seven-transmembrane receptors. *Trends Pharmacol Sci* 2007;**28**(8):416–22.
82. Schmid CL, Raehal KM, Bohn LM. Agonist-directed signaling of the serotonin 2A receptor depends on beta-arrestin-2 interactions *in vivo*. *Proc Natl Acad Sci USA* 2008;**105**(3):1079–84.
83. Swarthout JT, D'Alonzo RC, Selvamurugan N, Partridge NC. Parathyroid hormone-dependent signaling pathways regulating genes in bone cells. *Gene* 2002;**282**(1–2):1–17.
84. Ferrari SL, Pierroz DD, Glatt V, Goddard DS, Bianchi EN, Lin FT, et al. Bone response to intermittent parathyroid hormone is altered in mice null for β-Arrestin2. *Endocrinology* 2005;**146**(4):1854–62.
85. Gesty-Palmer D, Flannery P, Yuan L, Corsino L, Spurney R, Lefkowitz RJ, et al. A beta-arrestin-biased agonist of the parathyroid hormone receptor (PTH1R) promotes bone formation independent of G protein activation. *Sci Transl Med* 2009;**1**(1):1ra1.
86. Gesty-Palmer D, Yuan L, Martin B, Wood III WH, Lee MH, Janech MG, et al. Beta-arrestin-selective G protein-coupled receptor agonists engender unique biological efficacy *in vivo*. *Mol Endocrinol* 2013;**27**(2):296–314.
87. Appleton KM, Lee MH, Alele C, Alele C, Luttrell DK, Peterson YK, et al. Biasing the parathyroid hormone receptor: relating *in vitro* ligand efficacy to *in vivo* biological activity. *Methods Enzymol* 2013;**522**:229–62.
88. Sonoda N, Imamura T, Yoshizaki T, Babendure JL, Lu JC, Olefsky JM. Beta-Arrestin-1 mediates glucagon-like peptide-1 signaling to insulin secretion in cultured pancreatic beta cells. *Proc Natl Acad Sci USA* 2008;**105**(18):6614–9.
89. Wang Q, Zhao J, Brady AE, Feng J, Allen PB, Lefkowitz RJ, et al. Spinophilin blocks arrestin actions *in vitro* and *in vivo* at G protein-coupled receptors. *Science* 2004;**304**(5679):1940–4.
90. Cottingham C, Wang Q. alpha2 adrenergic receptor dysregulation in depressive disorders: implications for the neurobiology of depression and antidepressant therapy. *Neurosci Biobehav Rev* 2012;**36**(10):2214–25.
91. Gilon P, Henquin JC. Mechanisms and physiological significance of the cholinergic control of pancreatic beta-cell function. *Endocr Rev* 2001;**22**(5):565–604.
92. Gromada J, Hughes TE. Ringing the dinner bell for insulin: muscarinic M3 receptor activity in the control of pancreatic beta cell function. *Cell Metab* 2006;**3**(6):390–2.
93. Swayne LA, Mezghrani A, Varrault A, Chemin J, Bertrand G, Dalle S, et al. The NALCN ion channel is activated by M3 muscarinic receptors in a pancreatic beta-cell line. *EMBO Rep* 2009;**10**(8):873–80.
94. Kong KC, Butcher AJ, McWilliams P, Jones D, Wess J, Hamdan FF, et al. M3-muscarinic receptor promotes insulin release via receptor phosphorylation/arrestin-dependent activation of protein kinase D1. *Proc Natl Acad Sci USA* 2010;**107**(49):21181–6.
95. Bruchas MR, Yang T, Schreiber S, Defino M, Kwan SC, Li S, et al. Long-acting kappa opioid antagonists disrupt receptor signaling and produce noncompetitive effects by activating c-Jun N-terminal kinase. *J Biol Chem* 2007;**282**(41):29803–11.
96. Bohn LM, Lefkowitz RJ, Gainetdinov RR, Peppel K, Caron MG, Lin FT. Enhanced morphine analgesia in mice lacking beta-arrestin 2. *Science* 1999;**286**(5449):2495–8.
97. Bohn LM, Gainetdinov RR, Lin FT, Lefkowitz RJ, Caron MG. Mu-opioid receptor desensitization by beta-arrestin-2 determines morphine tolerance but not dependence. *Nature* 2000;**408**(6813):720–3.
98. Ryman-Rasmussen JP, Griffith A, Oloff S, Vaidehi N, Brown JT, Goddard III WA, et al. Functional selectivity of dopamine D1 receptor agonists in regulating the fate of internalized receptors. *Neuropharmacology* 2007;**52**(2):562–75.
99. Baba M, Nishimura O, Kanzaki N, Okamoto M, Sawada H, Iizawa Y, et al. A small-molecule, nonpeptide CCR5 antagonist with highly potent and selective anti-HIV-1 activity. *Proc Natl Acad Sci USA* 1999;**96**(10):5698–703.
100. Mack M, Luckow B, Nelson PJ, Cihak J, Simmons G, Clapham PR, et al. Aminooxypentane-RANTES induces CCR5 internalization but inhibits recycling: a novel inhibitory mechanism of HIV infectivity. *J Exp Med* 1998;**187**(8):1215–24.
101. Hartley O, Offord RE. Engineering chemokines to develop optimized HIV inhibitors. *Curr Protein Pept Sci* 2005;**6**(3):207–19.

102. Deshpande DA, Theriot BS, Penn RB, Walker JK. Beta-arrestins specifically constrain beta2-adrenergic receptor signaling and function in airway smooth muscle. *FASEB J* 2008;**22**(7):2134–41.
103. Wang WC, Mihlbachler KA, Brunnett AC, Liggett SB. Targeted transgenesis reveals discrete attenuator functions of GRK and PKA in airway beta2-adrenergic receptor physiologic signaling. *Proc Natl Acad Sci USA* 2009;**106**(35):15007–12.
104. Breivogel CS, Lambert JM, Gerfin S, Huffman JW, Razdan RK. Sensitivity to delta9-tetrahydrocannabinol is selectively enhanced in beta-arrestin2 -/- mice. *Behav Pharmacol* 2008;**19**(4):298–307.
105. Shenoy SK, Lefkowitz RJ. Receptor-specific ubiquitination of beta-arrestin directs assembly and targeting of seven-transmembrane receptor signalosomes. *J Biol Chem* 2005;**280**(15):15315–24.
106. Shenoy SK, Barak LS, Xiao K, Ahn S, Berthouze M, Shukla AK, et al. Ubiquitination of beta-arrestin links seven-transmembrane receptor endocytosis and ERK activation. *J Biol Chem* 2007;**282**(40):29549–62.
107. Shukla AK, Violin JD, Whalen EJ, Gesty-Palmer D, Shenoy SK, Lefkowitz RJ. Distinct conformational changes in beta-arrestin report biased agonism at seven-transmembrane receptors. *Proc Natl Acad Sci USA* 2008;**105**(29):9988–93.
108. Ren XR, Reiter E, Ahn S, Kim J, Chen W, Lefkowitz RJ. Different G protein-coupled receptor kinases govern G protein and beta-arrestin-mediated signaling of V2 vasopressin receptor. *Proc Natl Acad Sci USA* 2005;**102**(5):1448–53.
109. Zhang J, Ferguson SS, Barak LS, Bodduluri SR, Laporte SA, Law PY, et al. Role for G protein-coupled receptor kinase in agonist-specific regulation of mu-opioid receptor responsiveness. *Proc Natl Acad Sci USA* 1998;**95**(12):7157–62.
110. Groer CE, Schmid CL, Jaeger AM, Bohn LM. Agonist-directed interactions with specific beta-arrestins determine mu-opioid receptor trafficking, ubiquitination, and dephosphorylation. *J Biol Chem* 2011;**286**(36):31731–41.
111. Schulz R, Eisinger DA, Wehmeyer A. Opioid control of MAP kinase cascade. *Eur J Pharmacol* 2004;**500**(1–3):487–97.
112. Finn AK, Whistler JL. Endocytosis of the mu opioid receptor reduces tolerance and a cellular hallmark of opiate withdrawal. *Neuron* 2001;**32**(5):829–39.
113. He L, Whistler JL. An opiate cocktail that reduces morphine tolerance and dependence. *Curr Biol* 2005;**15**(11):1028–33.
114. Bockaert J, Marin P, Dumuis A, Fagni L. The "magic tail" of G protein-coupled receptors: an anchorage for functional protein networks. *FEBS Lett* 2003;**546**(1):65–72.
115. Dev KK. Making protein interactions druggable: targeting PDZ domains. *Nat Rev Drug Discov* 2004;**3**(12):1047–56.
116. Hall RA, Premont RT, Chow CW, Blitzer JT, Pitcher JA, Claing A, et al. The beta2-adrenergic receptor interacts with the Na + /H + -exchanger regulatory factor to control Na + /H + exchange. *Nature* 1998;**392**(6676):626–30.
117. Li JG, Chen C, Liu-Chen LY. Ezrin-radixin-moesin-binding phosphoprotein-50/Na + /H + exchanger regulatory factor (EBP50/NHERF) blocks U50,488H-induced down-regulation of the human kappa opioid receptor by enhancing its recycling rate. *J Biol Chem* 2002;**277**(30):27545–52.
118. Huang P, Steplock D, Weinman EJ, Hall RA, Ding Z, Li J, et al. kappa Opioid receptor interacts with Na(+)/H(+)-exchanger regulatory factor-1/Ezrin-radixin-moesin-binding phosphoprotein-50 (NHERF-1/EBP50) to stimulate Na(+)/H(+) exchange independent of G(i)/G(o) proteins. *J Biol Chem* 2004;**279**(24):25002–9.
119. Ritter SL, Hall RA. Fine-tuning of GPCR activity by receptor-interacting proteins. *Nat Rev Mol Cell Biol* 2009;**10**(12):819–30.
120. Daaka Y, Luttrell LM, Lefkowitz RJ. Switching of the coupling of the beta2-adrenergic receptor to different G proteins by protein kinase A. *Nature* 1997;**390**(6655):88–91.
121. Lawler OA, Miggin SM, Kinsella BT. Protein kinase A-mediated phosphorylation of serine 357 of the mouse prostacyclin receptor regulates its coupling to G(s)-, to G(i)-, and to G(q)-coupled effector signaling. *J Biol Chem* 2001;**276**(36):33596–607.
122. Zamah AM, Delahunty M, Luttrell LM, Lefkowitz RJ. Protein kinase A-mediated phosphorylation of the beta 2-adrenergic receptor regulates its coupling to Gs and Gi. Demonstration in a reconstituted system. *J Biol Chem* 2002;**277**(34):31249–56.

123. Chakrabarti S, Regec A, Gintzler AR. Biochemical demonstration of mu-opioid receptor association with Gsalpha: enhancement following morphine exposure. *Brain Res Mol Brain Res* 2005;**135**(1–2):217–24.
124. Malmberg A, Strange PG. Site-directed mutations in the third intracellular loop of the serotonin 5-HT(1A) receptor alter G protein coupling from G(i) to G(s) in a ligand-dependent manner. *J Neurochem* 2000;**75**(3):1283–93.
125. Chesley A, Lundberg MS, Asai T, Xiao RP, Ohtani S, Lakatta EG, et al. The beta(2)-adrenergic receptor delivers an antiapoptotic signal to cardiac myocytes through G(i)-dependent coupling to phosphatidylinositol 3′-kinase. *Circ Res* 2000;**87**(12):1172–9.
126. Hasseldine AR, Harper EA, Black JW. Cardiac-specific overexpression of human beta2 adrenoceptors in mice exposes coupling to both Gs and Gi proteins. *Br J Pharmacol* 2003;**138**(7):1358–66.
127. Bosier B, Muccioli GG, Hermans E, Lambert DM. Functionally selective cannabinoid receptor signalling: therapeutic implications and opportunities. *Biochem Pharmacol* 2010;**80**(1):1–12.
128. González-Maeso J, Sealfon SC. Functional selectivity in GPCR heterocomplexes. *Mini Rev Med Chem* 2012;**12**(9):851–5.
129. Glass M, Felder CC. Concurrent stimulation of cannabinoid CB1 and dopamine D2 receptors augments cAMP accumulation in striatal neurons: evidence for a Gs linkage to the CB1 receptor. *J Neurosci* 1997;**17**(14):5327–33.
130. Rashid AJ, So CH, Kong MM, Furtak T, El-Ghundi M, Cheng R, et al. D1-D2 dopamine receptor heterooligomers with unique pharmacology are coupled to rapid activation of Gq/11 in the striatum. *Proc Natl Acad Sci USA* 2007;**104**(2):654–9.
131. Hasbi A, Nguyen T, Fan T, Cheng R, Rashid A, Alijaniaram M, et al. Trafficking of preassembled opioid mu-delta heterooligomer-Gz signaling complexes to the plasma membrane: coregulation by agonists. *Biochemistry* 2007;**46**(45):12997–3009.
132. Verma V, Hasbi A, O'Dowd BF, George SR. Dopamine D1-D2 receptor Heteromer-mediated calcium release is desensitized by D1 receptor occupancy with or without signal activation: dual functional regulation by G protein-coupled receptor kinase 2. *J Biol Chem* 2010;**285**(45):35092–103.
133. Breit A, Gagnidze K, Devi LA, Lagace M, Bouvier M. Simultaneous activation of the delta opioid receptor (deltaOR)/sensory neuron-specific receptor-4 (SNSR-4) hetero-oligomer by the mixed bivalent agonist bovine adrenal medulla peptide 22 activates SNSR-4 but inhibits deltaOR signaling. *Mol Pharmacol* 2006;**70**(2):686–96.
134. Pfeiffer M, Kirscht S, Stumm R, Koch T, Wu D, Laugsch M, et al. Heterodimerization of substance P and mu-opioid receptors regulates receptor trafficking and resensitization. *J Biol Chem* 2003;**278**(51):51630–7.
135. Terrillon S, Barberis C, Bouvier M. Heterodimerization of V1a and V2 vasopressin receptors determines the interaction with beta-arrestin and their trafficking patterns. *Proc Natl Acad Sci USA* 2004;**101**(6):1548–53.
136. Grant M, Alturaihi H, Jaquet P, Collier B, Kumar U. Cell growth inhibition and functioning of human somatostatin receptor type 2 are modulated by receptor heterodimerization. *Mol Endocrinol* 2008;**22**(10):2278–92.
137. Gupta A, Mulder J, Gomes I, Rozenfeld R, Bushlin I, Ong E, et al. Increased abundance of opioid receptor heteromers after chronic morphine administration. *Sci Signal* 2010;**3**(131):ra54.
138. Rozenfeld R, Devi LA. Receptor heteromerization and drug discovery. *Trends Pharmacol Sci* 2010;**31**(3):124–30.
139. Bosier B, Hermans E, Lambert DM. Concomitant activation of adenylyl cyclase suppresses the opposite influences of CB(1) cannabinoid receptor agonists on tyrosine hydroxylase expression. *Biochem Pharmacol* 2009;**77**(2):216–27.
140. Koener B, Focant MC, Bosier B, Maloteaux JM, Hermans E. Increasing the density of the D2L receptor and manipulating the receptor environment are required to evidence the partial agonist properties of aripiprazole. *Prog Neuropsychopharmacol Biol Psychiatry* 2012;**36**(1):60–70.
141. Kenakin T. Functional selectivity through protean and biased agonism: who steers the ship? *Mol Pharmacol* 2007;**72**(6):1393–401.
142. Lohse MJ. Molecular mechanisms of membrane receptor desensitization. *Biochim Biophys Acta* 1993;**1179**(2):171–88.
143. Menard L, Ferguson SS, Zhang J, Lin FT, Lefkowitz RJ, Caron MG, et al. Synergistic regulation of beta2-adrenergic receptor sequestration: intracellular complement of beta-adrenergic receptor kinase and beta-arrestin determine kinetics of internalization. *Mol Pharmacol* 1997;**51**(5):800–8.

144. Tobin AB, Butcher AJ, Kong KC. Location, location, location...site-specific GPCR phosphorylation offers a mechanism for cell-type-specific signalling. *Trends Pharmacol Sci* 2008;**29**(8):413–20.
145. Philip F, Sengupta P, Scarlata S. Signaling through a G Protein-coupled receptor and its corresponding G protein follows a stoichiometrically limited model. *J Biol Chem* 2007;**282**(26):19203–16.
146. Reubi JC. Peptide receptors as molecular targets for cancer diagnosis and therapy. *Endocr Rev* 2003;**24**(4):389–427.
147. Cescato R, Loesch KA, Waser B, Macke HR, Rivier JE, Reubi JC, et al. Agonist-biased signaling at the sst2A receptor: the multi-somatostatin analogs KE108 and SOM230 activate and antagonize distinct signaling pathways. *Mol Endocrinol* 2010;**24**(1):240–9.
148. Mary S, Damian M, Louet M, Floquet N, Fehrentz JA, Marie J, et al. Ligands and signaling proteins govern the conformational landscape explored by a G protein-coupled receptor. *Proc Natl Acad Sci USA* 2012;**109**(21):8304–9.
149. van der Laan S, Meijer OC. Pharmacology of glucocorticoids: beyond receptors. *Eur J Pharmacol* 2008;**585**(2–3):483–91.
150. Gburcik V, Picard D. The cell-specific activity of the estrogen receptor alpha may be fine-tuned by phosphorylation-induced structural gymnastics. *Nucl Recept Signal* 2006;**4**(e005):1–4.
151. Gronemeyer H, Gustafsson JA, Laudet V. Principles for modulation of the nuclear receptor superfamily. *Nat Rev Drug Discov* 2004;**3**(11):950–64.
152. Arey BJ, Stevis PE, Deecher DC, Shen ES, Frail DE, Negro-Vilar A, et al. Induction of promiscuous G protein coupling of the follicle-stimulating hormone (FSH) receptor: a novel mechanism for transducing pleiotropic actions of FSH isoforms. *Mol Endocrinol* 1997;**11**(5):517–26.
153. Logsdon NJ, Jones BC, Allman JC, Izotova L, Schwartz B, Pestka S, et al. The IL-10R2 binding hot spot on IL-22 is located on the N-terminal helix and is dependent on N-linked glycosylation. *J Mol Biol* 2004;**342**(2):503–14.
154. Saremba S, Nickel J, Seher A, Kotzsch A, Sebald W, Mueller TD. Type I receptor binding of bone morphogenetic protein 6 is dependent on N-glycosylation of the ligand. *FEBS J* 2008;**275**(1):172–83.
155. Matzuk MM, Keene JL, Boime I. Site specificity of the chorionic gonadotropin N-linked oligosaccharides in signal transduction. *J Biol Chem* 1989;**264**(5):2409–14.
156. Nguyen VT, Singh V, Butnev VY, Gray CM, Westfall S, Davis JS, et al. Inositol phosphate stimulation by LH requires the entire alpha Asn56 oligosaccharide. *Mol Cell Endocrinol* 2003;**199**(1–2):73–86.
157. Sullivan SK, McGrath DA, Grigoriadis D, Bacon KB. Pharmacological and signaling analysis of human chemokine receptor CCR-7 stably expressed in HEK-293 cells: high-affinity binding of recombinant ligands MIP-3beta and SLC stimulates multiple signaling cascades. *Biochem Biophys Res Commun* 1999;**263**(3):685–90.
158. Kohout TA, Nicholas SL, Perry SJ, Reinhart G, Junger S, Struthers RS. Differential desensitization, receptor phosphorylation, beta-arrestin recruitment, and ERK1/2 activation by the two endogenous ligands for the CC chemokine receptor 7. *J Biol Chem* 2004;**279**(22):23214–22.
159. Palanche T, Ilien B, Zoffmann S, Reck MP, Bucher B, Edelstein SJ, et al. The neurokinin A receptor activates calcium and cAMP responses through distinct conformational states. *J Biol Chem* 2001;**276**(37):34853–61.
160. Shoemaker JL, Ruckle MB, Mayeux PR, Prather PL. Agonist-directed trafficking of response by endocannabinoids acting at CB2 receptors. *J Pharmacol Exp Ther* 2005;**315**(2):828–38.
161. Peverelli E, Lania AG, Mantovani G, Beck-Peccoz P, Spada A. Characterization of intracellular signaling mediated by human somatostatin receptor 5: role of the DRY motif and the third intracellular loop. *Endocrinology* 2009;**150**(7):3169–76.
162. Hu S, Wang D, Wu J, Jin J, Wei W, Sun W. Involvement of beta-arrestins in cancer progression. *Mol Biol Rep* 2013;**40**(2):1065–71.
163. Li TT, Alemayehu M, Aziziyeh AI, Pape C, Pampillo M, Postovit LM, et al. Beta-arrestin/Ral signaling regulates lysophosphatidic acid-mediated migration and invasion of human breast tumor cells. *Mol Cancer Res* 2009;**7**(7):1064–77.
164. Zou L, Yang R, Chai J, Pei G. Rapid xenograft tumor progression in beta-arrestin1 transgenic mice due to enhanced tumor angiogenesis. *FASEB J* 2008;**22**(2):355–64.
165. Fong AM, Premont RT, Richardson RM, Yu YR, Lefkowitz RJ, Patel DD. Defective lymphocyte chemotaxis in beta-arrestin2- and GRK6-deficient mice. *Proc Natl Acad Sci USA* 2002;**99**(11):7478–83.
166. Raghuwanshi SK, Nasser MW, Chen X, Strieter RM, Richardson RM. Depletion of beta-arrestin-2 promotes tumor growth and angiogenesis in a murine model of lung cancer. *J Immunol* 2008;**180**(8):5699–706.

167. Balabanian K, Lagane B, Pablos JL, Laurent L, Planchenault T, Verola O, et al. WHIM syndromes with different genetic anomalies are accounted for by impaired CXCR4 desensitization to CXCL12. *Blood* 2005;**105**(6):2449–57.
168. Yamashita F, Hashida M. Pharmacokinetic considerations for targeted drug delivery. *Adv Drug Deliv Rev* 2013;**65**(1):139–47.
169. Allen TM, Cullis PR. Liposomal drug delivery systems: from concept to clinical applications. *Adv Drug Deliv Rev* 2013;**65**(1):36–48.
170. Kumar A, Zhang X, Liang XJ. Gold nanoparticles: emerging paradigm for targeted drug delivery system. *Biotechnol Adv* 2013;**31**(5):593–606.
171. Gaillard PJ, Visser CC, Appeldoorn CC, Rip J. Targeted blood-to-brain drug delivery —10 key development criteria. *Curr Pharm Biotechnol* 2012;**13**(12):2328–39.
172. Levine TB, Francis GS, Goldsmith SR, Simon AB, Cohn JN. Activity of the sympathetic nervous system and renin-angiotensin system assessed by plasma hormone levels and their relation to hemodynamic abnormalities in congestive heart failure. *Am J Cardiol* 1982;**49**(7):1659–66.
173. Teerlink JR, Pfeffer JM, Pfeffer MA. Progressive ventricular remodeling in response to diffuse isoproterenol-induced myocardial necrosis in rats. *Circ Res* 1994;**75**(1):105–13.
174. Blaufarb IS, Sonnenblick EH. The renin-angiotensin system in left ventricular remodeling. *Am J Cardiol* 1996;**77**(13):8C–16C.
175. Woo AY, Xiao RP. beta-Adrenergic receptor subtype signaling in heart: from bench to bedside. *Acta Pharmacol Sin* 2012;**33**(3):335–41.
176. Rockman HA, Koch WJ, Lefkowitz RJ. Seven-transmembrane-spanning receptors and heart function. *Nature* 2002;**415**(6868):206–12.
177. Bisognano JD, Weinberger HD, Bohlmeyer TJ, Pende A, Raynolds MV, Sastravaha A, et al. Myocardial-directed overexpression of the human beta(1)-adrenergic receptor in transgenic mice. *J Mol Cell Cardiol* 2000;**32**(5):817–30.
178. Engelhardt S, Hein L, Wiesmann F, Lohse MJ. Progressive hypertrophy and heart failure in beta1-adrenergic receptor transgenic mice. *Proc Natl Acad Sci USA* 1999;**96**(12):7059–64.
179. Zhu W, Zeng X, Zheng M, Xiao RP. The enigma of beta2-adrenergic receptor Gi signaling in the heart: the good, the bad, and the ugly. *Circ Res* 2005;**97**(6):507–9.
180. Bristow MR, Ginsburg R, Umans V, Fowler M, Minobe W, Rasmussen R, et al. Beta 1- and beta 2-adrenergic-receptor subpopulations in nonfailing and failing human ventricular myocardium: coupling of both receptor subtypes to muscle contraction and selective beta 1-receptor down-regulation in heart failure. *Circ Res* 1986;**59**(3):297–309.
181. Bristow MR. Changes in myocardial and vascular receptors in heart failure. *J Am Coll Cardiol* 1993;**22**(4 Suppl. A):61A–71A.
182. Ahmet I, Krawczyk M, Heller P, Moon C, Lakatta EG, Talan MI. Beneficial effects of chronic pharmacological manipulation of beta-adrenoreceptor subtype signaling in rodent dilated ischemic cardiomyopathy. *Circulation* 2004;**110**(9):1083–90.
183. Ahmet I, Lakatta EG, Talan MI. Pharmacological stimulation of beta2-adrenergic receptors (beta2AR) enhances therapeutic effectiveness of beta1AR blockade in rodent dilated ischemic cardiomyopathy. *Heart Fail Rev* 2005;**10**(4):289–96.
184. Ahmet I, Krawczyk M, Zhu W, Woo AY, Morrell C, Poosala S, et al. Cardioprotective and survival benefits of long-term combined therapy with beta2 adrenoreceptor (AR) agonist and beta1 AR blocker in dilated cardiomyopathy postmyocardial infarction. *J Pharmacol Exp Ther* 2008;**325**(2):491–9.
185. Maudsley S, Pierce KL, Zamah AM, Miller WE, Ahn S, Daaka Y, et al. The beta(2)-adrenergic receptor mediates extracellular signal-regulated kinase activation via assembly of a multi-receptor complex with the epidermal growth factor receptor. *J Biol Chem* 2000;**275**(13):9572–80.
186. Kim J, Ahn S, Rajagopal K, Lefkowitz RJ. Independent beta-arrestin2 and Gq/protein kinase Czeta pathways for ERK stimulated by angiotensin type 1A receptors in vascular smooth muscle cells converge on transactivation of the epidermal growth factor receptor. *J Biol Chem* 2009;**284**(18):11953–62.
187. Jarpe MB, Widmann C, Knall C, Schlesinger TK, Gibson S, Yujiri T, et al. Anti-apoptotic versus pro-apoptotic signal transduction: checkpoints and stop signs along the road to death. *Oncogene* 1998;**17**(11 Reviews):1475–82.

188. McKay MM, Morrison DK. Integrating signals from RTKs to ERK/MAPK. *Oncogene* 2007;**26**(22):3113–21.
189. Wisler JW, DeWire SM, Whalen EJ, Violin JD, Drake MT, Ahn S, et al. A unique mechanism of beta-blocker action: carvedilol stimulates beta-arrestin signaling. *Proc Natl Acad Sci USA* 2007;**104**(42):16657–62.
190. Lo HW, Hsu SC, Ali-Seyed M, Gunduz M, Xia W, Wei Y, et al. Nuclear interaction of EGFR and STAT3 in the activation of the iNOS/NO pathway. *Cancer Cell* 2005;**7**(6):575–89.
191. Afonso RA, Patarrao RS, Macedo MP, Carmo MM. Carvedilol's actions are largely mediated by endogenous nitric oxide. *Rev Port Cardiol* 2006;**25**(10):911–7.
192. Abrams J. Beneficial actions of nitrates in cardiovascular disease. *Am J Cardiol* 1996;**77**(13):31C–7C.
193. Xiao K, McClatchy DB, Shukla AK, Zhao Y, Chen M, Shenoy SK, et al. Functional specialization of beta-arrestin interactions revealed by proteomic analysis. *Proc Natl Acad Sci USA* 2007;**104**(29):12011–6.
194. Christensen GL, Aplin M, Hansen JL. Therapeutic potential of functional selectivity in the treatment of heart failure. *Trends Cardiovasc Med* 2010;**20**(7):221–7.
195. Bohn LM, Dykstra LA, Lefkowitz RJ, Caron MG, Barak LS. Relative opioid efficacy is determined by the complements of the G protein-coupled receptor desensitization machinery. *Mol Pharmacol* 2004;**66**(1):106–12.
196. Raehal KM, Schmid CL, Groer CE, Bohn LM. Functional selectivity at the mu-opioid receptor: implications for understanding opioid analgesia and tolerance. *Pharmacol Rev* 2011;**63**(4):1001–19.
197. Kieffer BL. Opioids: first lessons from knockout mice. *Trends Pharmacol Sci* 1999;**20**(1):19–26.
198. Kieffer BL, Gaveriaux-Ruff C. Exploring the opioid system by gene knockout. *Prog Neurobiol* 2002;**66** (5):285–306.
199. Raehal KM, Bohn LM. The role of beta-arrestin2 in the severity of antinociceptive tolerance and physical dependence induced by different opioid pain therapeutics. *Neuropharmacology* 2011;**60**(1):58–65.
200. Melief EJ, Miyatake M, Bruchas MR, Chavkin C. Ligand-directed c-Jun N-terminal kinase activation disrupts opioid receptor signaling. *Proc Natl Acad Sci USA* 2010;**107**(25):11608–13.
201. Whistler JL, von ZM. Morphine-activated opioid receptors elude desensitization by beta-arrestin. *Proc Natl Acad Sci USA* 1998;**95**(17):9914–9.
202. Whistler JL, Chuang HH, Chu P, Jan LY, von ZM. Functional dissociation of mu opioid receptor signaling and endocytosis: implications for the biology of opiate tolerance and addiction. *Neuron* 1999;**23**(4):737–46.
203. Groer CE, Tidgewell K, Moyer RA, Harding WW, Rothman RB, Prisinzano TE, et al. An opioid agonist that does not induce mu-opioid receptor—arrestin interactions or receptor internalization. *Mol Pharmacol* 2007;**71** (2):549–57.
204. Harding WW, Tidgewell K, Byrd N, Cobb H, Dersch CM, Butelman ER, et al. Neoclerodane diterpenes as a novel scaffold for mu opioid receptor ligands. *J Med Chem* 2005;**48**(15):4765–71.
205. Xu H, Partilla JS, Wang X, Rutherford JM, Tidgewell K, Prisinzano TE, et al. A comparison of noninternalizing (herkinorin) and internalizing (DAMGO) mu-opioid agonists on cellular markers related to opioid tolerance and dependence. *Synapse* 2007;**61**(3):166–75.
206. Lamb K, Tidgewell K, Simpson DS, Bohn LM, Prisinzano TE. Antinociceptive effects of herkinorin, a MOP receptor agonist derived from salvinorin A in the formalin test in rats: new concepts in mu opioid receptor pharmacology: from a symposium on new concepts in mu-opioid pharmacology. *Drug Alcohol Depend* 2012;**121**(3):181–8.
207. Raehal KM, Walker JK, Bohn LM. Morphine side effects in beta-arrestin 2 knockout mice. *J Pharmacol Exp Ther* 2005;**314**(3):1195–201.
208. Ling GS, Spiegel K, Lockhart SH, Pasternak GW. Separation of opioid analgesia from respiratory depression: evidence for different receptor mechanisms. *J Pharmacol Exp Ther* 1985;**232**(1):149–55.
209. Ling GS, Paul D, Simantov R, Pasternak GW. Differential development of acute tolerance to analgesia, respiratory depression, gastrointestinal transit and hormone release in a morphine infusion model. *Life Sci* 1989;**45**(18):1627–36.
210. Raport CJ, Gosling J, Schweickart VL, Gray PW, Charo IF. Molecular cloning and functional characterization of a novel human CC chemokine receptor (CCR5) for RANTES, MIP-1beta, and MIP-1alpha. *J Biol Chem* 1996;**271**(29):17161–6.
211. Milligan G. G protein-coupled receptor hetero-dimerization: contribution to pharmacology and function. *Br J Pharmacol* 2009;**158**(1):5–14.
212. Rozenfeld R, Devi LA. Exploring a role for heteromerization in GPCR signalling specificity. *Biochem J* 2011;**433**(1):11–8.

213. Somvanshi RK, Kumar U. Pathophysiology of GPCR homo- and heterodimerization: special emphasis on somatostatin receptors. *Pharmaceuticals (Basel)* 2012;**5**(5):417–46.
214. Rabin RA, Regina M, Doat M, Winter JC. 5-HT2A receptor-stimulated phosphoinositide hydrolysis in the stimulus effects of hallucinogens. *Pharmacol Biochem Behav* 2002;**72**(1-2):29–37.
215. González-Maeso J, Yuen T, Ebersole BJ, Wurmbach E, Lira A, Zhou M, et al. Transcriptome fingerprints distinguish hallucinogenic and nonhallucinogenic 5-hydroxytryptamine 2A receptor agonist effects in mouse somatosensory cortex. *J Neurosci* 2003;**23**(26):8836–43.
216. González-Maeso J, Ang RL, Yuen T, Chan P, Weisstaub NV, López-Gimenez JF, et al. Identification of a serotonin/glutamate receptor complex implicated in psychosis. *Nature* 2008;**452**(7183):93–7.
217. Moreno JL, Holloway T, Albizu L, Sealfon SC, González-Maeso J. Metabotropic glutamate mGlu2 receptor is necessary for the pharmacological and behavioral effects induced by hallucinogenic 5-HT2A receptor agonists. *Neurosci Lett* 2011;**493**(3):76–9.
218. Pei L, Li S, Wang M, Diwan M, Anisman H, Fletcher PJ, et al. Uncoupling the dopamine D1–D2 receptor complex exerts antidepressant-like effects. *Nat Med* 2010;**16**(12):1393–5.
219. Conner DA, Mathier MA, Mortensen RM, Christe M, Vatner SF, Seidman CE, et al. beta-Arrestin1 knockout mice appear normal but demonstrate altered cardiac responses to beta-adrenergic stimulation. *Circ Res* 1997;**81**(6):1021–6.
220. Wettschureck N, Moers A, Offermanns S. Mouse models to study G protein-mediated signaling. *Pharmacol Ther* 2004;**101**(1):75–89.
221. Wettschureck N, Offermanns S. Mammalian G proteins and their cell type specific functions. *Physiol Rev* 2005;**85**(4):1159–204.
222. Robishaw JD, Berlot CH. Translating G protein subunit diversity into functional specificity. *Curr Opin Cell Biol* 2004;**16**(2):206–9.
223. Hansen CA, Schwindinger WF, Robishaw JD. Specificity of G-protein βγ dimer signaling. In: Dennis EA, Bradshaw RA, editors. *Transduction Mechanisms in Cellular Signaling*. Elsevier/Academic Press, London. 2011. p. 501.
224. Rudolph U, Spicher K, Birnbaumer L. Adenylyl cyclase inhibition and altered G protein subunit expression and ADP-ribosylation patterns in tissues and cells from Gi2 alpha-/-mice. *Proc Natl Acad Sci USA* 1996;**93**(8):3209–14.
225. Schwindinger WF, Giger KE, Betz KS, Stauffer AM, Sunderlin EM, Sim-Selley LJ, et al. Mice with deficiency of G protein gamma3 are lean and have seizures. *Mol Cell Biol* 2004;**24**(17):7758–68.
226. Metra M, Cas LD, Di LA, Poole-Wilson P. Beta-blockers in heart failure: are pharmacological differences clinically important? *Heart Fail Rev* 2004;**9**(2):123–30.
227. Kenakin T, Christopoulos A. Signalling bias in new drug discovery: detection, quantification and therapeutic impact. *Nat Rev Drug Discov* 2013;**12**(3):205–16.
228. Kenakin T, Christopoulos A. Measurements of ligand bias and functional affinity. *Nat Rev Drug Discov* 2013;**12**(6):483.
229. Rajagopal S, Rajagopal K, Lefkowitz RJ. Teaching old receptors new tricks: biasing seven-transmembrane receptors. *Nat Rev Drug Discov* 2010;**9**(5):373–86.
230. Chen X, Sassano MF, Zheng L, Setola V, Chen M, Bai X, et al. Structure-functional selectivity relationship studies of beta-arrestin-biased dopamine D(2) receptor agonists. *J Med Chem* 2012;**55**(16):7141–53.
231. Allen JA, Yost JM, Setola V, Chen X, Sassano MF, Chen M, et al. Discovery of beta-arrestin-biased dopamine D2 ligands for probing signal transduction pathways essential for antipsychotic efficacy. *Proc Natl Acad Sci USA* 2011;**108**(45):18488–93.
232. Cheng Z, Garvin D, Paguio A, Stecha P, Wood K, Fan F. Luciferase reporter assay system for deciphering GPCR pathways. *Curr Chem Genomics* 2010;**4**:84–91.
233. Bertrand L, Parent S, Caron M, Legault M, Joly E, Angers S, et al. The BRET2/arrestin assay in stable recombinant cells: a platform to screen for compounds that interact with G protein-coupled receptors (GPCRS). *J Recept Signal Transduct Res* 2002;**22**(1–4):533–41.
234. Zhao X, Jones A, Olson KR, Peng K, Wehrman T, Park A, et al. A homogeneous enzyme fragment complementation-based beta-arrestin translocation assay for high-throughput screening of G protein-coupled receptors. *J Biomol Screen* 2008;**13**(8):737–47.
235. Yin H, Chu A, Li W, Wang B, Shelton F, Otero F, et al. Lipid G protein-coupled receptor ligand identification using beta-arrestin PathHunter assay. *J Biol Chem* 2009;**284**(18):12328–38.

236. Barnea G, Strapps W, Herrada G, Berman Y, Ong J, Kloss B, et al. The genetic design of signaling cascades to record receptor activation. *Proc Natl Acad Sci USA* 2008;**105**(1):64–9.
237. Barnes WG, Reiter E, Violin JD, Ren XR, Milligan G, Lefkowitz RJ. beta-Arrestin 1 and Galphaq/11 coordinately activate RhoA and stress fiber formation following receptor stimulation. *J Biol Chem* 2005;**280**(9):8041–50.
238. Hunton DL, Barnes WG, Kim J, Ren XR, Violin JD, Reiter E, et al. Beta-arrestin 2-dependent angiotensin II type 1A receptor-mediated pathway of chemotaxis. *Mol Pharmacol* 2005;**67**(4):1229–36.
239. Charest PG, Terrillon S, Bouvier M. Monitoring agonist-promoted conformational changes of beta-arrestin in living cells by intramolecular BRET. *EMBO Rep* 2005;**6**(4):334–40.
240. Giaever I, Keese CR. Monitoring fibroblast behavior in tissue culture with an applied electric field. *Proc Natl Acad Sci USA* 1984;**81**(12):3761–4.
241. Fang Y. Label-free cell-based assays with optical biosensors in drug discovery. *Assay Drug Dev Technol* 2006;**4**(5):583–95.
242. Zanella F, Lorens JB, Link W. High content screening: seeing is believing. *Trends Biotechnol* 2010;**28**(5):237–45.
243. Fang Y, Li GG, Peng J. Optical biosensor provides insights for bradykinin B(2) receptor signaling in A431 cells. *FEBS Lett* 2005;**579**(28):6365–74.
244. Fang Y, Li G, Ferrie AM. Non-invasive optical biosensor for assaying endogenous G protein-coupled receptors in adherent cells. *J Pharmacol Toxicol Methods* 2007;**55**(3):314–22.
245. Peters MF, Vaillancourt F, Heroux M, Valiquette M, Scott CW. Comparing label-free biosensors for pharmacological screening with cell-based functional assays. *Assay Drug Dev Technol* 2010;**8**(2):219–27.
246. Scott CW, Peters MF. Label-free whole-cell assays: expanding the scope of GPCR screening. *Drug Discov Today* 2010;**15**(17–18):704–16.
247. Achour L, Kamal M, Jockers R, Marullo S. Using quantitative BRET to assess G protein-coupled receptor homo- and hetero-dimerization. *Methods Mol Biol* 2011;**756**:183–200.
248. Cottet M, Albizu L, Comps-Agrar L, Trinquet E, Pin JP, Mouillac B, et al. Time resolved FRET strategy with fluorescent ligands to analyze receptor interactions in native tissues: application to GPCR oligomerization. *Methods Mol Biol* 2011;**746**:373–87.
249. Comps-Agrar L, Maurel D, Rondard P, Pin JP, Trinquet E, Prezeau L. Cell-surface protein–protein interaction analysis with time-resolved FRET and snap-tag technologies: application to G protein-coupled receptor oligomerization. *Methods Mol Biol* 2011;**756**:201–14.
250. Maurel D, Comps-Agrar L, Brock C, Rives ML, Bourrier E, Ayoub MA, et al. Cell-surface protein–protein interaction analysis with time-resolved FRET and snap-tag technologies: application to GPCR oligomerization. *Nat Methods* 2008;**5**(6):561–7.
251. Venkatakrishnan AJ, Deupi X, Lebon G, Tate CG, Schertler GF, Babu MM. Molecular signatures of G protein-coupled receptors. *Nature* 2013;**494**(7436):185–94.
252. Deupi X, Standfuss J. Structural insights into agonist-induced activation of G protein-coupled receptors. *Curr Opin Struct Biol* 2011;**21**(4):541–51.
253. Audet M, Bouvier M. Insights into signaling from the beta2-adrenergic receptor structure. *Nat Chem Biol* 2008;**4**(7):397–403.
254. Rasmussen SG, DeVree BT, Zou Y, Kruse AC, Chung KY, Kobilka TS, et al. Crystal structure of the beta2 adrenergic receptor-Gs protein complex. *Nature* 2011;**477**(7366):549–55.
255. Warne T, Edwards PC, Leslie AG, Tate CG. Crystal structures of a stabilized beta1-adrenoceptor bound to the biased agonists bucindolol and carvedilol. *Structure* 2012;**20**(5):841–9.
256. Wacker D, Wang C, Katritch V, Han GW, Huang XP, Vardy E, et al. Structural features for functional selectivity at serotonin receptors. *Science* 2013;**340**(6132):615–9.
257. Kubon C, Mistry NB, Grundvold I, Halvorsen S, Kjeldsen SE, Westheim AS. The role of beta-blockers in the treatment of chronic heart failure. *Trends Pharmacol Sci* 2011;**32**(4):206–12.
258. Waagstein F. Beta blockers in heart failure. *Cardiology* 1993;**82**(Suppl. 3):13–8.
259. McMurray JJ, Pfeffer MA. Heart failure. *Lancet* 2005;**365**(9474):1877–89.
260. Sigmund M, Jakob H, Becker H, Hanrath P, Schumacher C, Eschenhagen T, et al. Effects of metoprolol on myocardial beta-adrenoceptors and Gi alpha-proteins in patients with congestive heart failure. *Eur J Clin Pharmacol* 1996;**51**(2):127–32.

261. Iaccarino G, Rockman HA, Shotwell KF, Tomhave ED, Koch WJ. Myocardial overexpression of GRK3 in transgenic mice: evidence for *in vivo* selectivity of GRKs. *Am J Physiol* 1998;**275**(4 Pt 2):H1298–306.
262. Packer M. Current role of beta-adrenergic blockers in the management of chronic heart failure. *Am J Med* 2001;**110**(Suppl. 7A):81S–94S.
263. Poole-Wilson PA, Swedberg K, Cleland JG, Di LA, Hanrath P, Komajda M, et al. Comparison of carvedilol and metoprolol on clinical outcomes in patients with chronic heart failure in the Carvedilol Or Metoprolol European Trial (COMET): randomised controlled trial. *Lancet* 2003;**362**(9377):7–13.
264. Effect of metoprolol CR/XL in chronic heart failure: Metoprolol CR/XL Randomised Intervention Trial in Congestive Heart Failure (MERIT-HF). *Lancet* 1999;353(9169):2001–7.
265. The Cardiac Insufficiency Bisoprolol Study II (CIBIS-II): a randomised trial. *Lancet* 1999;353(9146):9–13.
266. Kovacs K, Hanto K, Bognar Z, Tapodi A, Bognar E, Kiss GN, et al. Prevalent role of Akt and ERK activation in cardioprotective effect of Ca(2+) channel- and beta-adrenergic receptor blockers. *Mol Cell Biochem* 2009;**321** (1–2):155–64.
267. Baker JG, Hill SJ, Summers RJ. Evolution of beta-blockers: from anti-anginal drugs to ligand-directed signalling. *Trends Pharmacol Sci* 2011;**32**(4):227–34.
268. Nakaya M, Chikura S, Watari K, Mizuno N, Mochinaga K, Mangmool S, et al. Induction of cardiac fibrosis by beta-blocker in G protein-independent and G protein-coupled receptor kinase 5/beta-arrestin2-dependent Signaling pathways. *J Biol Chem* 2012;**287**(42):35669–77.
269. Burris KD, Molski TF, Xu C, Ryan E, Tottori K, Kikuchi T, et al. Aripiprazole, a novel antipsychotic, is a high-affinity partial agonist at human dopamine D2 receptors. *J Pharmacol Exp Ther* 2002;**302**(1):381–9.
270. Lawler CP, Prioleau C, Lewis MM, Mak C, Jiang D, Schetz JA, et al. Interactions of the novel antipsychotic aripiprazole (OPC-14597) with dopamine and serotonin receptor subtypes. *Neuropsychopharmacology* 1999;**20** (6):612–7.
271. Shapiro DA, Renock S, Arrington E, Chiodo LA, Liu LX, Sibley DR, et al. Aripiprazole, a novel atypical antipsychotic drug with a unique and robust pharmacology. *Neuropsychopharmacology* 2003;**28**(8):1400–11.
272. Urban JD, Vargas GA, von ZM, Mailman RB. Aripiprazole has functionally selective actions at dopamine D2 receptor-mediated signaling pathways. *Neuropsychopharmacology* 2007;**32**(1):67–77.
273. Kikuchi T, Tottori K, Uwahodo Y, Hirose T, Miwa T, Oshiro Y, et al. 7-(4-[4-(2,3-Dichlorophenyl)-1-piperazinyl]butyloxy)-3,4-dihydro-2(1H)-quinolinon e (OPC-14597), a new putative antipsychotic drug with both presynaptic dopamine autoreceptor agonistic activity and postsynaptic D2 receptor antagonistic activity. *J Pharmacol Exp Ther* 1995;**274**(1):329–36.
274. Koener B, Goursaud S, Van De Stadt M, Calas AG, Jeanjean AP, Maloteaux JM, et al. Pharmacological blockade of dopamine D2 receptors by aripiprazole is not associated with striatal sensitization. *Naunyn Schmiedebergs Arch Pharmacol* 2011;**383**(1):65–77.
275. Emamian ES, Hall D, Birnbaum MJ, Karayiorgou M, Gogos JA. Convergent evidence for impaired AKT1-GSK3beta signaling in schizophrenia. *Nat Genet* 2004;**36**(2):131–7.
276. Ikeda M, Iwata N, Suzuki T, Kitajima T, Yamanouchi Y, Kinoshita Y, et al. Association of AKT1 with schizophrenia confirmed in a Japanese population. *Biol Psychiatry* 2004;**56**(9):698–700.
277. Emamian ES. AKT/GSK3 signaling pathway and schizophrenia. *Front Mol Neurosci* 2012;**5**:33.
278. Meltzer HY, Massey BW. The role of serotonin receptors in the action of atypical antipsychotic drugs. *Curr Opin Pharmacol* 2011;**11**(1):59–67.
279. Chen XT, Pitis P, Liu G, Yuan C, Gotchev D, Cowan CL, et al. Structure-activity relationships and discovery of a G protein biased mu opioid receptor ligand, [(3-methoxythiophen-2-yl)methyl]({2-[(9R)-9-(pyridin-2-yl)-6-oxaspiro-[4.5]decan- 9-yl]ethyl})amine (TRV130), for the treatment of acute severe pain. *J Med Chem* 2013;**56** (20):8019–31.

Index

Note: Page numbers followed by "*f*" and "*t*" refers to figures and tables respectively.

R

S

Kolibri 360 GmbH
Ferdinand-Jühlke-Straße 7
99095 Erfurt, Deutschland
produktsicherheit@kolibri360.de